Structure and Function
of the
MUSCULOSKELETAL SYSTEM

James Watkins, PhD

University of Strathclyde
Glasgow, Scotland

Human Kinetics

Library of Congress Cataloging-in-Publication Data

Watkins, James, 1946-
    Structure and function of the musculoskeletal system / James Watkins.
        p.   cm.
    Includes bibliographical references and index.
    ISBN 0-88011-686-2
    1. Musculoskeletal system—Physiology.   2. Musculoskeletal system—Anatomy.
    3. Human   mechanics.   I. Title.
    QP301.W34   1999
    612.7--dc21   98-27930
                                                                                        CIP

ISBN: 0-88011-686-2

Acquisitions Editors: Kathy Riggen and Loarn Robertson; **Developmental Editor:** Julie Rhoda; **Assistant Editors:** Sandra Merz Bott and Cassandra Mitchell; **Copyeditor:** Judy Peterson; **Proofreader:** Pam Johnson; **Indexer:** Marie Rizzo; **Graphic Designer:** Nancy Rasmus, **Graphic Artist:** Kimberly Maxey; **Cover Designer:** Jack Davis; **Printer:** Edwards Brothers, Inc.

All medical illustrations were drawn by the author.

Printed in the United States of America   10 9 8 7 6 5 4 3 2 1

Human Kinetics
Web site: http://www.humankinetics.com/

United States: Human Kinetics
P.O. Box 5076
Champaign, IL 61825-5076
1-800-747-4457
e-mail: humank@hkusa.com

Canada: Human Kinetics
475 Devonshire Road Unit 100
Windsor, ON N8Y 2L5
1-800-465-7301 (in Canada only)
e-mail: humank@hkcanada.com

Europe: Human Kinetics
P.O. Box IW14
Leeds LS16 6TR, United Kingdom
(44) 1132 781708
e-mail: humank@hkeurope.com

Australia: Human Kinetics
57A Price Avenue
Lower Mitcham, South Australia 5062
(088) 277 1555
e-mail: humank@hkaustralia.com

New Zealand: Human Kinetics
P.O. Box 105-231
Auckland 1
(09) 523 3462
e-mail: humank@hknewz.com

Contents

Part III   Musculoskeletal Response and Adaptation to Loading

Preface

Human movement is brought about by the musculoskeletal system under the control of the nervous system. By coordinating activity among the various muscle groups, forces generated by the muscles are transmitted by the bones and joints to enable an individual to maintain an upright or partially-upright body posture and to voluntarily move. Thus, the musculoskeletal system, also referred to as the locomotor system, is essentially a machine—a powered mechanism for generating and transmitting forces to counteract the effects of gravity and to bring about desirable movements of the body.

The open-chain arrangement of the skeleton enables the body to adopt and perform a wide range of postures and movements. However, the body pays a price for this ability: the muscles, bones, and joints are subjected to very high forces in all postures and movements other than lying down. In response to the forces exerted on them, the musculoskeletal components experience physical strain, and the greater the force, the greater the strain. Under normal circumstances, the musculoskeletal components adapt their size, shape, and structure to readily withstand the strain of everyday physical activity. However, excessive strain results in injury. Consequently, there is an intimate relationship between the structure and function of the musculoskeletal system; this book's purpose is to develop understanding of this relationship.

Structure and Function of the Musculoskeletal System is primarily a textbook for undergraduate students of kinesiology, exercise science, sport science, and physical education; however, students of physical therapy, occupational therapy, and podiatric medicine will also find the book useful.

The book organizes 12 chapters into three parts. Part I (chapters 1 and 2) develops the concept of the musculoskeletal system as a machine capable of generating and transmitting forces to counteract the effects of gravity and bring about desirable movements. Since the same mechanical laws govern human movement as all other forms of movement, chapter 1 describes the elementary mechanical concepts and principles that underlie movement. These concepts are applied throughout the rest of the book starting with chapter 2, which describes the basic structure and function of the musculoskeletal system. Part II (chapters 3 through 9) describes the functional anatomy of the musculoskeletal system. Chapters 3 through 7 describe the skeletal system starting with the skeleton (chapter 3), followed by connective tissues (chapter 4), the articular system (chapter 5), joints of the axial skeleton (chapter 6), and joints of the appendicular skeleton (chapter 7). Chapter 8 describes the neuromuscular system, and chapter 9 describes the effects of the open-chain arrangement of the skeleton on the forces exerted in muscles and joints. Part III (chapters 10 through 12) describes the response and adaptation of the musculoskeletal system to loading, particularly in impact and nonimpact situations (chapter 10), how the musculoskeletal system adapts its structure to time-averaged changes in loading (chapter 11), and the mechanical risk factors associated with etiology of injuries and other disorders of the musculoskeletal system (chapter 12).

A student does not necessarily need previous knowledge of anatomy or mechanics to use this book. Anatomical and mechanical concepts and principles are introduced, clearly explained, and subsequently reinforced throughout the book. Using the open learning model approach, the book features chapter overviews to summarize the content of each chapter, key terms highlighted and defined in a running glossary throughout the text, learning objectives outlining what each student should glean from the text, key points set off from the text to make them easy for students to find and remember, applied examples, extensive use of illustrations and conceptual models, review questions, references to guide further reading, and an extensive index.

Whether you read this text as part of a course or for general interest, I hope it will encourage you to learn more about the structure and function of the musculoskeletal system. Adequate musculoskeletal functioning is an important determinant of an individual's quality of life and the information contained in this book is essential for anyone wishing to maintain, or help others to maintain, a healthy musculoskeletal system.

Acknowledgments

I thank my dear wife, Shelagh, for her continuous help and encouragement throughout the various stages of the writing of this book. I also thank the staff at Human Kinetics, in particular, Loarn Robertson and Julie Rhoda, and Sandra Merz Bott, for their constant support and expert editorial work. Lastly, I thank my academic colleagues and the large number of undergraduate and graduate students who have helped me, directly and indirectly, over a long period, to conceptualize much of the work presented in this book.

Part I
THE HUMAN MACHINE

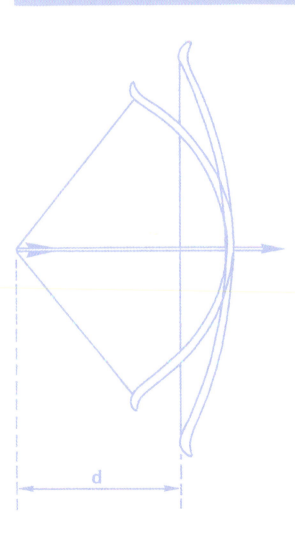

The musculoskeletal system under the control of the nervous system bring about human movement. By coordinated activity between the various muscle groups, bones and joints transmit muscle forces to enable us to maintain an upright posture and bring about voluntary controlled movements. Consequently, the musculoskeletal system is essentially a machine; it is a powered mechanism for generating and transmitting forces to bring about desirable movements. The same mechanical laws govern all movement, including human movement. Part I describes the basic mechanics of musculoskeletal function. Chapter 1—Elementary Mechanics—describes the concepts and principles that underlie all types of movement. Chapter 2—The Musculoskeletal System—describes the basic structure and function of the musculoskeletal system.

Chapter 1

Elementary Mechanics

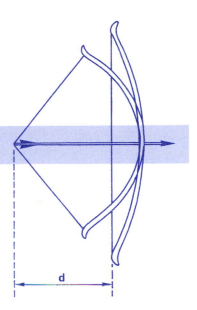

All movements and changes in movements are brought about by the action of forces. To move an object from a stationary position or to change its speed or direction of movement it is necessary to change the force acting on the object. An object's amount of change in speed or direction of movement resulting from the action of a force depends upon the magnitude and direction of the force; that is, there is a clear relationship between change of force and change in movement. Isaac Newton (1642–1727) described this relationship in what has come to be known as Newton's three laws of motion. The laws of motion apply to all forms of movement, including human movement.

Human movement is brought about by the musculoskeletal system—the muscles, bones, and joints—under the control of the nervous system. By coordinated activity between the various muscle groups, forces generated by the muscles are transmitted by the bones and joints to enable the individual to voluntarily control movements. To understand the relationship between structure and function of the musculoskeletal system it is necessary to understand Newton's laws of motion and other basic mechanical concepts and principles, in particular, kinematics and kinetics. The purpose of chapter 1 is to describe these concepts and principles before going on to consider the structure and function of the musculoskeletal system.

Objectives

After reading this chapter you should be able to do the following:

1. Define or describe the key terms.
2. Describe the different forms of motion.
3. Differentiate between kinematics and kinetics.
4. Differentiate between vector and scalar quantities.
5. Resolve a vector into two components at right angles to each other.
6. Differentiate between mass, inertia, and linear momentum.
7. Differentiate between contact forces and forces of attraction.
8. Differentiate between mass and weight.
9. Describe the three basic types of loads.
10. Distinguish between load, strain, and stress.
11. Differentiate between work, energy, and power.
12. Differentiate between the impulse-momentum relationship and the work-energy relationship.
13. Differentiate between stability and equilibrium.
14. Describe the three types of lever systems.
15. Explain the difference in mechanical advantage among the three types of lever systems.

Force and Mechanics

Force: that which alters or tends to alter an object's state of rest or type of movement

Mechnics: the study of the forces acting on objects and their effects on the objects' movement, size, shape, and structure

Biomechanics: the study of the forces that act on and within the human body and their effects on the body's movement, size, shape, and structure

Kinematics: the branch of mechanics that describes the motion of objects

Skeletal muscles pull on bones in order to control joint movements. The action of pulling, like pushing, is an example of the application of a **force**. Other examples of forces include the following:

- Pushing or pulling open a door
- Pushing the pedals and pulling the hand brake in a motor vehicle
- Kicking a ball (a form of pushing force)
- Throwing a javelin (the arm action involves a pulling phase followed by a pushing phase)

Objects move the way they do as a result of the forces acting on them. Consequently, a change in the pattern of forces acting on an object tends to change the way the object is moving (or the physical dimensions of the object). For example, when you kick a ball, your foot exerts a force on the ball that temporarily deforms it (while it is in contact with your foot) and moves it in the direction of the force. **Mechanics** is the study of the forces acting on objects and their effects on the objects' movement, size, shape, and structure. **Biomechanics** refers to the study of the forces that act on and within the human body and their effects on the body's movement, size, shape, and structure. For example, in kicking a ball, your muscles pull on your bones to control the movement in your joints so that you can remain upright and swing the kicking leg at the ball. **Kinematics** is the branch of mechanics that describes the motion of objects. A kinematic analysis describes the movement of an object in terms of form of motion (linear or angular), how far the object moves (distance), how fast the object moves (speed), and how consistently the object moves (acceleration). For example, a kinematic analysis of kicking a ball describes how body parts move in relation to each other during the kicking action, and how the ball moves while in contact with your foot and after leaving your foot. **Kinetics** is the branch of mechanics that describes the forces acting on objects, that is, the study of why objects move the way that they do. A kinetic analysis describes the cause of the observed kinematics. For example, a kinetic analysis of kicking a ball describes the forces exerted in your muscles and joints during the kicking action and those exerted between your foot and the ball.

> *All objects move the way that they do as a result of the forces acting on them. A kinematic analysis describes the way an object moves and a kinetic analysis describes the forces acting on the object that are responsible for the observed kinematics.*

Forms of Motion

There two basic forms of motion: linear and angular. **Linear motion**, also referred to as **translation**, occurs when all parts of an object or person move the same distance, in the same direction, in the same time. When the direction of movement is a straight line, as when a skater glides across the ice, the type of motion is referred to as **rectilinear translation** (figure 1.1). When the direction of movement follows a curved path, as experienced by a long jumper during the middle of the flight phase of the jump, the type of motion is referred to as **curvilinear translation** (figure 1.2).

Angular motion, or **rotation**, occurs when an object or part of an object moves along a circular path about some line in space called the **axis of rotation** such that all parts of the object move through the same angle, in the same direction, in the same time. The axis of rotation may be stationary or it may experience rectilinear or curvilinear motion. For example, when an individual bends an elbow with the upper arm stationary the forearm rotates about a more or less stationary axis through the elbow joint (figure 1.3). When rowing, the thighs rotate about an axis through the hip joints and the lower legs rotate about an axis through the knee joints (figure 1.4). During the propulsion (*a* to *b*) and recovery (*b* to *a*) phases of the stroke the hip joint axis undergoes rectilinear motion and the knee joint axis undergoes curvilinear motion.

Kinetics: the branch of mechanics that describes the forces acting on objects

Linear motion: motion of an object during which all parts of the object move the same distance, in the same direction, in the same time

Translation: another name for linear motion

Rectilinear translation: linear motion in which the direction of movement is a straight line

Curvilinear translation: linear motion in which the direction of movement follows a curved path

Angular motion: motion during which an object or part of an object moves along a circular path about a line in space such that all parts of the object move through the same angle, in the same direction, in the same time

Rotation: another name for angular motion

Axis of rotation: the line in space about which rotation of an object takes place

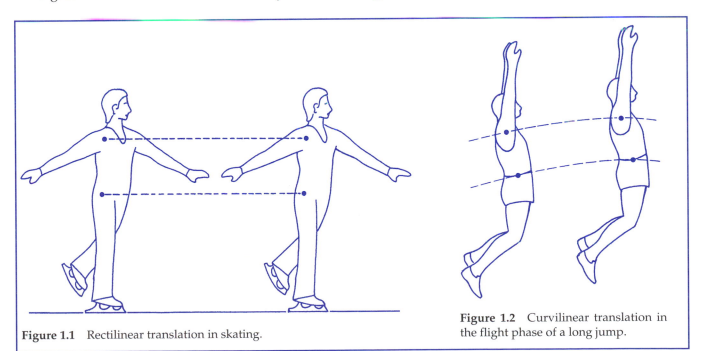

Figure 1.1   Rectilinear translation in skating.

Figure 1.2   Curvilinear translation in the flight phase of a long jump.

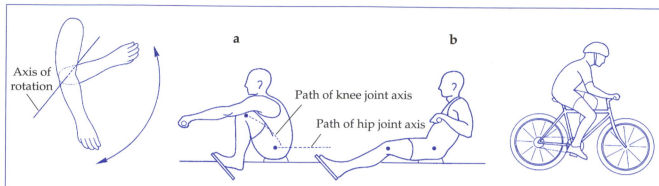

Figure 1.3   Rotation of the forearm and hand about the elbow joint

Figure 1.4   Rotation at the hip and knee joints in rowing: the axis through the hip joints undergoes rectilinear translation and the axis through the knee joints undergoes curvilinear translation.

Figure 1.5   General motion: simultaneous linear motion of the head, trunk, and arms and angular motion of the legs in cycling.

General motion: motion that involves a combination of linear motion and angular motion

Most movements of the human body involve a combination of linear and angular movements. For example, when cycling, the head, trunk, and arms experience more or less continuous rectilinear motion brought about by angular motion of the legs (figure 1.5). Motion involving a combination of linear and angular motion is usually referred to as **general motion.**

The musculoskeletal system is responsible for human movement: the muscles pull on the bones to control joint movements and thereby control the movement of the body as a whole. Most human movements involve general motion, that is, a combination of linear motion and angular motion.

Kinematics

Kinematics describes motion in terms of the position and change in position with time of an object with respect to its spatial environment. **Linear kinematics** describes linear motion and **angular kinematics** describes angular motion.

Linear Kinematics

Linear kinetics: the description of linear motion in terms of position and change in position with time

Angular kinetics: the description of angular motion in terms of position and change in position with time

Distance: the length of the path between two points

Displacement: the straight-line distance and direction from one point to another

Speed: the rate of change of distance

If a runner completes a 10 km cross-country run (1 km = 1000 m), then the length of the course is the linear distance (also simply called **distance**) covered by the runner (figure 1.6). However, the distance gives no indication of direction of travel during the run and, therefore, no indication of the location of the starting point with respect to the finish point. The location of the finish with respect to the start is specified by the **displacement** from the start to the finish, that is, the straight-line distance between the start and finish and the direction of the finish with respect to the start. For example, if the straight-line distance between the start and the finish is 100 m and the direction of the finish is due north of the start, then the displacement of the finish from the start is 100 m due north. Consequently, even though the runner travels a distance of 10 km his displacement is 100 m due north.

The speed of an object indicates how fast the object moves over a certain distance. **Speed** is defined as distance traveled per unit of time or rate of change of distance:

$$\text{average speed} = \frac{\text{distance}}{\text{time}}.$$

Thus,

$$\bar{s} = \frac{l}{t},$$

where $\bar{s}$ = "s bar" = average speed, l = distance, and t = time taken to cover l. For example, if the cross-country runner completed the 10 km course in 40 min, then his average speed during the run is given by

$$\bar{s} = \frac{10 \text{ km}}{40 \text{ min}}.$$

Converting minutes (min) to seconds (s) and substituting

$$40 \text{ min} \times 60 \text{ s} = 2400 \text{ s}$$

$$\bar{s} = \frac{10000 \text{ m}}{2400 \text{ s}}$$

$$\bar{s} = 4.2 \text{ m/s}.$$

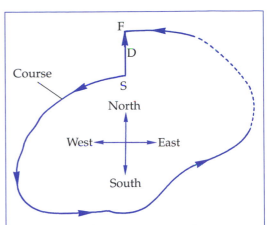

Figure 1.6   Cross country running course illustrating linear distance and linear displacement. The length of the course = linear distance traveled by the runner. The location of the finish F to the start S = linear displacement of the runner = D.

Whereas speed refers to rate of change of distance, **velocity** refers to rate of change of displacement, that is, rate of change of distance in a specific direction, or

Velocity: the rate of change of displacement

Acceleration: the rate of change of velocity

$$\text{average velocity } (\bar{v}) = \frac{\text{displacement (d)}}{\text{time (t)}},$$

where $\bar{v}$ = "v bar" = average velocity, d = displacement, t = time taken to cover d. When displacement is zero, as in a running event that starts and finishes at the same place, the average velocity is zero. When distance and displacement are the same, as in a 100 m sprint run, the average speed and average velocity are the same.

If a sprinter runs a 100 m race (on a straight track) in 11 s, her average velocity is 9.09 m/s (100 m divided by 11 s). However, her velocity varies considerably during the race. For example, her velocity increases from 0 m/s at the start to maximum velocity after about 50 m. She then attempts to maintain maximum velocity, but her velocity probably decreases toward the end of the race. Variation in velocity is measured in terms of **acceleration**, which is defined as rate of change of velocity.

$$\text{average acceleration } (\bar{a}) = \frac{\text{change in velocity } (v_2 - v_1),}{\text{time } (t_2 - t_1)}$$

where $\bar{a}$ = "a bar" = average acceleration, t_1 = time at the start of the period under consideration, t_2 = time at the end of the period under consideration, v_1 = velocity at t_1, and v_2 = velocity at t_2. For example, if the velocity of the sprinter changes from 0 m/s at the start to 9 m/s after 4 s, then her average acceleration over the first 4 s of the race is given by

$$\bar{a} = \frac{v_2 - v_1,}{t_2 - t_1}$$

where $t_1 = 0$, $t_2 = 4$ s, $v_1 = 0$, $v_2 = 9$ m/s. Thus,

$$\bar{a} = \frac{9 - 0 \text{ m/s}}{4 - 0 \text{ s}}$$

$$\bar{a} = \frac{9 \text{ m/s}}{4 \text{ s}}$$

$$\bar{a} = 2.25 \text{ m/s/s or m/s}^2.$$

The result of 2.25 m/s^2 indicates that the velocity of the sprinter increased during the first 4 s of the race by an average of 2.25 m/s every second. Similarly, if the sprinter completes the race in 11 s, and her velocity changes during the latter part of the race from 10.5 m/s after 6.5 s to 9.5 m/s at the finish, then her average acceleration over the last 4.5 s of the race is given by

$$\bar{a} = \frac{v_2 - v_1}{t_2 - t_1},$$

where $t_1 = 6.5$ s, $t_2 = 11$ s, $v_1 = 10.5$ m/s, and $v_2 = 9.5$ m/s. Thus,

$$\bar{a} = \frac{9.5 \text{ m/s} - 10.5 \text{ m/s}}{11 \text{ s} - 6.5 \text{ s}}$$

$$\bar{a} = \frac{-1 \text{ m/s}}{4.5 \text{ s}}$$

$$\bar{a} = -0.22 \text{ m/s}^2.$$

In this case the acceleration has a negative value indicating that the velocity of the sprinter decreased during the last 4.5 s of the race by an average of 0.22 m/s every second. It is usual to describe an object as being accelerated when its velocity is increasing and decelerated when its velocity is decreasing.

Vector and Scalar Quantities

All quantities within the physical and life sciences can be categorized as either scalar or vector quantities. Quantities that can be completely specified in terms of their magnitude (size) are called **scalar quantities**; these include area, volume, temperature, distance, and speed. Quantities that require specification in both magnitude and direction are called **vector quantities**; these include displacement, velocity, acceleration, and force. A vector quantity can be represented diagrammatically by a straight line with an arrow head; the length of the line corresponds (with respect to an appropriate scale) to the magnitude of the quantity and the orientation of the line and arrow head (with respect to an appropriate reference axis) indicates the direction.

Scalar quantity: a quantity that can be completely specified by its magnitude

Vector quantity: a quantity that must be specified in both magnitude and direction

> All quantities within the physical and life sciences can be categorized as scalar or vector quantities. A scalar quantity can be specified completely by magnitude, but a vector quantity has to be specified in both magnitude and direction.

Displacement Vectors If a man ran 4 mi from point A to point B and then walked 3 mi from point B to point C, it is clear that he traveled a total distance of 7 mi, but it is not possible to determine the position of C in relation to A since no information is available concerning the directions in which he ran and walked. However, if we are told that the man ran 4 mi due south from A to B and then walked 3 mi due east from B to C, the position of C in relation to A can be determined by considering the displacement vectors $\overline{AB}$ and $\overline{BC}$. The line above the letters denotes a vector. The displacements $\overline{AB}$ and $\overline{BC}$ are shown in figure 1.7a. The position of C in relation to A is specified by the vector $\overline{AC}$. The distance between A and C can be determined by measuring the length of the line $\overline{AC}$ and converting this to miles by using the distance scale. The direction of C with respect to A is specified by the angle θ.

The vector $\overline{AC}$ is referred to as the resultant of $\overline{AB}$ and $\overline{BC}$, and $\overline{AB}$ and $\overline{BC}$ are referred to as component vectors. It is clear that vector addition is not the same as arithmetic (or scalar) addition. If, as in figure 1.7b, the man walked 2 mi north from C to D, the position of D with respect to A is specified by the vector $\overline{AD}$. In this case the resultant vector $\overline{AD}$ is the resultant of the three component vectors $\overline{AB}$, $\overline{BC}$, and

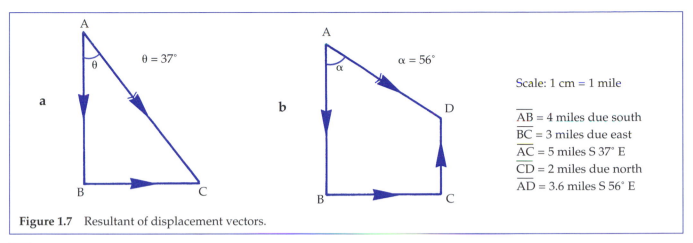

Figure 1.7   Resultant of displacement vectors.

$\overline{CD}$. This example illustrates that irrespective of the number of component displacement vectors used to describe the movement of an object, the net effect of all the component vectors is a single resultant vector. This general principle applies to all vector quantities. The method used in figure 1.7 to determine the resultant of the component vectors is called the **vector chain** method, that is, the component vectors are linked together in a chain (in any order) and the resultant vector runs from the starting point of the first component vector to the end point of the last component vector.

Velocity Vectors   In addition to the vector chain method of determining the resultant of a number of component vectors, there is another method, the **parallelogram of vectors**, that is useful when there are only two component vectors, but somewhat laborious when there are three or more component vectors. In this method two component vectors extend from the same point to form adjacent sides of a parallelogram. The resultant of the two component vectors is given by the diagonal of the completed parallelogram. For example, if a ship without a keel starts to sail due north in a wind blowing S 60° W (south 60° west) the resultant velocity of the ship is specified by the resultant of

1. the velocity $\overline{V}_1$ of the ship resulting from the drive of the engines, and
2. the velocity $\overline{V}_2$ of the ship resulting from the wind.

The vector chain and parallelogram of vectors methods of determining the resultant velocity of the ship are shown in figure 1.8. If $\overline{V}_1 = 15$ knots due north and $\overline{V}_2 = 6$ knots S 60° W, the resultant velocity of the ship $\overline{V}_R = 13.05$ knots N 23.5° W.

When the parallelogram of vectors method is used to determine the resultant of three or more component vectors the first step is to find the resultant $\overline{R}$ of any two component vectors. The resultant of $\overline{R}$ and another component vector is then found and the process is repeated until the resultant of all the component vectors is found.

Resolution of a Vector

Just as the resultant of any number of component vectors can be determined, any single vector can be replaced by any number of component vectors that have the same effect as the single vector. The process of replacing a vector by two or more component vectors is referred to as the **resolution of a vector**. In analyzing the effects of forces on human movement it can be necessary to resolve a force into two components at right angles to each other. This procedure requires an understanding of basic trigonometry.

Trigonometry of a Right-Angled Triangle   Trigonometry is the branch of mathematics that deals with the relationships among the lengths of the sides and the sizes of the angles in a triangle (Enoka 1994). Figure 1.9 shows a right-angled triangle, in

Vector chain: the method of determining the resultant of a number of component vectors by joining the component vectors together in a chain

Parallelogram of vectors: the method of determining the resultant of two component vectors by constructing a parallelogram in which the two component vectors arise from the same point and form two adjacent sides of the parallelogram

Resolution of a vector: the process of replacing a vector by two or more component vectors

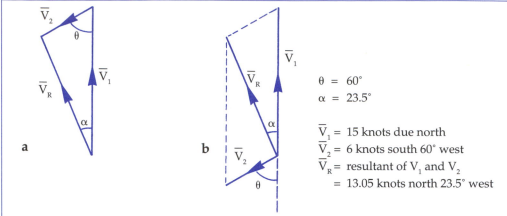

θ = 60°
α = 23.5°

$\overline{V}_1$ = 15 knots due north
$\overline{V}_2$ = 6 knots south 60° west
$\overline{V}_R$ = resultant of V_1 and V_2
   = 13.05 knots north 23.5° west

Figure 1.8 Resultant velocity of a ship; (*a*) vector chain method and (*b*) parallelogram of vectors method.

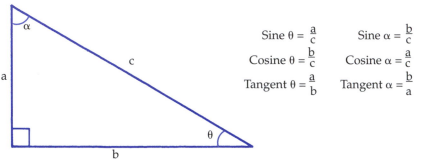

Sine θ = $\frac{a}{c}$ Sine α = $\frac{b}{c}$

Cosine θ = $\frac{b}{c}$ Cosine α = $\frac{a}{c}$

Tangent θ = $\frac{a}{b}$ Tangent α = $\frac{b}{a}$

Figure 1.9 Definitions of sine, cosine, and tangent in a right-angled triangle.

which one angle (between sides a and b) is 90°. The angles between sides a and c, and sides b and c, are denoted α and θ, respectively.

In a right-angled triangle the two angles less than 90°, namely, α and θ in figure 1.9, can be specified by the ratio between the lengths of any two sides of the triangle. The three most common ratios are sine, cosine, and tangent, and they are defined in relation to the particular angle under consideration. For example, in relation to angle θ in figure 1.9, a is referred to as the opposite side, b is referred to as the adjacent side, and c is the hypotenuse, the side of the triangle opposite the right angle.

The sine of θ is defined as the ratio of the opposite side to the hypotenuse, such that,

$$\text{sine } \theta = \frac{\text{opposite side}}{\text{hypotenuse}} = \frac{a}{c}.$$

The cosine of θ is defined as the ratio of the adjacent side to the hypotenuse, such that,

$$\text{cosine } \theta = \frac{\text{adjacent side}}{\text{hypotenuse}} = \frac{b}{c}.$$

The tangent of θ is defined as the ratio of the opposite side to the adjacent side, such that,

$$\text{tangent } \theta = \frac{\text{opposite side}}{\text{adjacent side}} = \frac{a}{b}.$$

Most pocket calculators provide a range of trigonometric ratios including sine (sin), cosine (cos), and tangent (tan). Alternatively, tables of sine, cosine, and tangent (for angles between 0° and 90°) can be obtained in publications such as Castle (1969). The lengths of sides and sizes of angles in right-angled triangles can be calculated using sine, cosine, and tangent functions provided that two sides or one side and one other angle are known. With reference to figure 1.9, for example, if c = 10 cm and θ = 30°, the lengths of sides a and b and the size of angle α can be determined as follows.

1. Calculation of the length of side a:

$$\frac{a}{c} \;=\; \sin\theta$$

$$a \;=\; c \times \sin\theta \text{ (i.e., c multiplied by } \sin\theta\text{)}$$

$$a \;=\; c \times \sin 30°.$$

From sine tables it can be found that sin 30° = 0.5 (i.e., the ratio of the length of side a to the length of side c is 0.5). Since c = 10 cm and sin 30° = 0.5, it follows that

$$a \;=\; 10 \text{ cm} \times 0.5$$

$$a \;=\; 5 \text{ cm}.$$

2. Calculation of the length of side b:

$$\frac{b}{c} \;=\; \cos\theta$$

$$b \;=\; c \times \cos\theta \text{ (i.e., c multiplied by } \cos\theta\text{)}$$

$$b \;=\; c \times \cos 30°.$$

From cosine tables it can be found that cos 30° = 0.866 (i.e., the ratio of the length of side b to the length of side c is 0.866). Since c = 10 cm and cos 30° = 0.866, it follows that

$$b \;=\; 10 \text{ cm} \times 0.866$$

$$b \;=\; 8.66 \text{ cm}.$$

3. Calculation of angle α:

Angle α can be determined a number of ways:

a. The sum of the three angles in any triangle (with or without a right angle) is 180°. Here, since the sum of the two known angles is 120°, it follows that α = 180° − 120° = 60°.

b. The lengths of all three sides of the triangle are known, namely, a = 5 cm, b = 8.66 cm and c = 10 cm. Consequently, α can be determined by calculating the sine, cosine, or tangent of the angle:

$$\sin\alpha \;=\; \frac{b}{c} \;=\; \frac{8.66 \text{ cm}}{10 \text{ cm}} = 0.866 \text{ (i.e., } \alpha = 60°\text{).}$$

$$\cos\alpha \;=\; \frac{a}{c} \;=\; \frac{5 \text{ cm}}{10 \text{ cm}} = 0.5 \text{ (i.e., } \alpha = 60°\text{).}$$

$$\tan\alpha \;=\; \frac{b}{a} \;=\; \frac{8.66 \text{ cm}}{5 \text{ cm}} = 1.732 \text{ (i.e., } \alpha = 60°\text{).}$$

Pythagoras's Theorem   Pythagoras, a Greek mathematician (572–497 B.C.), showed that in a right-angled triangle the square of the hypotenuse is equal to the sum of the squares of the other two sides. Therefore, with respect to figure 1.9

$$c^2 \;=\; a^2 + b^2$$

$$c \;=\; \sqrt{(a^2 + b^2)}.$$

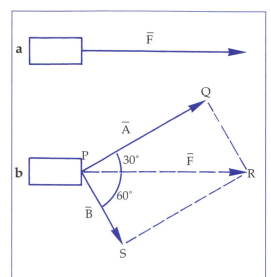

Figure 1.10 Resolution of a vector: replacement of a single force acting on a sled by two forces at right angles to each other.

This can be demonstrated with the data from the above example, where a = 5 cm, b = 8.66 cm, and c = 10 cm: a² = 25, b² = 75, and c² = 100.

Resolution of a Force Figure 1.10 shows a sled being pulled across a frozen pond; assume that the sled slides smoothly over the ice. In figure 1.10*a* the sled is being pulled by a single force $\bar{F}$. In figure 1.10*b* the force $\bar{F}$ has been replaced by two components $\bar{A}$ and $\bar{B}$ at right angles to each other such that $\bar{A}$ makes an angle of 30° with $\bar{F}$ and $\bar{B}$ makes an angle of 60° with $\bar{F}$. The parallelogram PQRS is a rectangle and the triangles PQR and PSR are right-angled triangles. Consequently, if $\bar{F}$ = 100 N, then $\bar{A}$ and $\bar{B}$ can be calculated as follows:

$$\frac{\bar{A}}{\bar{F}} = \cos 30°$$

$$\bar{A} = \bar{F} \times \cos 30°$$

$$\bar{A} = 100 \text{ N} \times 0.866$$

$$\bar{A} = 86.6 \text{ N}.$$

Similarly,

$$\frac{\bar{B}}{\bar{F}} = \cos 60°$$

$$\bar{B} = \bar{F} \times \cos 60°$$

$$\bar{B} = 100 \text{ N} \times 0.5$$

$$\bar{B} = 50 \text{ N}.$$

Angular Kinematics

Figure 1.11 shows a gymnast performing a giant circle on a high bar. In rotating from position *a* to position *d* the gymnast moves through an **angular distance** and an **angular displacement** (angular distance in a specified direction with respect to reference position *a*) of 180°. In moving from *a* through *d* and back to *a* the gymnast moves through an angular distance of 360°, but the angular displacement is zero. **Angular speed** is the rate of change of angular distance:

$$\text{average angular speed } (\bar{\sigma} = \text{"sigma bar"}) = \frac{\text{angular distance } (\phi)}{\text{time taken to cover } \phi \text{ (t)}}$$

$$\bar{\sigma} = \frac{\phi}{t}.$$

For example, if it takes the gymnast 0.6 s to move from position *a* to position *d*, his average angular speed during this period is given by

$$\bar{\sigma} = \frac{180°}{0.6 \text{ s}}$$

$$\bar{\sigma} = 300°/\text{s}.$$

Angular velocity is the rate of change of angular displacement:

$$\text{average angular velocity } (\bar{\omega} = \text{"omega bar"}) = \frac{\text{angular displacement } (\theta)}{\text{time (t)}}.$$

Angular distance: the angle through which an object rotates from one position to another about a particular axis of rotation

Angular displacement: the smaller of the two angles (with respect to a scale of 360°) between the initial and final positions of an object following a period of rotation; the displacement may be positive or negative depending on the relationship of the initial position to the final position

Angular speed: the rate of change of angular distance

Angular velocity: the rate of angular displacement

Angular acceleration is the rate of change of angular velocity:

$$\text{average angular acceleration } (\bar{\alpha}) = \frac{\text{change in angular velocity}}{\text{time}}.$$

Therefore,

$$\bar{\alpha} = \frac{\omega_2 - \omega_1}{t_2 - t_1},$$

where t_1 = time at the start of the period under consideration, t_2 = time at the end of the period under consideration, ω_1 = angular velocity at t_1, and ω_2 = angular velocity at t_2. For example, if the angular velocity of the gymnast changes from $210°/s$ at position e to $450°/s$ at position f in 0.1 s then his average acceleration between e and f is given by

$$\bar{\alpha} = \frac{\omega_2 - \omega_1}{t_2 - t_1},$$

where $\omega_1 = 210°/s$, $\omega_2 = 450°/s$, and $t_2 - t_1 = 0.1$ s. That is

$$\bar{\alpha} = \frac{450°/s - 210°/s}{0.1 \text{ s}}$$

$$\bar{\alpha} = \frac{240°/s}{0.1 \text{ s}},$$

$$\bar{\alpha} = 2400°/s/s \text{ or } 2400°/s^2.$$

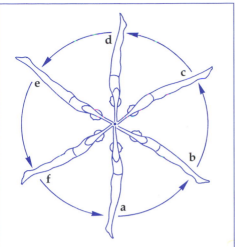

Figure 1.11   A gymnast performing a giant circle on the high bar.

Kinetics

While linear kinetics is the study of the causes and changes of linear motion, angular kinetics studies the causes and changes of angular motion.

Linear Kinetics

Linear kinetics is concerned with the relationship between change in force and change in linear kinematics. This relationship is summarized in Newton's three laws of motion as they apply to linear motion. Before considering the laws of motion it is necessary to introduce the concepts of inertia, mass, and linear momentum.

The **mass** of an object is the quantity of matter comprising the object. Mass is measured in units such as pounds (lb) and kilograms (kg). An object's mass depends on its density—the concentration of mass per unit volume of the object—and its volume. If two objects have the same density, the one with the greater volume will have the greater mass.

An object that is stationary—or "at rest"—such as a book lying on a table, exhibits a certain reluctance to move; that is, a certain amount of effort is needed to move the book, and the greater the mass of the book the greater the effort needed to move it. The reluctance of a stationary object to move is referred to as its **inertia** and the mass of the object is a measure of its inertia.

Just as a stationary object exhibits a reluctance to start moving, a moving object exhibits a reluctance to change the way it is moving. In this case the inertia of the object depends not only on its mass, but also on its linear velocity. The product of an object's mass and its linear velocity is referred to as the **linear momentum** of the object:

$$\text{linear momentum (M)} = \text{mass (m)} \times \text{linear velocity (v)}.$$

Angular acceleration: the rate of change of angular velocity

Mass: the quantity of matter comprising an object

Inertia: the reluctance of an object to change its state of rest or the way it is moving

Linear momentum: the product of the mass of an object and its linear velocity

That is

$$M = mv.$$

For example, the linear momentum of a football player of mass 85 kg and linear velocity 8 m/s is given by

$$M = 85 \text{ kg} \times 8 \text{ m/s}$$

$$M = 680 \text{ kg m/s}.$$

Newton's First Law of Motion

To move an object from rest or to change the way it is moving, it is necessary to change the pattern of forces acting on the object. This phenomenon is the basis of **Newton's first law of motion**: an object remains at rest or continues to move with constant velocity in a straight line unless compelled to move or change the way it is moving by a change in the pattern of forces acting upon it. For example, a book resting on a table will remain at rest until someone moves it. A passenger traveling in a bus moves at the same velocity as the bus. If the bus suddenly brakes the passenger will be thrown forward, especially a standing passenger, since he will tend to move forward with the velocity that the bus had immediately before braking. Passengers in vehicles often wear seat belts to prevent getting hurt due to sudden changes in velocity.

Newton's Second Law of Motion

Any change in the velocity of an object results in a change in the linear momentum of the object. **Newton's second law of motion** describes the relationship between the change in linear momentum experienced by an object and the force responsible for the change—when a force acts on an object the change in linear momentum experienced by the object takes place in the direction of the force and is proportional to the magnitude and duration of the force. This statement can be expressed algebraically as

(1)
$$Ft = mv_2 - mv_1,$$

where F = magnitude of the force, t = duration of the force, m = mass of object, v_1 = velocity of object immediately prior to application of the force, and v_2 = velocity of object immediately after removal of the force. The product of force and time, Ft, is referred to as the **impulse of the force**. From *(1)*, it follows that

(2)
$$F = \frac{m(v_2 - v_1)}{t}.$$

Since $\frac{v_2 - v_1}{t}$ = a = acceleration, it follows that

(3)
$$F = ma$$

From *(3)*, it follows that when the resultant force acting on an object is zero the acceleration is also zero, that is, the object is either at rest or moving with constant velocity. It also follows from *(3)* that

(4)
$$a = \frac{F}{m}.$$

Newton's second law is often expressed in terms of *(4)*: when a force acts on an object the acceleration experienced by the object is inversely proportional to the mass of the object, directly proportional to the magnitude of the force, and takes place in the direction of the force. Consequently, for a constant mass, the larger the force the higher the acceleration, and for a constant force, the larger the mass the lower the acceleration.

Newton's first law of motion: an object remains at rest or continues to move with constant velocity in a straight line unless compelled to move or change the way it is moving by a change in the pattern of forces acting upon it

Newton's second law of motion: when a force acts on an object the change in linear momentum experienced by the object takes place in the direction of the force and is proportional to the magnitude of the force and the duration of the force

Impulse of a force: the product of the magnitude of a force and the duration of the force

Newton's Third Law of Motion

Objects in contact exert equal and opposite forces on each other. This is **Newton's third law of motion**: when one object exerts a force on another there is an equal and opposite force exerted by the second object on the first. Alternatively, for every action there is an equal and opposite reaction.

Contact Forces and Forces of Attraction

There are two types of forces: contact forces and forces of attraction. **Contact forces** result from physical contact between different objects. Forces of attraction (or **attraction forces**) tend to make objects move toward each other whether or not the objects are in contact. A good example of a force of attraction is magnetism in iron. The human body is constantly subjected (unless in orbit around the earth) to one very considerable attraction force—**body weight**. To understand body weight it is necessary to consider **Newton's law of gravitation**.

> *Contact forces result from physical contact between objects, and forces of attraction tend to make objects move toward each other whether or not the objects are in contact.*

Newton's Law of Gravitation

There is a well-known story about Sir Isaac Newton, who, it is said, was sitting under an apple tree one day when he saw an apple fall to the ground. This observation led him to formulate his law of gravitation (law of attraction): every object attracts every other object with a force that is directly proportional to the product of the masses of the two objects and inversely proportional to the square of the distance between the two objects. In algebraic terms the law can be represented as

$$(5) \qquad F = \frac{Gm_1 m_2}{d^2},$$

where F = force of attraction, G = constant of gravitation, m_1 and m_2 = the masses of the two objects, and d = the distance between the centers of mass of the two objects. It is difficult to appreciate that a force of attraction exists between any two objects, but the magnitude of such forces is usually extremely small and of no practical consequence. However, there is one object, the earth, that exerts a significant force of attraction on every other object. This force of attraction is referred to as the force of gravity or body weight. The earth has a huge mass; though there is a large distance between its center and an object on its surface, the force of gravity is strong enough to keep objects in contact with or bring them back to the surface of the earth. For example, a ball thrown up in the air quickly falls.

From equation *(5)*, it follows that the weight W of an object of mass m is given by

$$(6) \qquad W = \frac{GmM}{d^2},$$

where M = the mass of the earth and d = the distance between the center of the earth and the object on its surface. It follows from (6) that

$$(7) \qquad W = mg,$$

where

$$g = \frac{GM}{d^2}.$$

From Newton's second law of motion (F = ma), it follows that g represents acceleration (due to gravity). The distance d varies slightly at different points on the earth's surface. Consequently g also varies slightly, with an average value of 9.81 m/s². From *(7)*, it follows that the weight of an object is the product of its mass and its acceleration due to gravity.

Newton's third law of motion: when one object exerts a force on another there is an equal and opposite force exerted by the second object on the first

Contact force: a force exerted by one object on another due to physical contact

Attraction force: a force that tends to make objects move toward each other whether or not the objects are in contact

Body Weight: the force of attraction the earth exerts on an object

Newton's law of gravitation: every object attracts every other object with a force directly proportional to the product of the masses of the two objects and inversely proportional to the square of the distance between the two objects

Center of Gravity   The human body is composed of a number of segments linked by joints. Each segment contributes to the body's total weight (figure 1.12*a*). Movement of the body segments relative to each other alters the weight distribution of the body. However, in any particular body posture the body behaves (in terms of the effect of body weight on the movement of the body) as if the total weight of the body is concentrated at a single point called the **center of gravity** (figure 1.12*b*). The concept of a center of gravity applies to all objects, animate and inanimate.

The position of an object's center of gravity depends on the distribution of the weight of the object. For a regular-shaped object such as a cube or oblong the center of gravity is located at the object's geometric center (figure 1.13). For an irregular-shaped object the center of gravity may be inside or outside the object. For example, consider a triangle-shaped card with sides 15 cm, 20 cm, and 25 cm in length, with vertices A, B, and C. Figure 1.14*a* shows the card suspended from a freely moveable pin joint close to vertex A. The line of action of the card's weight, that is, the direction of the weight of the card through its center of gravity, coincides with the vertical line through the point of suspension. By suspending a plumb line in front of the card from the same pin joint, the line of action of the card's weight can be determined. Figure 1.14, *b* and *c*, shows this process repeated from points of suspension close to vertices B and C, respectively. In figure 1.14*d* the lines of action of the card's weight in positions 2 and 3 are shown superimposed on the line of action in position 1. The lines of action of the weight of the card in the three positions intersect at a single point—the card's center of gravity. In this example, the card's center of gravity lies inside the body of the card.

Figure 1.15 shows the same process carried out with an L-shaped card with arms of length 15 cm and 20 cm and of width 4 cm. In this case the card's center of gravity is found to lie outside the body of the card (figure 1.15*d*).

The human body is an irregular shape. In upright standing an adult male's center of gravity is located inside his body close to the level of his navel, that is, at about 54% of his stature when measured from the floor, and midway between the front and back of his body (figure 1.16*a*) (Watkins 1983). With respect to upright standing, movements of the arms, which comprise about 11% of body weight, result in a fairly slight change in the position of the center of gravity. However, movement of the trunk, which comprises approximately 50% of body weight, likely results in a relatively large change in the center of gravity; for example, full flexion of the trunk can result in the center of gravity being located outside the body (figure 1.16*b*). This position is similar to the position adopted by a pole vaulter when clearing the bar (figure 1.16*c*).

The body's center of gravity may also be located outside the body during postures involving full extension of the trunk, as in clearing the bar using the Fosbury flop technique in high jumping (figure 1.16*d*). Movements involving continuous change in the orientation of body segments to each other, such as walking and running, result in continuous change in the body's center of gravity.

Movements that involve continuous change in the orientation of the body segments to each other result in continuous change in the body's center of gravity.

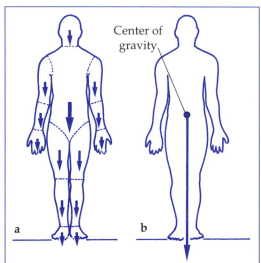

Figure 1.12   Center of gravity; frontal view of the human body (*a*) divided into 14 segments; (*b*) showing the position of the center of gravity and the line of action of body weight.

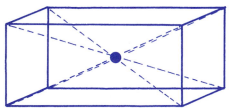

Figure 1.13   Location of the center of gravity of a regular-shaped object.

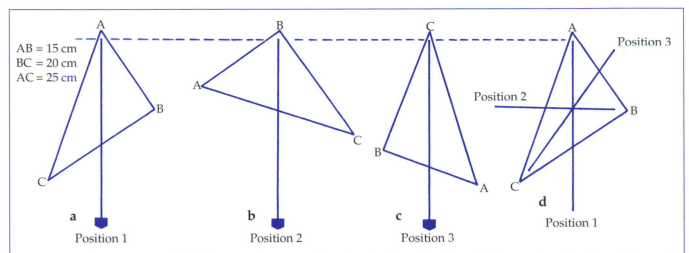

AB = 15 cm
BC = 20 cm
AC = 25 cm

a Position 1

b Position 2

c Position 3

d Position 1, Position 2, Position 3

Figure 1.14   Suspension method of locating the center of gravity of an irregular-shaped object: a triangular-shaped piece of card.

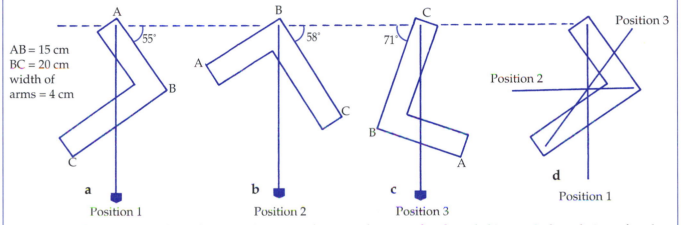

AB = 15 cm
BC = 20 cm
width of
arms = 4 cm

55°   58°   71°

a Position 1

b Position 2

c Position 3

d Position 1, Position 2, Position 3

Figure 1.15   Suspension method of locating the center of gravity of an irregular-shaped object: an L-shaped piece of card.

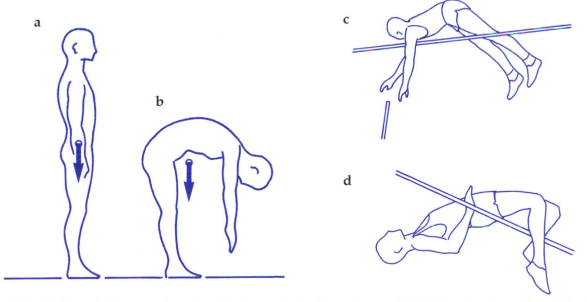

a

b

c

d

Figure 1.16   Position of the center of gravity of the human body; (*a*)standing upright; (*b*) bending forward; and clearing the bar in the (*c*) pole vault and (*d*) high jump (Fosbury flop).

W = body weight

R = ground reaction force

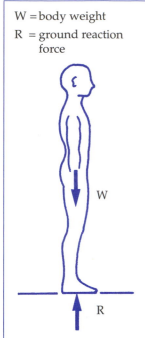

Figure 1.17 Free body diagram of the forces acting on a man standing upright.

Free body diagram: a diagram showing all of the forces acting on an object

Newton: the unit of force in the meter-kilogram-second (m-kg-s) system of units—the force acting on a mass of 1 kg that gives it an acceleration of 1 m/s²

English system: the foot-pound-second (ft-lb-s) and foot-slug-second (ft-sl-s) systems of units of mechanical variables

Metric system: the centimeter-gram-second (cm-g-s) and meter-kilogram-second (m-kg-s) systems of units of mechanical variables

Free Body Diagram A diagram showing all of the forces acting on an object is referred to as a **free body diagram**. If the magnitude and direction of the various forces are known, the forces may be shown on the diagram as vectors. However, if the magnitude and directions of some (or all) of the forces are not known, then it is usual to indicate the points of application and the assumed direction of the forces. The forces represented are a combination of attraction forces such as weight and contact forces such as pushes, pulls, support forces (including buoyancy effects), and wind resistance. For example, figure 1.17 shows a free body diagram of a person standing upright. The only forces acting on him (in the absence of any wind resistance) are his body weight W and the support force (ground reaction force) R exerted by the floor. Since he is stationary the resultant force acting on him must be zero, that is, R is equal and opposite to W.

Units of Force In the metric system the unit of force is the **newton (N)**. A newton is defined, in accordance with Newton's second law of motion, as the force acting on a mass of 1 kg that gives it an acceleration of 1 m/s², that is,

$$1 \text{ N} = 1 \text{ kg} \times 1 \text{ m/s}^2.$$

Since the acceleration due to gravity is 9.81 m/s², it follows that the weight of a mass of 1 kg, which may be referred to as 1 kg wt (kilogram weight), is given by

$$1 \text{ kg wt} = 1 \text{ kg} \times 9.81 \text{ m/s}^2 = 9.81 \text{ N}.$$

A mass of 1 kg is equivalent to 2.2046 lb. That is, 1 kg wt is equivalent to 2.046 lb wt (pound weight). The lb wt and kg wt are gravitational units of force and most weighing machines in everyday use, for example, kitchen scales and bathroom scales are graduated in lb wt and kg wt. Body weight is often recorded in lb or kg in health records and elsewhere. Although this makes no practical difference, the correct units are the lb wt and the kg wt.

Most kitchen scales and bathroom scales are graduated in lb wt and kg wt—gravitational units of force.

Systems of Units

There are two systems of units for mechanical variables, namely, the **English system** and the **metric system** (table 1.1). The International System of Units is an expanded version of the metric system, and is also referred to as the Système internationale d' unites (SI system). Both systems are based on variables of length, mass, and time, and each system has two most commonly used versions. In both versions of the English system the base units of length and time are the foot (ft) and second (s), respectively. However, in one version the base unit of mass is the pound (lb) and in the other version the base unit of mass is the slug (sl), where 1 sl = 32.2 lb. The two versions of the English system are usually referred to as the foot-pound-second (ft-lb-s) system and the foot-slug-second (ft-sl-s) system.

In both versions of the metric system the base unit of time is the second (s). However, in one version the base units of length and mass are the centimeter (cm) and the gram (g), and in the other version the base units of length and mass are the meter (m) and the kilogram (kg); 1 kg = 1000 g and 1 m = 100 cm. The two versions of the metric system are usually referred to as the centimeter-gram-second (cm-g-s) system and the meter-kilogram-second (m-kg-s) system. The English system was established before the metric system, but the metric system is now more widely used. In this book the m-kg-s version of the metric system is preferred.

Table 1.1 Systems of Mechanical Units

	English system		Metric system	
	foot-pound-second version	**foot-slug-second version**	**centimeter-gram-second version**	**meter-kilogram-second version**
Distance	foot (ft)	foot (ft)	centimeter (cm)	meter (m)
Speed	foot per second (ft/s)	foot per second (ft/s)	centimeter per second (cm/s)	meter per second (m/s)
Acceleration	foot per second per second (ft/s^2)	foot per second per second (ft/s^2)	centimeter per second per second (cm/s^2)	meter per second per second (m/s^2)
Mass	pound (lb)	slug (sl)	gram (g)	kilogram (kg)
Linear momentum	pound foot per second (lb ft/s)	slug foot per second (sl ft/s)	gram centimeter per second (g cm/s)	kilogram meter per second (kg m/s)
Force	poundal (pdl) 1pdl=1lbx1ft/s^2	pound force (lb f) 1lb f=1slx1ft/s^2	dyne (dyn) 1dyn=1gx1cm/s^2	newton (N) 1N=1kgx1m/s^2
Weight	pound weight (lb wt) 1lb wt=32.2 pdl	slug weight (sl wt) 1sl wt=32.2lb f	gram weight (g wt) 1g wt=981dyn	kilogram weight 1kg wt=9.81N
Angular distance	radian (rad)	radian (rad)	radian (rad)	radian (rad)
Angular speed	radian per second (rad/s)	radian per second (rad/s)	radian per second (rad/s)	radian per second (rad/s)
Angular acceleration	radian per second per second (rad/s^2)	radian per second per second (rad/s^2)	radian per second per second (rad/s^2)	radian per second per second (rad/s^2)
Moment of inertia	pound foot squared (lb ft^2)	slug foot squared (sl ft^2)	gram centimeter squared (g cm^2)	kilogram meter squared (kg m^2)
Angular momentum	pound foot squared per second (lb ft^2/s)	slug foot squared per second (sl ft^2/s)	gram centimeter squared per second (g cm^2/s)	kilogram meter squared per second (kg m^2/s)
Turning moment	poundal foot (pdl ft)	pound force foot (lb f ft)	dyne centimeter (dyn cm)	newton meter (N m)

Load, Strain, and Stress

A **load** is any force or combination of forces applied to an object. There are three types of loads: **tension**, **compression**, and **shear** (figure 1.18). Loads tend to deform the objects on which they act. Tension is a pulling (stretching) load that tends to make an object longer and thinner along the line of the force (figure 1.18, *a* and *b*). Compression is a pushing or pressing load that tends to make an object shorter and thicker along the line of the force (figure 1.18, *a* and *c*). A shear load is comprised of two equal (in magnitude), opposite (in direction), parallel forces that tend to displace one part of an object with respect to an adjacent part along a plane parallel to and between the lines of force (figure 1.18, *a* and *d*). The cutting load produced by scissors and garden shears is a shear load, while the cutting load produced by a knife is a compression load. The three types of loads frequently occur in combination with each other, especially in the forms of **bending** and **torsion** (figure 1.18, *a*, *e*, and *f*). Bending involves both tension and compression. An object subjected to bending experiences tension on one side and compression on the other. Torsion involves all three types of load. An object subjected to torsion simultaneously experiences tension, compression, and shear.

In mechanical terms, the deformation of an object that occurs in response to a load is referred to as **strain**. In a tug-of-war, the rope is subjected to a tension load and as a result experiences tension strain. Similarly, an object subjected to a compression load experiences compression strain and an object subjected to a shear load experiences shear strain. Strain denotes deformation of the intermolecular bonds making up the structure of an object. When an object experiences strain, the intermolecular bonds exert forces that tend to restore the original (unloaded) size and shape of the object. The forces exerted by the intermolecular bonds of an object under strain are referred to as **stress**. Stress is the resistance of the intermolecular bonds to the strain caused by the load.

The stress on an object resulting from a particular load is distributed throughout the whole of the material sustaining the load. However, the level of stress in different regions of the material varies depending upon the amount of material sustaining the load in the different regions; the more material sustaining the load, the lower the stress. Consequently, stress is measured in terms of the average load on the plane of material sustaining the load at the point of interest.

An object subjected to a load experiences strain; stress is the resistance of the intermolecular bonds of the object to the strain.

Load: a force or combination of forces applied to an object

Tension: a pulling load that tends to make an object longer and thinner along the line of the force

Compression: a pushing or pressing load that tends to make an object shorter and thicker along the line of the force

Shear: a load comprised of two equal, opposite, parallel forces that tends to displace one part of an object with respect to an adjacent part along a plane parallel to and between the lines of force

Bending: a load involving both tension and compression; an object subjected to bending will experience tension on one side and compression on the other

Torsion: a load involving tension, compression and shear; an object subjected to torsion will experience tension, compression, and shear simultaneously

Strain: the amount of deformation experienced by an object in response to a load

Stress: the resistance of the intermolecular bonds of an object to the strain caused by a load

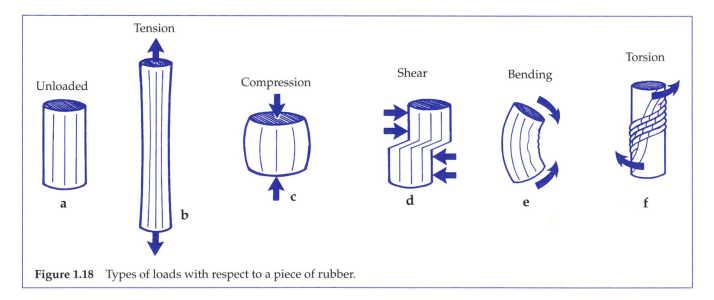

Figure 1.18   Types of loads with respect to a piece of rubber.

Tension Stress   Figure 1.19 shows a section of rope from a tug-of-war. If the tension load on the rope is 5000 N and the cross-sectional area of the rope at the point P is 6 cm² (square centimeters) then the tension stress on the rope at P is given by

$$\text{Tension stress at P} = \frac{5000 \text{ N}}{6 \text{ cm}^2}$$

$$= 833 \text{ N/cm}^2.$$

Compression Stress   When standing upright on both feet the weight of the body is transmitted to the floor by the soles of the feet (figure 1.20*a*). In accordance with Newton's third law of motion, the supporting area of the feet experience a compression load, the ground reaction force, equal and opposite to the weight of the body. This compression load is distributed over the whole of the supporting area of the feet, which in an adult male is approximately 200 cm² and in an adult female about 150 cm². For a man weighing 686 N (70 kg wt), the compression stress on the supporting area of his feet (on a level floor, supporting area perpendicular to the compression load) is given by

$$\text{Compression stress} = \frac{686 \text{ N}}{200 \text{ cm}^2}$$

$$= 3.43 \text{ N/cm}^2.$$

By raising the heels off the floor the size of the supporting area is approximately halved, as shown in figure 1.20*b*. Since the compression load (body weight) is the same as before, it follows that the compression stress on the reduced supporting area is approximately doubled.

Shear Stress   In a bookcase the weight of each shelf and any books on the shelf is supported by the two ends of the shelf. Figure 1.21*a* represents a section from a bookcase, showing part of one of the sides with a shelf jointed into it. The downward force L exerted on the joint together with the reaction force R exert a shear load on the plane of the joint parallel to the shear load (figure 1.21, *b* and *c*). If the shear load is 150 N and the area of the plane of the joint

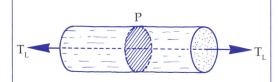

T_L = tension load on rope

A_P = cross-sectional area of the rope at P

$\dfrac{T_L}{A_P}$ = tension stress at P

Figure 1.19   Tension stress on a rope.

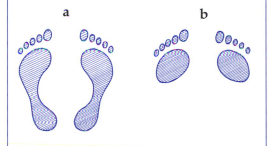

Figure 1.20   Supporting area of the feet; (*a*) normal upright standing posture; (*b*) standing with heels raised off the floor.

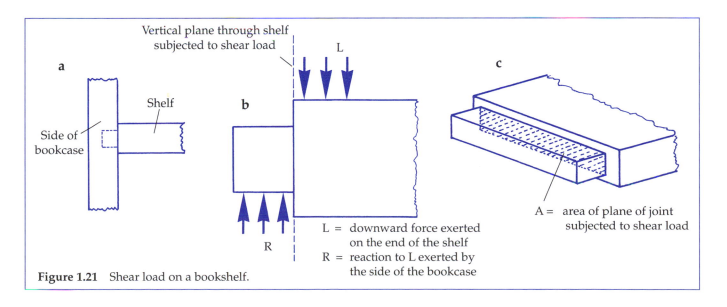

Figure 1.21   Shear load on a bookshelf.

sustaining the shear load is 30 cm², then the shear stress on the joint is given by the following:

$$\text{Shear stress} = \frac{150 \text{ N}}{30 \text{ cm}^2}$$

$$= 5 \text{ N/cm}^2.$$

Work, Energy, and Power

Work: the product of a force and the distance moved by the point of application of the force in the direction of the force

Newton-meter (Nm) or joule (J): the work done by a force of 1 newton (N) when it moves its point of application a distance of 1 meter (m) in the direction of the force

In mechanics a force does **work** when it moves its point of application in the direction of the force and the amount of work done by a force f is the product of the force and the distance d moved by the point of application of the force. For example, in lifting a box from the floor as shown in figure 1.22*a* the work done by the man on the box is given by

$$\text{work} = \text{force (f)} \times \text{distance (d)},$$

where f = weight of box and d = vertical displacement of the box. In reality the upward force acting on the box must be greater than the weight of the box, otherwise it would not move, that is, there must be a resultant upward force acting on the box. However, the resultant upward force need only be just greater than zero and, as such, when speed of movement is not a consideration, it is reasonable to consider the work done in lifting the box to be equal to f × d.

If f = 20 kg wt and d = 0.65 m, then the work done by the man on the box is given by

$$\text{work} = \text{f} \times \text{d}$$

$$= 20 \text{ kg wt} \times 0.65 \text{ m}$$
(and since 1 kg wt = 9.81 N)

$$= 196.2 \text{ N} \times 0.65 \text{ m}$$

$$= 127.5 \text{ Nm.}$$

One **newton-meter** (Nm) is the work done by a force of one newton when it moves its point of application a distance of one meter in the direction of the force. The newton-meter is usually referred to as the **joule** (J), that is, 1 J of work is the same as 1 Nm of work.

Similarly, when a man moves from a sitting to a standing position, as shown in figure 1.22*b*, the work done by his muscles is equivalent to the amount of work done in displacing his center of gravity (CG) vertically upward through the distance D. If the weight of the man is 70 kg wt and the vertical displacement of his CG is 0.43 m then the work done by his muscles is given by

$$\text{work} = \text{force} \times \text{distance}$$

$$= 70 \text{ kg wt} \times 0.43 \text{ m}$$

$$= 686.7 \text{ N} \times 0.43 \text{ m}$$

$$= 295.3 \text{ Nm}$$

$$= 295.3 \text{ J.}$$

In figure 1.22*a* the work done by the muscles in moving the body from a squat to a standing position while lifting the box is equivalent to the amount of work done in displacing the combined CG of the person and box vertically upward from its location in the squat position to its location in the standing position. In both of the above

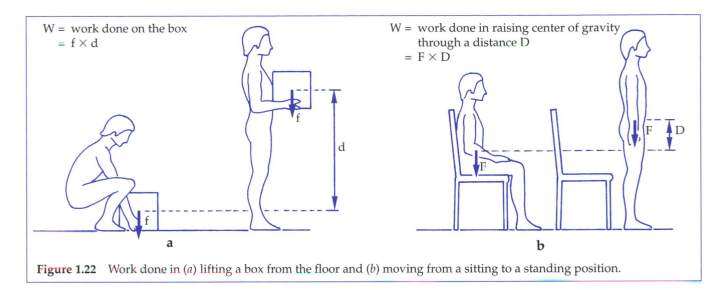

Figure 1.22   Work done in (*a*) lifting a box from the floor and (*b*) moving from a sitting to a standing position.

examples the force responsible for moving the CG of the person and box, as in figure 1.22*a*, and the person, as in figure 1.22*b*, is produced by the muscles, that is, the muscles do the work. In doing the work the muscles expend **energy**, that is, chemical energy in the muscles is converted to mechanical energy in the form of work. A body (or object) is said to have energy if it can do work. Apart from when the body is in a recumbent position the muscles of the body are constantly expending energy to counteract the effects of gravity—maintain an upright (e.g., standing) or partly upright (e.g., sitting) posture—and bring about purposeful movements.

> *In mechanics a force does work when it moves its point of application in the direction of the force. A body (or object) is said to have energy if it can do work.*

Conservation of Energy   In drawing a bow, as in figure 1.23, an archer deforms (strains) the bow; some of the energy expended by the archer is stored in the bow. When the bowstring is released the bow recoils and the stored energy propels the arrow toward the target. The energy stored in the bow is referred to as **strain energy**. Many materials store strain energy in response to loading, for example, a stretched elastic band. Strain energy is a form of **potential energy**—stored energy that, given appropriate conditions, may be used to do work. With regard to the arrow, it follows from Newton's second law of motion that

(8) $$ft = mv - mu,$$

where f = average force exerted on the arrow by the bowstring from the instant of release to the instant that the arrow separates from the bowstring, t = duration of the impulse of f, m = mass of the arrow, u = velocity of the arrow at release = 0, and v = velocity of the arrow at the instant it separates from the bowstring. Since u = 0, it follows from (8) that

(9) $$ft = mv.$$

The impulse of the force exerting on the arrow is given by ft. However, the work done by f on the arrow is given by fd, where d is the distance over which f acts (figure 1.23). Consequently, the average velocity of the arrow during the impulse of f is given by

(10) $$\text{average velocity of arrow} = \frac{d}{t}.$$

Energy: the capacity to do work

Strain energy: the energy stored in an object as a result of being strained

Potential energy: stored energy

The average velocity of an object during the impulse of a constant force is given by

(11)   average velocity of object $= \dfrac{v_1 + v_2}{2}$,

where v_1 and v_2 are the initial and final velocity of the object. Since the force f acting on the arrow is an average force and, as such, can be considered a constant force, and the initial velocity of the arrow is zero, it follows from equation (11) that

(12)   average velocity of arrow $= \dfrac{v}{2}$.

From equations (10) and (12) it follows that

$$\dfrac{d}{t} = \dfrac{v}{2}$$

(13) $$t = \dfrac{2d}{v}.$$

By substituting t from equation (13) into equation (9), it follows that

$$f \times \dfrac{2d}{v} = mv$$

(14) $$fd = \dfrac{mv^2}{2}.$$

Kinetic energy: the energy an object possesses by virtue of its movement

Principle of conservation of energy: energy cannot be created or destroyed, it can only be converted from one form of energy to another

Impulse-momentum relationship: the relationship between the impulse of a force and the resulting change in linear momentum

Work-energy relationship: the relationship between the work done by a force and the resulting change in kinetic energy

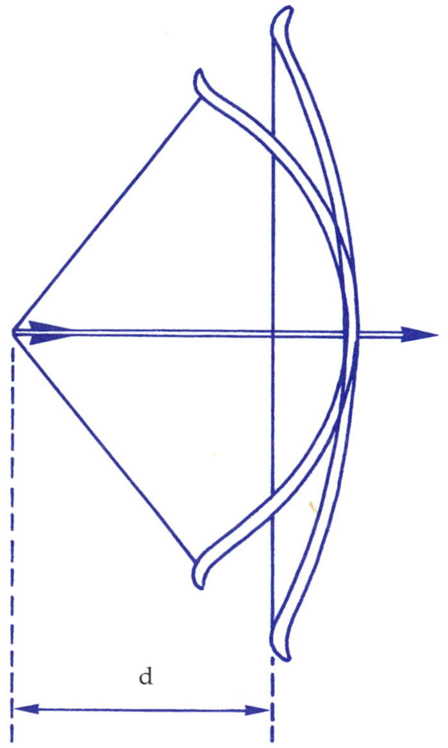

Figure 1.23 Storage of strain energy in a bow; d = the distance over which the arrow is pulled backward in drawing the bow.

The term $mv^2/2$ is referred to as the **kinetic energy** of the arrow—the energy of the arrow by virtue of its movement. Clearly, the work done (fd) on the arrow by the recoil of the bow is equivalent to the kinetic energy of the arrow; that is, the strain energy stored in the bow has been transformed into kinetic (movement) energy in the arrow. Furthermore, the total amount of energy in the system (i.e., the bow and arrow) before and after recoil of the bow is the same. This illustrates the **principle of conservation of energy**—energy cannot be created or destroyed, it can only be converted from one form of energy to another (Tricker and Tricker 1967).

There are many different forms of energy, including sound, light, electricity, magnetism, kinetic energy, heat, and various forms of potential energy. In the human body the energy contained in foodstuffs is converted into various forms of chemical energy that can be used to carry out the various life processes including, for example, the production of heat to maintain an appropriate body temperature. The muscles convert chemical energy into kinetic energy contained in moving body segments and this kinetic energy may in turn be converted to strain energy in musculoskeletal components.

In mechanics, the relationship between force, time, and linear momentum—the relationship described by Newton's second law of motion—is sometimes referred to as the **impulse-momentum relationship** (equation 8). Similarly, the relationship between force, distance, and kinetic energy is sometimes referred to as the **work-energy relationship** (equation 14).

Many materials store strain energy in response to loading. Strain energy is a form of potential energy.

Gravitational Potential Energy One of the most common forms of potential energy is **gravitational potential energy**—the energy possessed by an object by virtue of its height (vertical displacement of CG) above any particular reference position, usually ground level. If an object is held above ground level and then released it will fall to the ground under its own weight. The work done on an object by the force of its own weight mg when it falls a distance h is given by mgh. Since mg is a constant force it follows (from equation 14) that

$$mgh = \frac{mv^2}{2},$$

where v and $mv^2/2$ are the velocity and kinetic energy of the object, respectively, after falling the distance h.

It follows that when an object of weight mg is held a distance h above ground level it possesses potential energy equivalent to mgh, which may be transformed into kinetic energy if it is allowed to fall. This form of potential energy is called gravitational potential energy because it is due to the effect of gravity; the greater the height of the CG of an object above the ground (or some other horizontal reference position), the greater its level of gravitational potential energy. Furthermore, when an object is allowed to fall freely, the decrease in gravitational potential energy it experiences (as the height of its CG above the ground decreases) is matched by an equivalent increase in kinetic energy (as its velocity increases).

Power Figure 1.24 shows a weightlifter lifting a barbell from the floor to shoulder height. If the barbell is at rest at the start and end of the lift the work done on the bar by the lifter during the lift is equivalent to the increase in gravitational potential energy fd of the bar, where f is the weight of the bar and d is the upward displacement of the bar. If the duration of the lift was t seconds then the rate at which work was done on the bar by the lifter during the lift is given by fd/t. Work rate is referred to as **power**. For example, if the barbell weighs 200 kg wt and it is lifted 1.2 m in 1.5 s, then the average power output of the lifter (on the bar) during the lift is given by

$$\text{power output} = \frac{fd}{t}$$
$$= \frac{200 \text{ kg wt} \times 1.2 \text{ m}}{1.5 \text{ s}}$$
$$= \frac{1962 \text{ N} \times 1.2 \text{ m}}{1.5 \text{ s}}$$
$$= 1570 \text{ J/s}.$$

The joule per second (J/s) is usually referred to as the **watt** (W).

Transformation of relatively small amounts of energy can have a considerable effect when the transformation involves sufficient power. Frost (1967) illustrates this well with respect to light energy. Exposure of a piece of steel to an ordinary beam of light for 10 s has no visible effect on the steel. However, if the same amount of light energy is discharged in one picosecond (one millionth of one millionth of one second) it will burn a hole in the steel; this is the basis of laser technology.

Angular Kinetics

Angular kinetics is the study of the causes of and changes in angular motion. Whereas a small change in the magnitude of force acting on an object tends to result in a small change in linear motion, a small

Gravitational potential energy: the energy possessed by an object by virtue of its height above any particular reference position, usually ground level

Power: work rate

Watt: work rate of 1 joule per second (J/s)

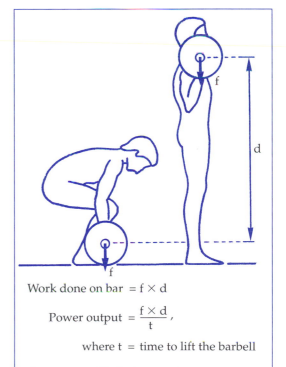

Work done on bar $= f \times d$

Power output $= \frac{f \times d}{t}$,

where t = time to lift the barbell

Figure 1.24 Work done and power output in lifting a barbell.

change in the direction of force acting on an object may result in a large change in angular motion.

Stability

Figure 1.25*a* shows a regular cube-shaped block of wood resting on a horizontal surface. The center of gravity of the block of wood is located at its geometric center and the line of action of the weight of the wood intersects the base of support ABCD. If the block of wood is tilted over on any of the edges of the base of support, AB, BC, CD, or AD, it returns to its original position provided that, at release, the line of action of its weight intersects the plane of the original base of support ABCD. This situation is shown in figure 1.25*b* with respect to the edge AB. However, if, at release, the line of action of its weight does not intersect the original base of support, the block of wood falls onto one of its other faces as shown in figure 1.25, *c* and *d*. With respect to a particular base of support, an object is said to be stable when the line of action of its weight intersects the plane of the base of support, and to be unstable when it does not. Consequently, the block of wood in figure 1.25 is stable with respect to the base of support ABCD in the positions shown in figure 1.25, *a* and *b*, and unstable with respect to the base of support ABCD in the position shown in figure 1.25*c*.

> *With respect to a particular base of support an object is said to be stable when the line of action of its weight intersects the plane of the base of support, and unstable when it does not.*

Stability: the state of equilibrium to which an object returns after being disturbed

An object's degree of **stability**—risk of becoming unstable relative to its normal base of support—depends on the dimensions of the base of support in relation to the height of the object's center of gravity above its normal base of support. Figure 1.26 shows two blocks of wood, 10 cm and 20 cm long, respectively, with the same square cross section of 4 cm × 4 cm. When the two pieces of wood rest on any of their larger

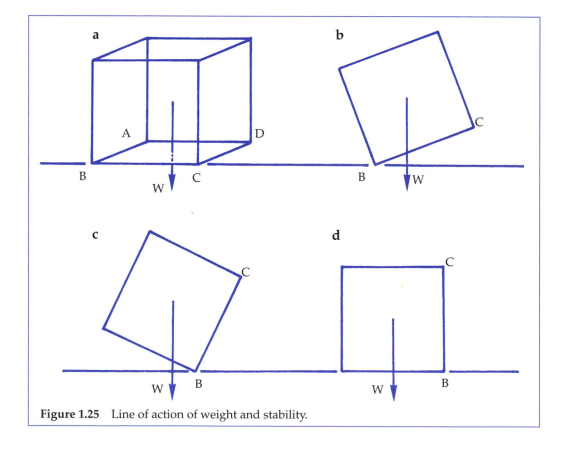

Figure 1.25   Line of action of weight and stability.

faces the heights of their centers of gravity above the base of support are the same, that is, 2 cm. With respect to the X axis (CD in the longer block of wood and RS in the shorter block of wood) the stability of the longer block of wood is greater than that of the shorter block of wood since the longer one would need to be tilted through a larger angle (79°) than the shorter one (68°) before it became unstable (figure 1.26, c and d). However, with respect to the Y axis (AD in the longer block of wood and PS in the shorter block of wood) the stability of the two blocks is the same; each block would need to be tilted through an angle of 45° before it became unstable (figure 1.26e).

Clearly, for a given height of center of gravity, the broader the base of support with respect to a particular axis of tilt, the greater the object's stability. In figure 1.27a the longer block of wood is shown standing on one of its ends, and in figure 1.27c the shorter block is shown standing on one of its ends. In this situation the dimensions of

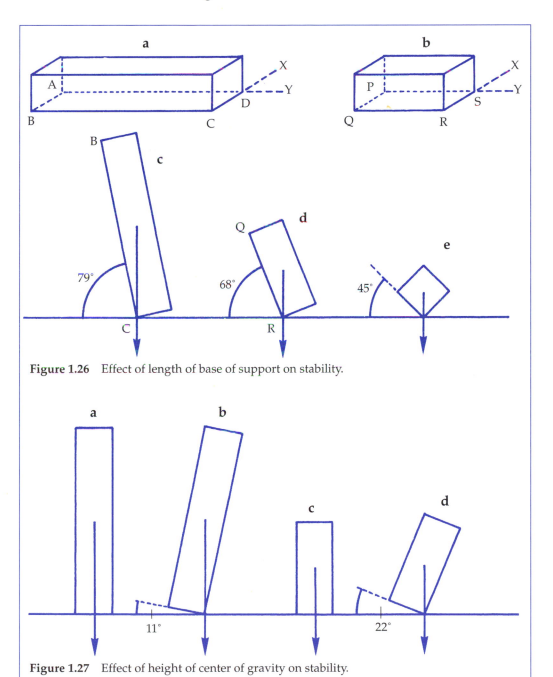

Figure 1.26 Effect of length of base of support on stability.

Figure 1.27 Effect of height of center of gravity on stability.

each block's base of support are the same, but the center of gravity of the taller block is 10 cm above its base, and that of the shorter block is 5 cm above its base. When both blocks are tilted on one edge, the taller becomes unstable after being tilted through a much smaller angle (11°) than the shorter (22°). Consequently, in this situation the taller block is less stable than the shorter block.

It follows from the above experiments that for any particular object the lower the ratio of the height of the center of gravity to the length of the base of support (with respect to each possible tilt axis), the greater the stability of the object. This principle is used, for example, in the design of vehicles in order to minimize their risk of overturning during normal use.

With regard to human movement, the terms stability and balance are often used synonymously. Maintaining stability of the human body is a fairly complex, albeit largely unconscious, process (Roberts 1995). In upright standing the line of action of body weight intersects the base of support formed by the area beneath and between the feet (figure 1.28, *a* and *b*). Moving the feet farther apart can increase the size of the base of support. For example, moving one foot in front of the other increases anteroposterior stability, and moving one foot laterally increases side-to-side stability (figure 1.28, *c* and *d*). Combining these movements with a degree of flexion of the hips, knees, and ankles, as in certain movements in wrestling and boxing, reduces the height of the center of gravity and, thereby, further increases stability.

Movement of the body from one base of support to another, such as in moving from standing to sitting, illustrates the unconscious way in which the balance systems of the body automatically redistribute body weight to maintain stability. Figure 1.29

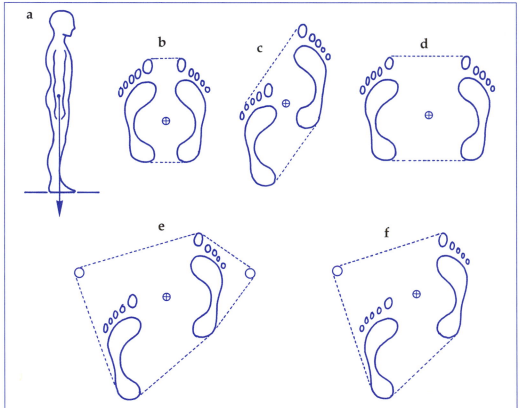

Figure 1.28   Line of action of body weight and the base of support; (*a* and *b*) standing; (*c*) increase in anteroposterior stability; (*d*) increase in side-to-side stability; (*e*) base of support when walking with the aid of crutches or two walking canes; and (*f*) base of support when walking while using a walking cane in the left hand. The symbol ⊕ denotes point of intersection of line of action of body weight with base of support.

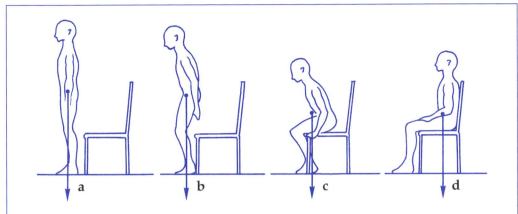

Figure 1.29 Line of action of body weight in relation to base of support when moving from standing to sitting.

shows a person moving from a standing position to sitting in a chair. The person moves his feet close to the front of the chair and then lowers his body by flexing his knees and bending his trunk forward while maintaining the same base of support, that is, the area beneath and between his feet (figure 1.29, *a* and *b*). He may or may not take hold of the sides of the chair as his thighs approach the seat of the chair. If he does take hold of the chair, his base of support immediately increases to include the area bounded by the legs of the chair as well as that beneath and between his feet, but the line of action of his body weight will still be over the area between his feet (figure 1.29*c*). When his thighs come close to the seat of the chair he begins to transfer his weight from over his feet to over the seat by gently rocking the trunk backward (figure 1.29*d*). These movements are reversed when moving from a sitting to a standing position.

In general, the lower the center of gravity and the larger the area of the base of support, the greater stability is likely to be. For example, by moving from a standing to a sitting position the body's center of gravity is lowered and the area of the base of support is increased (figure 1.29*d*). The recumbent position is the most stable position of the human body since it is the position in which the area of the base is greatest and the height of the center of gravity lowest. As the area of the base of support increases, the degree of muscular effort needed to maintain stability tends to decrease. For example, it is usually easier, in terms of muscular effort, to maintain stability when standing on both feet than when standing on one foot. Similarly, it is usually less tiring to sit than to stand, and less tiring to lie down than to sit. A person recovering from a leg injury may use crutches or a walking cane in order to relieve the load on the injured limb. The use of crutches also increases the area of the base of support and makes it easier for the user to maintain stability (figure 1.28, *e* and *f*).

In certain situations, especially in games and sports, body movements may depend on unstable rather than stable postures. For example, in the set position of a sprint start the sprinter tends to move her center of gravity as far forward as possible without overbalancing in order to obtain the best position from which to drive her body forward when the gun sounds. In this position the line of action of her body weight passes through the anterior limit of her base of support (figure 1.30).

As the area of the base of support decreases, if stability is to be maintained the degree of tolerance in the movement of the line of action of body weight also decreases. When the base of support

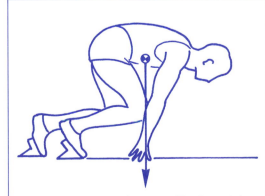

Figure 1.30 Line of action of body weight in the set position of a sprint start.

becomes a knife-edge or something similar such as a tightrope or very narrow beam, the amount of tolerance in the movement of the center of gravity is zero in any direction other than along the line of support. Consequently, when an object is in a balanced position on a knife-edge support the center of gravity of the object is located in the vertical plane through the line of support. By balancing an object in a number of different positions and noting the orientation of the vertical support plane to the object in each position, it may be possible to determine the position of the center of gravity of the object; the center of gravity is located at the point of intersection of the support planes. This method could be used with the irregular-shaped pieces of card referred to earlier in this chapter.

Moment of a Force

Moment of a force: the product of a force and the perpendicular distance between the line of action of the force and the axis of rotation

Fulcrum: axis of rotation

Moment arm: the perpendicular distance between the line of action of a force and the fulcrum

Figure 1.31*a* shows a block of wood resting on a table. If the block of wood is tilted about one of the edges of its base of support, for example PQ, the block of wood tends to rotate to its original position. This tendency to restore the original position is the result of the **moment** of the weight of the block of wood about the axis of rotation PQ. The magnitude of the moment is the product of the weight W of the block and the perpendicular distance D between the line of action of the weight of the block and the axis of rotation (figure 1.31*b*). In general, when a force acting on an object rotates or tends to rotate the object about a particular axis, the moment of the force (also referred to as turning moment, turning effect, or torque) is defined as the product of the force and the perpendicular distance between the line of action of the force and the axis of rotation. The axis of rotation is often referred to as the **fulcrum** and the perpendicular distance between the line of action of the force and the axis of rotation is usually referred to as the **moment arm** of the force.

In figure 1.31 the weight W of the block of wood is constant, but the moment arm of W about the fulcrum PQ varies with the angle of tilt; the greater the angle the smaller the moment arm of W and, therefore, the smaller the moment of W about PQ (figure 1.31, *b* and *c*). When the line of action of W passes through PQ the moment arm of W will be zero and, consequently, the moment of W about PQ will also be zero (figure 1.31*d*).

Resultant Moment

When an object is acted upon by two or more forces that rotate the object about a particular axis, the direction and speed of rotation of the object is determined by the resultant moment, that is, by the net effect of the moments exerted by the various forces. For example, consider figure 1.32*a*, which shows two children, A and B, sitting on a seesaw. If the seesaw is balanced with respect to the fulcrum, that is, if the line

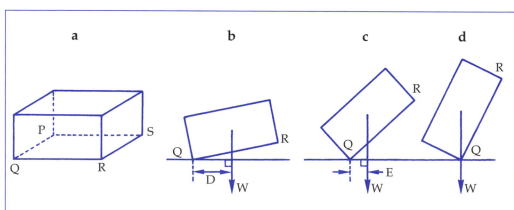

Figure 1.31   Moment of a force. In (*b*) the moment of W about PQ = W × D. In (*c*) the moment of W about PQ = W × E. The symbol □ denotes a right angle.

of action of the weight of the seesaw passes through the fulcrum in all positions of the seesaw, then the weights of the two children, W_A and W_B, are the only forces tending to rotate the seesaw about the fulcrum. With respect to figure 1.32a, W_A exerts a moment of $W_A \times M_A$ on the seesaw, where M_A is the moment arm of W_A about the fulcrum, tending to rotate the seesaw in a counterclockwise direction. Similarly, W_B exerts a moment of $W_B \times M_B$ on the seesaw, where M_B is the moment arm of W_B about the fulcrum, tending to rotate the seesaw clockwise. Using the convention of positive clockwise moments and negative counterclockwise moments, the resultant moment RM exerted by W_A and W_B is given by

$$RM = (W_B \times M_B) - (W_A \times M_A).$$

If RM is positive the seesaw rotates in a clockwise direction, and if RM is negative it rotates in a counterclockwise direction. If RM is zero—if the clockwise moment is equal and opposite to the counterclockwise moment—the seesaw is balanced and, as such, remains stationary. Consequently, if the weight of one of the children is known, but not the other, it is possible to determine the weight of the other child by balancing the seesaw with one child on each side of the fulcrum, both off the floor with the seesaw stationary, and equating the clockwise and counterclockwise moments. For example, if $W_B = 40$ kg wt, $M_B = 1.5$ m, and $M_A = 2.0$ m, then

$$W_A \times M_A = W_B \times M_B$$

$$W_A \times 2.0 \text{ m} = 40 \text{ kg wt} \times 1.5 \text{ m}$$

$$W_A = \frac{40 \text{ kg wt} \times 1.5 \text{ m}}{2.0 \text{ m}}$$

$$W_A = 30 \text{ kg wt.}$$

When an object is acted upon by two or more forces that tend to rotate the object about a particular axis, the resultant moment determines the object's direction and speed of rotation.

Equilibrium

In the above example there are three downward forces exerted on the seesaw: the weights of the two children and the weight of the seesaw itself. Since the seesaw is stationary the resultant force acting on the seesaw must be zero. Consequently, the downward forces must be counteracted by one or more upward forces. In this

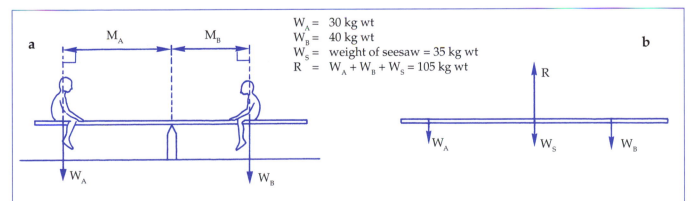

Figure 1.32 Moments exerted by two children sitting on a seesaw; (*a*) two children sitting on a seesaw where W_A and W_B are the weights of the two children and M_A and M_B are the moment arms of W_A and W_B, respectively; (*b*) free body diagram of the seesaw.

Equilibrium: when the resultant force and resultant moment acting on an object are both zero

Lever: a rigid or quasi-rigid object that can be made to rotate about a fulcrum to exert a force on another object

situation there is a single counteracting force R exerted by the fulcrum on the seesaw. Figure 1.32*b* shows a free body diagram of the seesaw. When the resultant force and resultant moment acting on an object (with respect to any reference axis of rotation) are both zero the object is said to be in a state of **equilibrium**.

The Lever

In figure 1.32 all of the forces acting on the seesaw are vertical forces. However, there are many situations where the forces tending to rotate an object are neither vertical nor parallel. For example, figure 1.33*a* shows a screwdriver being used to pry the lid off a metal can. The edge of the can forms a fulcrum about which the screwdriver can be rotated to apply a force to the underside of the lid. In response to a force E applied to the handle of the screwdriver, the lid of the can resists movement with a force R, which acts on the end of the screwdriver. Figure 1.33*b* shows a force-moment arm diagram, that is, the E and R forces and their moment arms are shown in relation to the fulcrum in order to more clearly show their turning effects on the screwdriver. The lid opens if the clockwise moment exerted by E is greater than the counterclockwise moment exerted by R.

In this situation the screwdriver is being used as a **lever**, a rigid or quasi-rigid object that can be made to rotate about a fulcrum to exert a force on another object. As in the example of the screwdriver, a lever encounters a resistance force R in response to an effort force E. Levers are classified into three systems depending on the position of the E and R forces in relation to the fulcrum (Watkins 1983). In a first class lever system the fulcrum is between the E and R forces (figure 1.34*a*). The use of a screwdriver to pry the lid off a metal can is an example of a first class lever system (figure 1.33). Scissors are a pair of first class levers that share the same fulcrum (figure 1.34*b*). In a second class lever system the R force is between the fulcrum and the E force as in, for example, a wheelbarrow (figure 1.34, *c* and *d*). In a third class lever system the E force is between the fulcrum and the R force as, for example, when holding a fishing rod (figure 1.34, *e* and *f*).

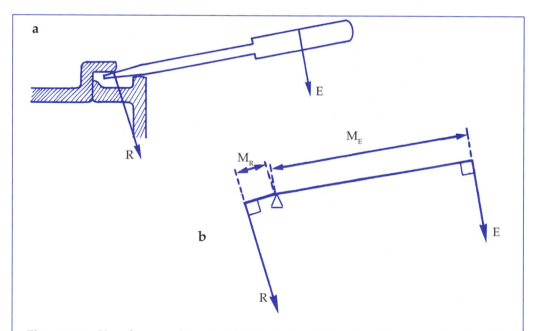

Figure 1.33   Use of a screwdriver to lift the lid of a metal can (E = force exerted on handle, R = resistance of lid, M_E and M_R = moment arms of E and R, respectively); (*a*) screwdriver under the edge of the lid of the can; (*b*) force-moment arm diagram corresponding to *a*. The symbol Δ denotes the fulcrum, □ denotes a right angle.

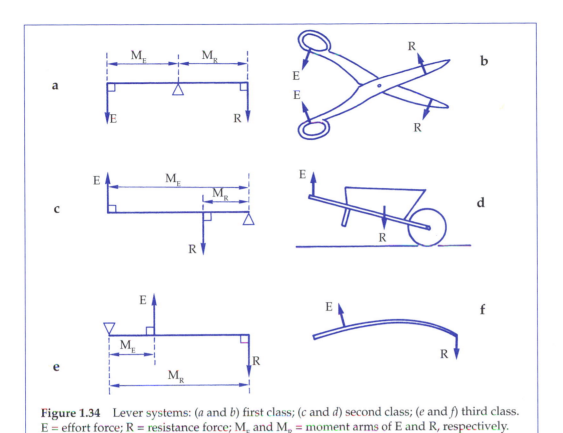

Figure 1.34 Lever systems: (*a* and *b*) first class; (*c* and *d*) second class; (*e* and *f*) third class. E = effort force; R = resistance force; M_E and M_R = moment arms of E and R, respectively.

Mechanical Advantage

The **mechanical advantage** (MA) of a lever system is a measure of its efficiency in terms of the amount of effort needed to overcome a particular resistance, that is,

$$MA = \frac{\text{magnitude of resistance (R)}}{\text{magnitude of effort (E)}}$$

$$= \frac{\text{length of moment arm of E } (M_E)}{\text{length of moment arm of R } (M_R)}.$$

Any machine with a mechanical advantage greater than 1.0 is regarded as very efficient. A first class lever may have a mechanical advantage greater than 1.0 or less than 1.0. The first class lever system in figure 1.33 has a mechanical advantage much greater than 1.0 since M_E is much greater than M_R. All second class lever systems have mechanical advantages greater than 1.0 since M_E is always greater than M_R. All third class lever systems have mechanical advantages less than 1.0 since M_E is always smaller than M_R. In all three lever systems, the greater the length of M_E in relation to the length of M_R the greater the leverage of the system.

Summary

The same mechanical principles that govern all other forms of movement govern human movement; these principles are reflected in the structure and function of the musculoskeletal system. This chapter described the elementary mechanical concepts and principles that underlie linear and angular motion. The next chapter provides an overview of the composition and function of the musculoskeletal system.

Mechanical advantage: the efficiency of a lever system in terms of the amount of effort needed to overcome a particular resistance

Review Questions

1. With regard to human movement, list three examples of rectilinear, curvilinear, and angular motion.
2. Differentiate between vector and scalar quantities.
3. Differentiate between linear distance and linear displacement.
4. If a runner completes the first and second laps of an 800 m race (on a 400 m track) in 56 s and 52 s, respectively, calculate the average speed of the runner in each lap and over the whole race.
5. If the angular velocity of the gymnast in figure 1.11 changes from 440°/s at b to 200°/s at c in 0.12 s, calculate the average angular acceleration of the gymnast during this period.
6. Define the terms sine, cosine, and tangent with respect to a right-angled triangle.
7. If the angle θ in figure 1.9 is 40° and the length of side c is 20 cm calculate the lengths of sides a and b given that cos 40° = 0.766 and sin 40° = 0.643.
8. In figure 1.10, if the angle between the component force $\bar{A}$ and the resultant force $\bar{F}$ is 40° and the angle between component forces $\bar{B}$ and $\bar{F}$ is 50°, calculate $\bar{A}$ and $\bar{B}$ given that $\bar{F}$ = 200 N, cos 40° = 0.766, and cos 50° = 0.643.
9. Differentiate between mass, inertia, and linear momentum.
10. Differentiate between mass and weight.
11. Describe the three basic types of loads.
12. Differentiate among work, energy, and power.
13. Differentiate between stability and equilibrium.
14. Describe the three types of lever systems.
15. Explain the difference in mechanical advantage among the three types of lever systems.

Chapter 2

The Musculoskeletal System

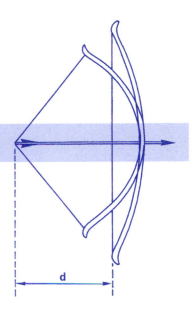

A ll living organisms are made up of one or more cells. The human body, like most organisms, is made up of billions of cells organized into complex functional groups including, for example, the musculoskeletal system. The first part of this chapter describes the four basic types of cells, called tissues, and the three main levels of cellular organization within functional groups. The second part of this chapter describes the composition and function of the musculoskeletal system in preparation for considering the functional anatomy of the musculoskeletal system in part II.

Objectives

After reading this chapter you should be able to do the following:

1. Define or describe the key terms.
2. Describe the four basic types of tissues.
3. Describe cellular organization in multicellular organisms.
4. Describe the composition and function of the musculoskeletal system.

Unicellular and Multicellular Organisms

Cell: the basic structural and functional unit of all organisms

Life processes: the activities a cell carries out in order to sustain the life of the cell

The basic structural and functional unit of life is the **cell**. All living organisms consist of one or more cells. All cells have four main components: cytosol, the nucleus, organelles, and the cell membrane (figure 2.1). Cytosol is a semitransparent fluid consisting of a complex solution of proteins, salts, and sugars. The nucleus and organelles are suspended in the cytosol. The nucleus contains genetic material and controls the **life processes** of the cell, which include movement, growth and development, respiration, circulation, digestion, excretion, and reproduction. The organelles, under the direction of the nucleus, carry out the life processes; each organelle carries out a specific function for the cell as a whole. The cytosol and organelles are usually referred to collectively as cytoplasm.

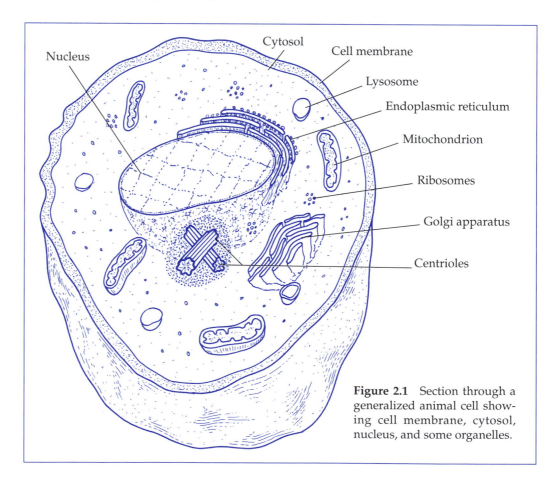

Nucleus

Cytosol

Cell membrane

Lysosome

Endoplasmic reticulum

Mitochondrion

Ribosomes

Golgi apparatus

Centrioles

Figure 2.1 Section through a generalized animal cell showing cell membrane, cytosol, nucleus, and some organelles.

The basic structural and functional unit of life is the cell, and all living organisms consist of one or more cells.

The cell membrane, also referred to as the plasma membrane or plasmalemma, encloses the cytoplasm and, as such, forms the external boundary of the cell. Like cytosol, the cell membrane consists of a viscous fluid, but it is usually much more viscous than the cytosol that it encloses. The viscous nature of the cell membrane and cytosol enables a cell to change its shape without losing its integrity. During normal functioning, all cells continuously change shape to a certain extent. The amount of change varies among different types of cells. For example, bone cells are likely to experience little change in shape, whereas in muscle cells the ability to change shape is highly specialized and an essential feature of normal function.

The number of cells that make up an organism reflect, to a certain extent, its level of evolution. For example, in the lowest forms of animal life such as in the amoeba and euglena the entire organism consists of a single cell. These organisms are referred to as **unicellular** organisms. In contrast, most animals consist of many cells and are referred to as **multicellular** organisms. For example, the adult human body consists of about 25 million billion cells, that is, 25×10^{15} cells (Basmajian 1970). The size and shape of cells varies considerably among different organisms and within the same organism. Differences in cell size and shape tend to reflect differences in function.

Unicellular organism: an organism that consists of a single cell

Multicellular organism: an organism that consists of many cells

There are basically two types of living organisms: unicellular and multicellular.

Cellular Organization in Multicellular Organisms

All living organisms are similar in that they are capable of carrying out all of the essential life processes. In unicellular organisms the life processes are, in biological terms, relatively simple. However, in multicellular organisms the cells are organized into complex functional groups that carry out the various life processes for the body as a whole. Multicellular organization can be divided into three structural levels: tissues, organs, and systems.

All living organisms are capable of carrying out a number of essential life processes. In multicellular organisms the life processes involve a high level of organization and integration between the cells. Cellular organization is on three levels: tissues, organs, and systems.

Tissues

In multicellular organisms all of the cells originate from a single cell formed by the fertilization of a female ovum by a male sperm. This cell undergoes rapid cell division to form a ball of cells. Soon after this, the cells begin to differentiate in size, shape, and structure to fulfill different functions in the body of the organism. This process of **cellular differentiation** results in the formation of four types of cells called tissues: epithelia, nerve, muscle, and connective. A **tissue** is a group of cells having the same specialized structure, enabling them to perform a particular function in the body (Freeman and Bracegirdle 1967). The word tissue is also used in a general sense to refer to any part of the body such as, for example, soft tissues, bone tissue, and skin tissue.

Cellular differentiation: the specialization of cells into tissues

Tissue: a group of cells having the same specialized structure enabling them to perform a particular function in the body

Epithelial Tissue

There are two types of epithelial tissue: covering and glandular. Covering epithelia form the surface layer(s) of cells of all the internal and external free surfaces of the body except the surfaces inside synovial joints. For example, the surfaces of the skin and the lining of the digestive tract, heart chambers, and blood vessels are all examples of covering epithelia.

All cells can secrete fluid to a greater or lesser extent, but glandular epithelial cells are specialized for this purpose and as such form the two types of glands found in the body: exocrine and endocrine. Many of the exocrine glands, such as the gastric glands in the lining of the stomach, secrete fluids containing enzymes necessary for the digestion of food. Endocrine glands, such as the pituitary at the base of the brain and the adrenals at the upper end of each kidney, secrete hormones that, in association with the nervous system, regulate and coordinate the various body functions.

Nerve Tissue

Nerve cells (neurons) are specialized to conduct electrochemical impulses throughout the body to regulate and coordinate the various body functions. The structure and function of nerve tissue is covered in detail in chapter 8, which deals with the neuromuscular system.

Muscle Tissue

Muscle cells are specialized to contract (i.e., create pulling forces to bring about movement). There are three types of muscle cells: skeletal, visceral, and cardiac.

Skeletal muscle forms the abdominal wall and is responsible for controlling movement in the joints between the bones of the skeleton. It is called skeletal muscle because it generally is attached to the skeleton. It is also called voluntary muscle because it is normally under the conscious control of the individual. The structure and functions of skeletal muscle are covered in detail in chapter 8.

Visceral or involuntary muscle is found in parts of the body that involve involuntary movement, that is, in body parts not under the conscious control of the individual. It is found, for example, in the walls of the digestive tract and the larger arteries.

Cardiac muscle is found only in the heart. It has characteristics of both skeletal and visceral muscle, but differs from them in that it contracts rhythmically throughout life even though the rate of contractions (heart rate) may alter frequently.

Connective Tissue

As its name suggests, one of the main functions of connective tissue is to support and bind other tissues together. The bones of the skeleton and the fibrous structures holding the bones together at joints are all forms of connective tissue. The structure and functions of connective tissue are covered in detail in chapter 4.

Organs and Systems

Organ: a combination of different tissues designed to carry out a specific bodily function

An **organ** is a combination of different tissues designed to carry out a specific bodily function. For example, the heart is designed to pump blood around the body. The structure of the heart consists of

- cardiac muscle cells,
- connective tissue, which binds the muscle cells together,
- epithelial tissue, which lines the chambers of the heart, and
- nerve tissue, which innervates the muscle cells.

Other examples of organs are the lungs, the stomach, and a skeletal muscle complete with its tendons, which attach the muscle to the skeletal system.

A **system** is a combination of different organs working together to carry out a particular function in the body. For example, the cardiovascular system consists of the heart and blood vessels; it is responsible for transporting blood around the body. There are basically 11 separate systems in the human body (Tortora and Anagnostakos 1984):

System: a combination of different organs working together to carry out a particular bodily function

Integumentary system: the external covering of the body, that is, the skin and associated structures such as nails

Skeletal system: the bones of the skeleton and the structures that form the joints between the bones

Muscular system: the skeletal muscles

Nervous system: the nerves, organized into central (brain and spinal cord) and peripheral (spinal nerves) components

Endocrine system: the glands that secrete hormones, which regulate and coordinate the various body functions in association with the nervous system

Cardiovascular system: the heart and blood vessels

Lymphatic system: the system of vessels and associated structures that drains and returns fluid leaked from the blood and protects against disease

Respiratory system: the lungs and associated passageways

Digestive system: the digestive tract and associated structures that break down food and eliminate solid waste

Urinary system: the kidneys, bladder, and associated structures that eliminate nitrogenous waste as urine

Reproductive system: the ovaries and associated structures (female) and testes and associated structures (male), which enable the body to produce offspring

These systems are responsible for carrying out the body's life processes. Whereas all of the life processes involve a certain degree of integration between different systems, some processes involve closer integration among different systems than others do. For example, the transport of oxygen from the air to all the cells of the body is carried out by the combined activity of the nervous, respiratory, and cardiovascular systems. Similarly, movement of the body is brought about by the combined activity of the nervous, muscular, and skeletal systems. Consequently, for descriptive purposes it is usual to refer to combinations of systems: for example, the cardiorespiratory system, the musculoskeletal system, and the neuromuscular system. Cellular differentiation and organization in multicellular organisms are illustrated in figure 2.2.

All of the cells of the body originate from a single cell—a fertilized ovum. This cell undergoes rapid cell division and, subsequently, cellular differentiation results in the formation of different types of tissues, which combine to form specialized organs. Organs then work together to form systems to carry out the life processes of multicellular organisms.

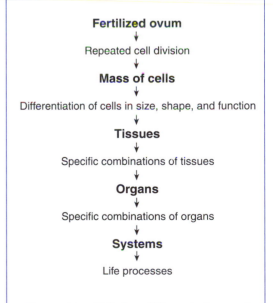

Figure 2.2 Cellular differentiation and organization in multicellular organisms.

Musculoskeletal System Structure

Musculoskeletal system:the combined system consisting of the skeletal system and the muscular system

Skeletal system: the bones and the structures that form the joints between the bones

Muscular system: the skeletal muscles

The **musculoskeletal system** consists of the **skeletal system** and the **muscular system** (figure 2.3). The skeletal system consists of the skeleton—the bones—and the various structures forming the joints between the bones. It is similar to the metal framework of a building in that it gives the body its basic shape and provides a supporting framework for all the other systems. Bone is an ideal support material since it is not only strong, but also fairly lightweight. The adult skeletal system normally has 206 bones, more than 200 joints, and accounts for between 12% and 15% of total body weight (McArdle, Katch, and Katch 1996).

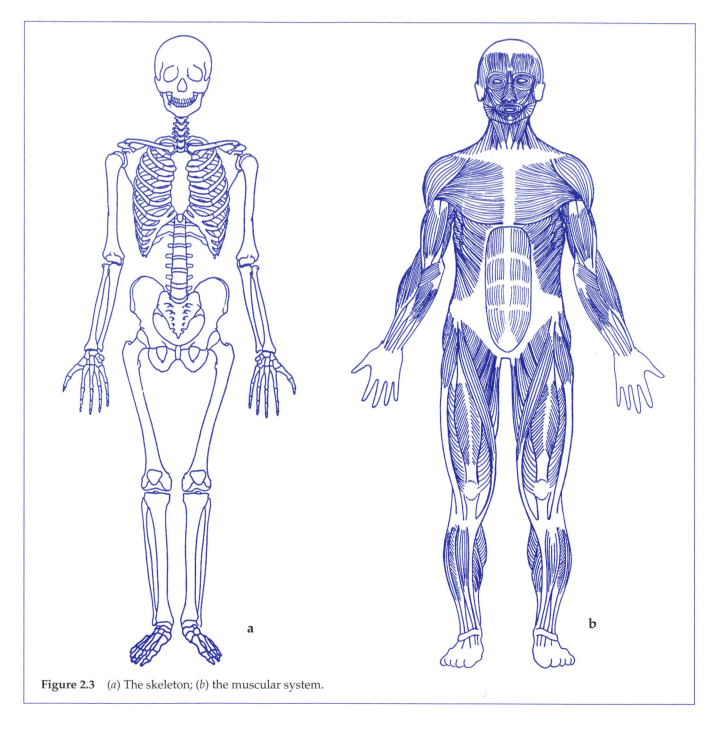

a   b

Figure 2.3   (*a*) The skeleton; (*b*) the muscular system.

The musculoskeletal system consists of the skeletal system (the bones and the various structures forming the joints between the bones) and the muscular system (the skeletal muscles).

As previously described, there are three types of muscle tissue: skeletal, visceral, and cardiac. The muscular system, however, refers only to the skeletal muscles. Visceral and cardiac muscle tissue belong to other systems (Tortora and Anagnostakos 1984). For example, cardiac muscle tissue, found only in the heart, is part of the cardiovascular system. Similarly, visceral muscle in the walls of the digestive tract is part of the digestive system. With regard to the muscular system, there are approximately 640 skeletal muscles, and they account for about 36% and 45% of total body weight in untrained adult females and males, respectively (McArdle, Katch, and Katch 1996).

The skeletal muscles are the powerhouses of the body; they are the organs that enable us to make voluntary, controlled movements of the body. The most important property of all types of muscle tissue is **contractility**—the ability to create a pulling force and, if necessary, change in length (increase or decrease) while maintaining a pulling force. This property is developed to a high level in skeletal muscle tissue. Each complete skeletal muscle consists of a large number of long muscle cells bound together by various layers of connective tissues (see chapter 4). In most muscles the muscle cells occupy the main belly of the muscle, but the ends of the muscle consist entirely of thickened cords or bands of virtually inextensible connective tissue that anchor the muscle onto the skeleton (figure 2.4). The shape of these connective tissue attachments depends on the shape of the muscle and the area of bone available for attachment. In general, there are two basic shapes: a cord or narrow band called a **tendon** and a broad band called an **aponeurosis** (figure 2.4). A skeletal muscle and its tendons and aponeuroses are usually referred to as a **musculotendinous unit** (Taylor et al. 1990).

The skeletal muscles are arranged in groups, and each group is attached to the skeleton such that it crosses over one or more joints. Since a muscle can only pull in one direction, the movement of each joint is controlled by one or more opposing pairs of muscle groups; one group is responsible for moving a joint in one direction and the other group is responsible for moving the joint in the opposite direction. For this reason the various pairs of muscle groups are referred to as **antagonistic pairs.** By coordinated activity between each antagonistic pair of muscles that cross a joint, the amount and rate of movement in the joint can be carefully controlled. For example, when kicking a ball, the quadriceps and hamstrings work together as an antagonistic pair of muscle groups to control movement at the hip and knee joints (figure 2.5).

The knee joint is basically designed to rotate in one plane about a transverse (side-to-side) axis as shown in figure 2.5. However, other joints such as the hip and shoulder are designed to move in more than one plane (see chapter 3). Consequently, two or more antagonistic pairs of muscle groups may simultaneously control movement in these joints.

Contractility: the ability of muscle tissue to create a pulling force

Tendon: a cord or narrow band of connective tissue that attaches skeletal muscle to bone

Aponeurosis: a broad band of connective tissue that attaches a skeletal muscle to bone

Musculotendinous unit: a skeletal muscle and its tendons or aponeuroses

Antagonistic pair: a pair of musculotendinous units that control a joint's movement in a particular plane

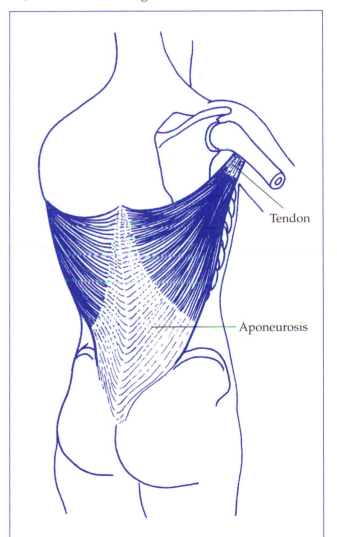

Tendon

Aponeurosis

Figure 2.4 Tendon and aponeurosis: The latissimus dorsi muscle on each side of the body is attached to the back by an aponeurosis and to the arm by a bandlike tendon.

Each skeletal muscle is attached to the skeleton such that it crosses one or more joints; when a muscle contracts it tends to bring about movement in the joints it crosses. The muscles are arranged in antagonistic pairs that coordinate action to control the amount and rate of movement in each joint.

Musculoskeletal System Function

Posture refers to the orientation of body segments to each other and is usually applied to sitting and standing body positions. In the upright standing position there are two forces acting on the body: body weight and the ground reaction force (figure 2.6a). The combined effect of body weight and the ground reaction force is a compression load that tends to collapse the body in a heap on the ground (compare figure 2.6a with figure 1.18c). This compression load increases with any additional weight carried by the body (figure 2.6b). Clearly, to prevent the body from collapsing while simultaneously bringing about desired movements, the movements of the

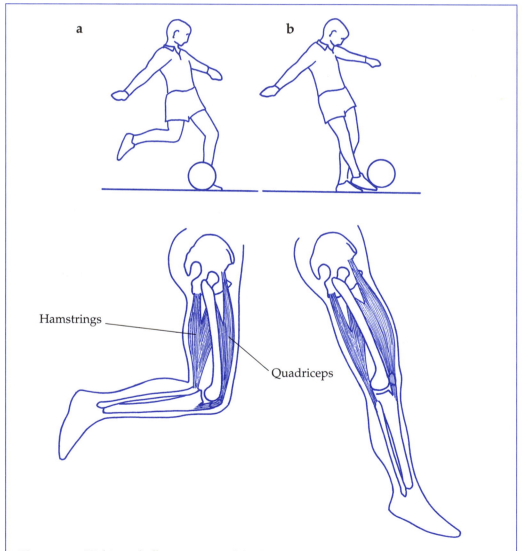

Figure 2.5   Kicking a ball; movement of the knee joint is controlled by the quadriceps and hamstrings.

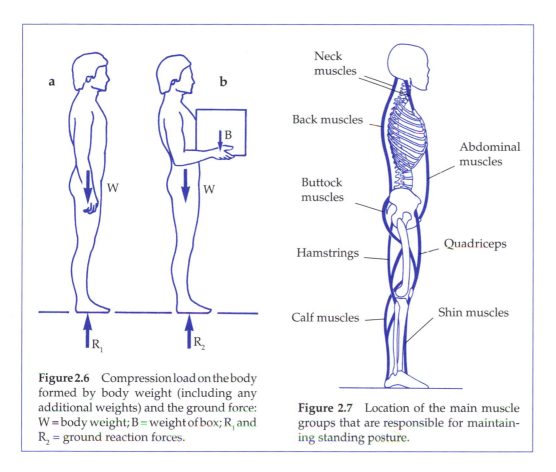

Figure 2.6 Compression load on the body formed by body weight (including any additional weights) and the ground force: W = body weight; B = weight of box; R_1 and R_2 = ground reaction forces.

Figure 2.7 Location of the main muscle groups that are responsible for maintaining standing posture.

various joints need to be carefully controlled by the musculoskeletal system under the control of the nervous system.

By coordinated activity between the various muscle groups, forces are generated by the muscles and are transmitted via the bones and joints to enable the individual to maintain an upright body position and bring about voluntary, controlled movements of the body. For example, when standing upright the joints of the neck, trunk, and legs must be prevented from moving by the muscles that control them; otherwise the body collapses (figure 2.7). Consequently, the weight of the whole body is transmitted to the floor by the feet, but the weight of individual body segments above the feet (head, arms, trunk, and legs) is transmitted indirectly to the floor by the skeletal chain formed by the bones and joints of the neck, trunk, and legs.

Transmitting body weight to the ground while maintaining an upright body posture illustrates the essential feature of musculoskeletal function—the generation (by the muscles) and transmission (by the bones and joints) of forces. Forces produced by muscles and transmitted by joints are referred to as **internal forces.** Body weight and the ground reaction force are referred to as **external forces.** Other external forces include water resistance, air resistance, and the forces arising from contact with other objects. The musculoskeletal system generates and transmits internal forces to counteract (in static postures) or overcome (during purposeful movements) the external forces acting on the body.

In any particular posture or movement, coordinated activity among various muscle groups results in control of joint movements so that forces can be transmitted throughout the skeleton. In this way the body can adopt postures that enable it to resist forces (such as body weight) tending to deform or move the body and apply forces to other objects, usually via the hands and feet, to transport the body and manipulate objects. Consequently, in terms of force transmission by the musculo-

Internal force: a force produced or transmitted by the musculoskeletal system

External force: a force exerted on the body by an external agent

skeletal system, there are three broad categories of movement: maintenance of upright posture, transport of the body, and manipulation of objects. Most movements involve a combination of two or all three basic categories of movement.

The musculoskeletal system generates and transmits internal forces to counteract or overcome external forces and thereby bring about controlled movement of the body.

Maintaining Upright Posture

Movements in this category involve fairly static postures in which the weight of one or more body segments is transmitted indirectly to the floor or other support surface, such as a chair, by other segments. These postures include standing, sitting, and other balancing activities (figures 2.8 and 2.9). Some upright postures may involve more than one support surface. For example, figure 2.8e shows a person leaning on a table. In this case his weight is supported partly by the table and partly by the floor. Similarly, figure 2.8c shows a person sitting in an armchair; his arms are supported by the armrests and his feet rest on the floor. In this case his weight is distributed over four support areas: the backrest, seat, armrests, and floor.

The degree of muscular activity required to maintain a particular body posture depends on the number and size of the support surfaces. Consequently, muscular activity is minimal in the recumbent posture since all of the body segments are supported directly by the support surface (figure 2.9a). In contrast, balancing on one hand involves a considerable amount of muscular activity to transmit the entire body weight through the small area beneath the hand (figure 2.9b). In this case the weight of each body segment above the grounded hand is transmitted indirectly to the floor, from segment to segment down through the skeletal chain.

Body Transport

To move from one place to another, the body must push or pull against something to provide the necessary force to drive it in the required direction. In walking and running, forward movement is combined with an upright body posture and movement is achieved by pushing a foot obliquely downward and backward against the ground. Provided the foot does not slip, the leg thrust results in a ground reaction force directed obliquely upward and forward. The effect of the ground reaction force is that the body moves forward with an upright posture (figure 2.10).

When swimming, the water largely supports body weight (buoyancy force). Consequently, the main function of the musculoskeletal system is to enable the arms

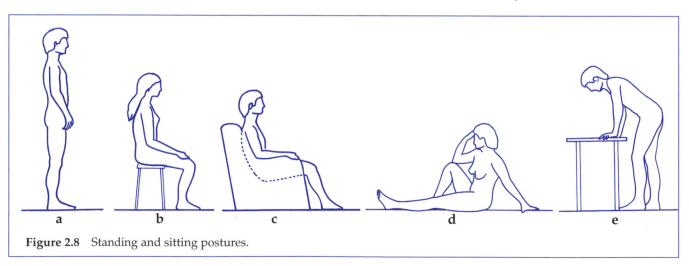

Figure 2.8 Standing and sitting postures.

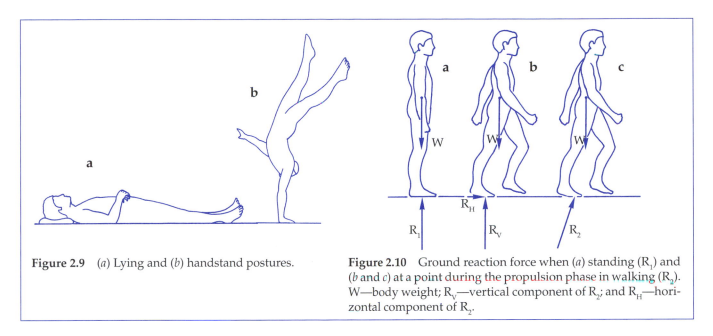

Figure 2.9   (*a*) Lying and (*b*) handstand postures.

Figure 2.10   Ground reaction force when (*a*) standing (R$_1$) and (*b* and *c*) at a point during the propulsion phase in walking (R$_2$). W—body weight; R$_V$—vertical component of R$_2$; and R$_H$—horizontal component of R$_2$.

and legs to pull and push backward against the water to move the body forward. Since the musculoskeletal system does not have to support the weight of the body in the water, the muscle forces and joint reaction forces that occur during swimming tend to be considerably less than in land-based activities. For this reason, swimming is often prescribed as part of a rehabilitation program to restore normal muscle function and joint flexibility following injury.

Manipulating Objects

Many movements involve manipulating objects with the hands or the feet. These manipulations may involve fairly forceful pushes and pulls or fine manual dexterity. For example, forceful pushes and pulls are required to push a wheelbarrow (figure 2.11*a*), pull on an upright post (figure 2.11*b*), carry a heavy suitcase, and throw or kick a ball. Activities involving manual dexterity are usually associated with a fairly static body posture (a stable base of support). Such activities include, for example, driving a car, writing or typing at a desk (figure 2.11*c*), turning a door handle, and taking a book off a shelf (figure 2.11*d*).

The Human Machine

The skeletal system consists basically of a number of fairly rigid components (bones) joined together to allow a certain amount of relative motion (linear or angular motion of one bone relative to the other) in each joint. The muscular system, under the control of the nervous system, provides the power to control the movement of the joints and, thereby, enables the body to apply forces on other objects. In this sense, the musculoskeletal system operates like a **machine**—a powered mechanism designed to move such that it is capable of applying forces to other objects (Dempster 1965). Figure 2.12 shows a simple machine that can apply a gripping force at one end by applying a compression load at the other end.

Machine: a powered mechanism designed to move such that it is capable of applying forces to other objects

The musculoskeletal system operates basically like a machine.

The movements of the components of an ordinary machine and the magnitude and direction of forces the machine exerts on other objects are very predictable, but this is not the case with the human musculoskeletal system. The human machine enables an

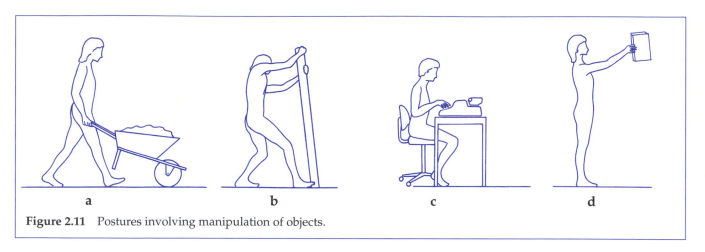

Figure 2.11   Postures involving manipulation of objects.

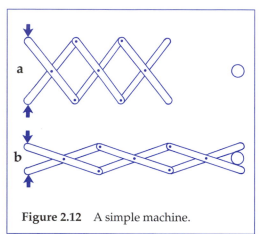

Figure 2.12   A simple machine.

individual to adopt the most suitable posture for applying forces in any given situation. The body's ability to adopt a wide range of postures is due to the skeletal system's open-chain arrangement of bones.

The arms and legs form four peripheral chains, free at their extremities, which attach onto a central chain (see figure 2.3a). This open-chain arrangement allows any part of the body to move more or less independently of the rest; movement of one part of the body does not necessarily result in movement in the rest of the body. For example, movements of the arms can accomplish a variety of tasks such as eating, writing, and typing while the rest of the body is in a more or less stationary sitting position.

In contrast to this open-chain arrangement, the components that make up inanimate machines tend to be in the form of closed chains. This arrangement produces simultaneous and predictable movements of the joints between the various components; the movement of one joint determines the movement in all the other joints.

The human machine has a large number of force-producing components (muscles). Muscle activity is carefully coordinated to utilize the muscles most efficiently. For example, in a throwing action such as pitching a baseball, the speed of the ball as it leaves the pitcher's hand is, in accordance with Newton's second law of motion, determined by the impulse of the force exerted on the ball during the pitching action (figure 2.13). The muscles generate the force, and the more muscles recruited during the pitching action, the greater the ball speed. In a well-coordinated pitching action, the forces produced by the individual muscle groups are summated to maximize the force exerted on the ball during the pitch (Fleisig et al. 1996). The pitching action starts with the rear leg driving the body forward into the step (figure 2.13, a and b). The next stage involves foot placement of the lead leg, which provides a firm base for the muscles of the legs and hips to drive the rear hip and, therefore, the trunk forward (figure 2.13, b and c). The muscles of the trunk then drive the throwing arm forward (figure 2.13, c and d). Finally, the shoulder, elbow, and wrist of the throwing arm drive the ball forward until release (figure 2.13, d and e).

In propulsive movements like throwing, the number of muscle groups recruited to drive the body and, therefore, the implement in the intended direction limits the amount of force applied to the implement. If all of an adult's skeletal muscles could contract simultaneously in the same direction, they would exert a force in the region of 22 tons (Elftman 1966). However, the arrangement of the muscles does not permit this theoretical maximum to be achieved. Even in a highly trained and highly motivated athlete, the amount of force generated in propulsive whole-body movements such as jumping and throwing is usually a small fraction of the theoretical maximum. Furthermore, injury to any part of the musculoskeletal system used in the movement reduces this force even further (Nicholas and Marino 1987). For example, injury to an ankle severely affects a pitcher's performance since he is unable to provide a strong, stable base of support for upper body action (figure 2.13). It has been shown that without the initial step, the speed of a baseball pitch is reduced to approximately 80% of normal, and without hip and trunk rotation, the speed of the pitch is reduced to about 50% of normal (Miller 1980).

> *If all of an adult's skeletal muscles could contract simultaneously in the same direction they would exert a force in the region of 22 tons. However, the arrangement of the muscles is such that the amount of force generated in propulsive whole-body movements is usually a small fraction of the theoretical maximum.*

Loading on the Musculoskeletal System

The open-chain arrangement of the bones of the skeleton and the arrangement of the muscles on the skeleton maximize the range of possible body postures. However, the body pays a price for this movement capability: many of the muscles operate in lever systems that have very low mechanical advantages and, as such, usually have to exert much larger forces than the weights of the body segments they control. Furthermore, the size of the forces exerted in joints, the joint reaction forces, are determined by the size of the muscle forces; the larger the muscle forces, the larger the joint reaction forces. For example, in walking, the peak hip, knee, and ankle joint reaction forces in an adult are normally in the range of 5 to 6, 3 to 8, and 3 to 5 times body weight, respectively (Nigg 1985). The more dynamic the activity, the greater the muscle forces and, therefore, the greater the joint reaction forces. For example, in fast running (800 m pace), the peak knee and ankle joint reaction forces in an adult are likely to be in the region of 20 and 8 times body weight, respectively (Nigg 1985).

In any body position other than the relaxed recumbent position, the musculoskeletal system is likely to be subjected to considerable loading. In response to the forces exerted on them the musculoskeletal components experience strain—they are deformed to a certain extent—and the greater the force, the greater the strain. Under normal circumstances the musculoskeletal components adapt their size, shape, and structure to the time-averaged forces exerted on them in order to more readily

a   b   c   d   e

Figure 2.13   Pitching a baseball.

withstand the strain. However, when the degree of strain experienced by a particular component exceeds its strength, it becomes injured. Consequently, there is an intimate relationship between the structure and function of the musculoskeletal system.

The open-chain arrangement of the skeleton enables an individual to perform a wide range of postures and movements. However, the body pays a price for this ability: the muscles, bones, and joints are subjected to very high forces in virtually all postures other than lying down.

Summary

This chapter discussed how the billions of cells that make up the body originate from a single fertilized ovum through a process involving cell division and cell differentiation. It also described how the cells are organized into functional groups for carrying out the essential life processes and discussed the composition and function of the musculoskeletal system. The musculoskeletal system mainly functions to generate and transmit forces to counter the external forces acting on the body and, in so doing, enable the body to move in a controlled manner. The open-chain arrangement of the skeleton allows an individual to adopt a wide range of postures and movements, but only at the expense of large forces in muscles, bones, and joints. The structure of the musculoskeletal system reflects the ability to sustain such loading. Part II describes the functional anatomy of the musculoskeletal system starting with the skeleton.

Review Questions

1. Differentiate between cellular differentiation and cellular organization in multicellular organisms.
2. Describe the four basic tissues.
3. Describe what is meant by a combined system.
4. With regard to musculoskeletal system function, describe the relationship between external and internal forces.
5. Briefly describe the three broad categories of movement brought about by the musculoskeletal system.
6. With reference to recumbent posture and standing posture:
   - explain the difference between direct and indirect transmission of the weight of body segments to the support surface, and
   - describe how direct and indirect transmission of the weight of body segments to the support surface likely affects the degree of activity in the muscles.
7. Describe the main advantage and disadvantage of the open-chain arrangement of the skeleton.

Part II

FUNCTIONAL ANATOMY OF THE MUSCULAR AND SKELETAL SYSTEMS

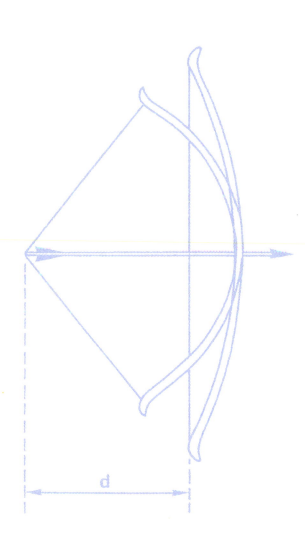

While part I described the basic structure and function of the musculoskeletal system, part II directs attention to how the system's function—to generate and transmit forces—is reflected in the structure of its components. Chapter 3—The Skeleton—describes the open-chain arrangement of the bones in the axial and appendicular skeletons and the features of the bones associated with force transmission and relative motion between bones. Chapter 4—Connective Tissues—differentiates the ordinary connective tissues, in particular, ligaments, tendons, and fascia, and the special connective tissues of cartilage and bone. Chapter 5—The Articular System—explains the differences in types of joints and, in particular, the way joint design reflects a trade-off between flexibility and stability. Chapter 6—Joints of the Axial Skeleton—describes the joints between the vertebrae and the joints of the pelvis, focusing on the association between abnormal load transmission and pain in these regions. Chapter 7—Joints of the Appendicular Skeleton—describes the joints and joint complexes of the appendicular skeleton and the functional interdependence between them with regard to load transmission. Chapter 8—The Neuromuscular System—discusses the interrelationship between the nervous and muscular systems in terms of force generation and proprioception. Finally, chapter 9—Forces in Muscles and Joints—focuses on how the open-chain arrangement of the skeleton affects the forces exerted in muscles and joints and, in particular, how this arrangement affects muscle and joint forces adopting different postures for the same tasks.

Chapter 3
The Skeleton

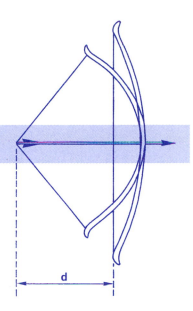

The skeleton gives the body its basic shape and provides a strong, protective, and supporting framework for all other systems of the body. With regard to movement, the bones of the skeleton act as levers operated by the skeletal muscles. The levers within the musculoskeletal system vary considerably in terms of mechanical advantage; this is reflected in the wide variety in size and shape of the bones. This chapter describes the bones of the skeleton and, in particular, the features of the bones associated with force transmission and relative motion between bones.

Objectives

After reading this chapter you should be able to do the following:
1. Define or describe the key terms.
2. Describe the composition and functions of the skeleton.
3. List the main bones of the skeleton.
4. Identify and describe the main features of the axial skeleton.
5. Distinguish vertebrae from different regions of the vertebral column.
6. Identify and describe the main features of the appendicular skeleton.
7. Distinguish between left and right with regard to the large bones of the appendicular skeleton.

Composition and Function of the Skeleton

At birth the human skeleton consists of approximately 270 bones. During growth and development of the skeleton (see chapter 4), some of the bones fuse together so that the adult skeleton normally consists of 206 bones. However, variations in this basic number do occur; for example, some adults have 11 or 13 pairs of ribs, whereas most adults have 12 pairs.

> While humans have approximately 270 bones at birth, some bones fuse together during growth so that in the adult the skeleton normally consists of 206 bones.

The skeleton performs three main mechanical functions:

1. It acts as a supporting framework for the rest of the body.
2. It acts as a system of levers on which the muscles can pull in order to stabilize and move the body.
3. It protects certain organs. For example, the skull protects the brain, the vertebral column protects the spinal cord, and the rib cage helps protect the heart and lungs.

Terminology

Axial skeleton: the bones of the skull, vertebral column, and ribcage

Appendicular skeleton: the bones of the upper and lower limbs

Aspect: the appearance of a particular bone (or any other part of the body) from a particular viewpoint

For descriptive purposes the bones are usually divided into two main groups: the **axial skeleton** and the **appendicular skeleton** (axial = axis, appendicular = appendage). The adult axial skeleton consists of 80 bones comprising the skull, vertebral column (backbone), and ribs. The adult appendicular skeleton consists of 126 bones that make up the upper limbs (arms and hands) and the lower limbs (legs and feet) (figure 3.1).

In anatomy, the term **aspect** refers to the appearance of a particular bone (or any other part of the body) from a particular viewpoint. For example, the anterior aspect of the skeleton refers to the features of the skeleton seen from an anterior (frontal) viewpoint (figure 3.1a). Similarly, lateral aspect (view from the side) (figure 3.1b), posterior aspect (view from the back), superior aspect (view from above), and inferior aspect (view from below) describe other viewpoints.

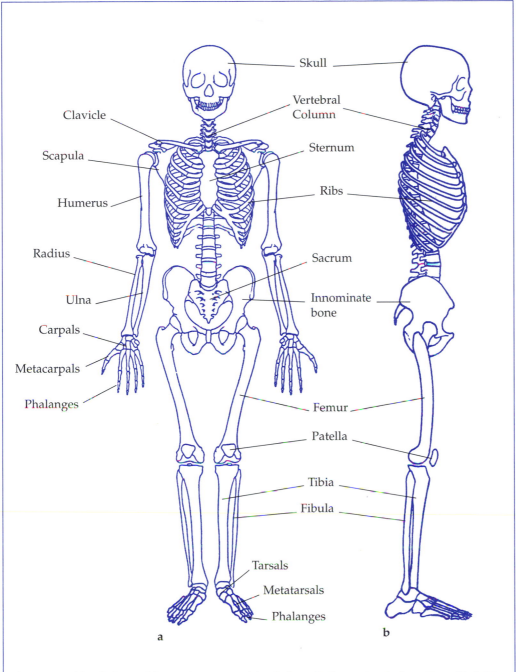

Figure 3.1   The skeleton; (*a*) anterior aspect of the skeleton in the anatomical position; (*b*) right lateral aspect of the trunk and lower limb.

The bones vary considerably in size and shape. There are four general categories of shape: long bones, short bones, flat bones, and irregular bones. Some bones fit into more than one category; for example, the small bones of the wrist are categorized as short and irregular. Whereas there are considerable differences in the size and shape of bones, there are a number of features such as articular surfaces and points of attachment of tendons that are common to many bones. These common features are frequently referred to in the description of bones and, therefore, it is useful to list them.

Common Bone Features

The common features of bones are illustrated in figure 3.2:

Articular surface: part of a bone that forms a joint with another bone.

Concave articular surface: a rounded depression.

Convex articular surface: a rounded elevation.

Facet: a small fairly flat articular surface. A convex facet on one bone usually articulates with a concave facet on an adjacent bone.

Condyle: a rounded projection of bone that provides the base for a rounded articular surface. A convex condyle on one bone usually articulates with a concave condyle on an adjacent bone.

Trochlea: a pulley-shaped condyle.

Fossa: an oval or circular depression or cavity that may also be an articular surface.

Notch: an oval depression that is often an articular surface. A notch may also take the form of a depressed region on the edge of a flat bone.

Groove or sulcus: an elongated depression (like a trench). One or more tendons usually occupy grooves.

Ridge or line: an elongated elevation. A ridge is usually the site of attachment of one or more aponeuroses.

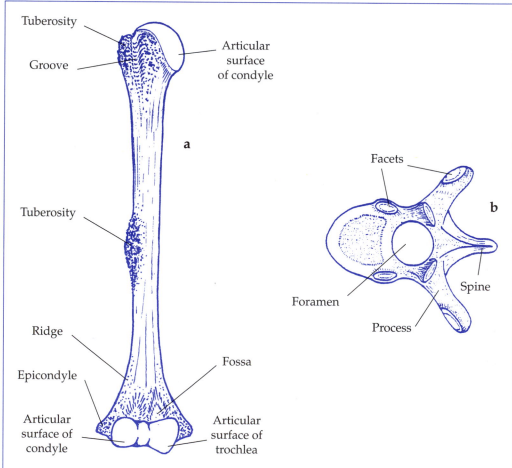

Figure 3.2 Common features of bones; (*a*) anterior aspect of the humerus; (*b*) superior aspect of a vertebra.

Crest: a broad ridge.

Process: a projection of bone from the main body usually providing attachment for tendons or ligaments.

Spine: a smooth process that may be slender or flat.

Epicondyle: a small process adjacent to a condyle.

Tubercle: a small roughened process.

Tuberosity: a large roughened process.

Trochanter: another name for a tuberosity used specifically in the description of the thigh bone (femur).

Foramen: a hole through a bone for the passage of blood vessels and nerves.

> *There is considerable variation in size and shape among the bones of the skeleton. In terms of shape there are four general categories: long, short, flat, and irregular. The four features common to many bones are*
>
> - *articular surfaces,*
> - *areas of attachment of tendons and ligaments,*
> - *bony grooves that guide tendons, and*
> - *holes for the passage of blood vessels and nerves.*

Those parts of the bone surface that do not have specific features such as articular surfaces are normally quite smooth. These smooth areas, usually fairly large, are regions where muscles attach directly to the bone. When a muscle is attached to a bone by a tendon, the area of attachment of the tendon to the bone is usually rough and, as such, is likely to be referred to as a tubercle, tuberosity, or trochanter.

Reference Planes and Spatial Terminology

In order to describe the spatial orientation of the particular features of a bone, or to describe the position of one bone (or body part) in relation to another, it is necessary to use standard terminology with reference to a standard body posture. In the standard posture, also called the **anatomical position** (see figure 3.1*a*), the body is upright with the arms by the sides and palms of the hands facing forward. There are three main reference planes: median, coronal, and transverse.

The **median plane** is a vertical plane that divides the body down the middle into more-or-less symmetrical left and right portions (figure 3.3). The median plane is also frequently referred to as the **sagittal plane;** the terms sagittal, paramedian, and parasagittal (para = beside or against) are also sometimes used to refer to any plane parallel to the median plane. In this book the term sagittal is used to refer to any plane parallel to the median plane.

The terms lateral and medial are used to describe the relationship of different parts of a bone (or body part) to the median plane. Lateral means farther away from the median plane and medial means closer to the median plane. For example, in figure 3.1*a* the lateral end of the clavicle (collarbone) articulates with the scapula (shoulder bone) and the medial end of the clavicle articulates with the sternum (breast bone). Similarly, in the anatomical position, the fingers of each hand are medial to the thumbs (and the thumbs are lateral to the fingers).

Anatomical position: reference body posture for descriptive purposes; the body is upright with the arms by the sides and the palms of the hands facing forward

Median plane: a vertical plane that divides the body down the middle into more or less symmetrical left and right portions

Sagittal plane: any plane parallel to the median plane

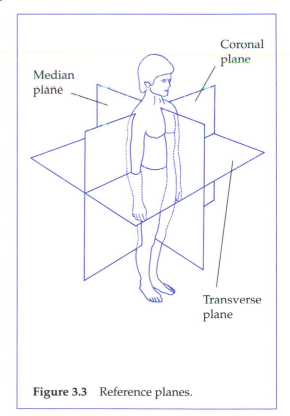

Figure 3.3 Reference planes.

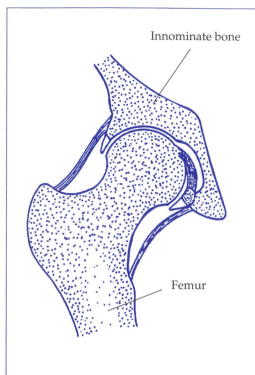

Innominate bone

Femur

Figure 3.4   Coronal section through the right hip joint.

Coronal plane: any vertical plane perpendicular to the median plane

Transverse plane: any plane perpendicular to both the median and coronal planes

Any vertical plane perpendicular to the median plane is called a **coronal plane** (or frontal plane) (see figure 3.3). The terms anterior (in front) and posterior (behind) describe the position of structures with respect to coronal planes. For example, the face forms the anterior part of the skull, the sternum is anterior to the vertebral column (backbone) and the patella (kneecap) is anterior to the lower end of the femur. Similarly, the calcaneus (heel bone) is posterior to the toes and the fibula (thin long bone in the lower leg) is posterior to the tibia (thicker of the two bones in the lower leg) (see figure 3.1b). The terms ventral and dorsal are synonymous with anterior and posterior, respectively.

Any plane perpendicular to both the median and coronal planes is called a **transverse plane.** All transverse planes are horizontal (see figure 3.3). The terms superior (above or upward) and inferior (below or downward) describe the position of structures with respect to transverse planes. For example, as seen in figure 3.1, the ribs are superior to the innominate bones (hip bones) and the patellae (kneecaps) are inferior to the innominate bones. Similarly, the superior end of the right femur (thigh bone) articulates with the right innominate bone to form the right hip joint, and the inferior end of the right femur articulates with the right patella and right tibia to form the right knee joint.

To describe the precise location and orientation of specific features of a particular bone, it is usually necessary to use combinations of the six spatial terms that are applicable to all bones: lateral, medial, anterior, posterior, superior, and inferior. For example, a particular feature may be described as being at the superior lateral part of a bone; another feature may be described as being at the anterior inferior lateral part of the bone.

However, there are some spatial terms that apply to some bones, but not to others. For example, the terms proximal and distal are normally only used in reference to the long bones of the limbs. Superior features of these bones (with respect to the anatomical position) are referred to as proximal, whereas inferior features of these bones are referred to as distal. For example, in each arm the proximal end of the humerus (upper arm bone) articulates with its corresponding scapula to form the shoulder joint. The distal end of the humerus articulates with the proximal ends of the radius and ulna (forearm bones) to form the elbow joint. The distal ends of the radius and ulna articulate with the carpals (wrist bones) to form the wrist joint.

The names of the three reference planes are often used to describe sectional views of bones. For example, figure 3.4 shows a coronal section through the right hip joint. The term *longitudinal section* normally refers to a vertical section, as in figure 3.4; a longitudinal section may be in the median plane, a paramedian plane, a coronal plane, or some other vertical plane. The term *cross section* is a general term that may refer to a section in one of the reference planes or to an oblique plane (relative to the reference planes).

The Axial Skeleton

The axial skeleton consists of the skull, the vertebral column, and the rib cage. The skull consists of 29 fairly flat or irregular bones that encase the brain, provide bases for the major sense organs, and form the upper and lower jaws. The vertebral column comprises 26 irregular bones stacked on top of each other. The vertebral column supports the weight of the head, arms, and trunk, and provides protection for the spinal cord. The rib cage consists of 25 bones—the sternum (breastbone) and 12 pairs of ribs. The sternum is a rather flat bone. Even though the ribs are considerably curved, they are fairly flat in cross section. The rib cage is a flexible structure that provides protection for the heart and lungs, and is also very important in the ventilation of the lungs during breathing.

The Skull

The 29 bones of the skull comprise 8 cranial bones (cranium), 13 facial bones (face), 6 ear ossicles (3 in each middle ear), the mandible (lower jaw), and the hyoid bone (part of the larynx). The bones of the cranium and face form a single unit that makes up most of the skull (figure 3.5). The cranium encloses the brain and is made up of 8 relatively flat irregular bones.

As figures 3.6 through 3.8 show, the frontal bone forms the anterior and anterior superior part of the cranium including the forehead. The two parietal bones form a large part of the superior and lateral aspects of the cranium. The two temporal bones form a large part of the superior and lateral aspects of the base and sides of the cranium. The sphenoid bone together with the ethmoid bone and the inferior aspects of the frontal bone forms the anterior half of the base of the cranium. The occipital bone forms the posterior inferior aspect of the cranium and the major portion of the posterior half of the base of the cranium.

The temporal and occipital bones have a number of important features. On the inferior lateral aspect of each of the temporal bones is a funnel-shaped opening that leads on to an open channel called the external acoustic meatus, the external part of the ear (see figure 3.7). Just behind the external acoustic meatus is a rounded process projecting downward called the mastoid process. The

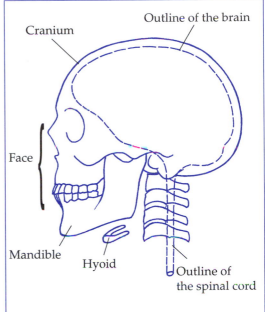

Figure 3.5 The main components of the skull in relation to the brain and spinal cord.

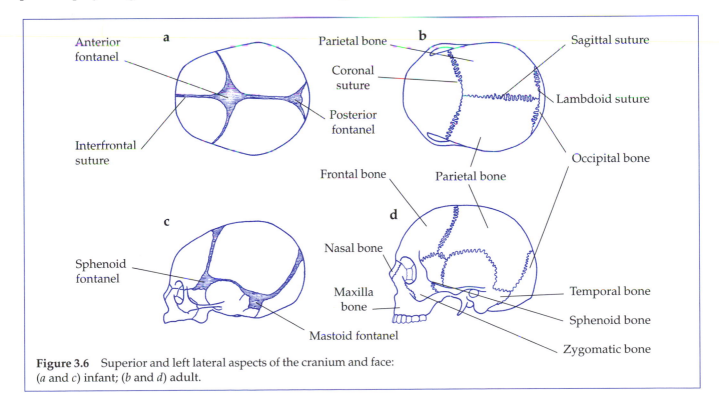

Figure 3.6 Superior and left lateral aspects of the cranium and face: (*a* and *c*) infant; (*b* and *d*) adult.

mastoid process can easily be felt as a bony bump beneath the skin just behind the ear. On the inferior aspect of each temporal bone is a relatively long slender process arising from the inferior aspect of the external acoustic meatus. This process is called the styloid process and projects medially, forward, and downward. Just in front of the external acoustic meatus there is a concave condyle called the mandibular fossa that articulates with the corresponding condyle of the mandible to form the temporoman-dibular joint. The styloid process provides part of the area of attachment for muscles that control the temporomandibular joint.

The occipital bone has a large hole, the foramen magnum, situated anteriorly. The foramen is occupied by the start of the spinal cord, which is continuous with the brain (see figure 3.5). Two occipital condyles articulate with the first vertebra (the atlas) and are situated on the anterior lateral aspects of the foramen magnum (see figure 3.7). A curved border between the inferior and posterior aspects of the occipital bone is the superior nuchal line. Anterior and concentric to the superior nuchal line is the inferior nuchal line. Running between the nuchal lines in the median plane is the external occipital crest. At the posterior end of the external occipital crest is a tuberosity called the external occipital protuberance. The muscles that hold the head upright (and tilt it backward) are attached to the area defined by these features.

The 13 bones of the face form the middle third of the anterior aspect of the skull (see figures 3.5 and 3.8a). The facial bones form the upper jaw, the inferior two-thirds of the eye sockets (orbits), and the anterior part of the nasal cavity. The two maxilla bones that join anteriorly in the median plane almost entirely form the upper jaw, which provides sockets for the upper teeth. The two palatine bones articulate with the posterior aspects of the maxillae to complete the upper jaw. The eye sockets are formed by parts of the frontal, sphenoid, and ethmoid bones, together with the two lacrimal bones and the two zygomatic (cheek) bones (see figures 3.6 and 3.8a).

The nasal cavity—the large cavity that stretches backward from the nose to the throat—is formed by bones of the cranium and face (see figure 3.8b). The anterior part of the nasal cavity, the bony part of the nose, is formed by the maxillae (floor and sides) and the two nasal bones (roof). The posterior part of the nasal cavity is much larger than the anterior part. The ethmoid bone forms the sides and part of the roof,

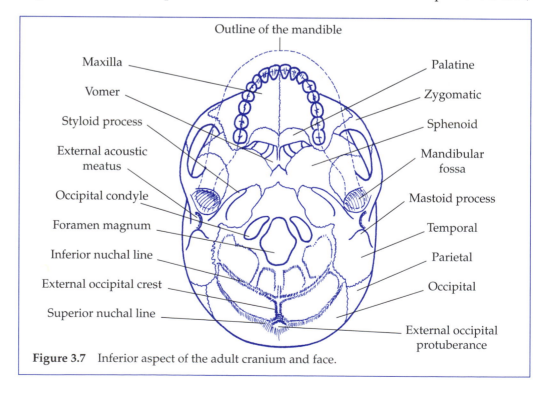

Figure 3.7   Inferior aspect of the adult cranium and face.

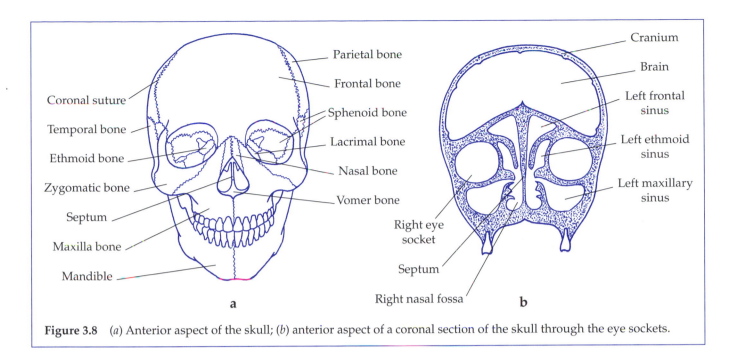

Figure 3.8   (*a*) Anterior aspect of the skull; (*b*) anterior aspect of a coronal section of the skull through the eye sockets.

while the frontal and sphenoid bones form the remainder of the roof. The sphenoid bone also forms the posterior wall. The vomer bone and the two inferior nasal conchae form the floor of the posterior part of the nasal cavity.

The nasal cavity is divided into left and right nasal fossae by a bony-cartilaginous septum in the median plane (see figure 3.8). The posterior part of the septum consists of a bony plate that projects downward from the ethmoid and articulates with the vomer. The anterior part of the septum consists of cartilage. The structure and functions of cartilage are described in detail in chapter 4.

The mandible, or lower jaw, basically consists of two L-shaped plates of bone that join anteriorly in the median plane. The upright part of each half of the mandible is called the ramus (branch) and the horizontal part, which provides sockets for the lower teeth, is called the body. At the posterior superior aspect of each ramus there is a convex condyle that articulates with the mandibular fossa on its corresponding temporal bone (see figure 3.7; figure 3.9). These joints enable the mandible to swing up and down, as in closing and opening the mouth, and to move from side to side. Chewing food involves a combination of these two types of movement.

The ear ossicles are tiny bones, all less than 1 cm in length, located in a chamber within each temporal bone known as the middle ear. The three ossicles—called the malleus, incus, and stapes—link the lateral and medial walls of the middle-ear chamber. They transmit sound waves from the outer part of the ear to the sound receptors in the inner part of the ear, which is medial to the middle ear.

The hyoid bone is not really part of the skull, but it is convenient to describe it in relation to the skull. The hyoid is a U-shaped bone suspended in front of the neck (in front of the fourth cervical vertebra) by ligaments from the styloid processes of the temporal bones (see figure 3.5). The hyoid forms part of the larynx (voice box) and provides attachment for some of the muscles that move the mouth and the tongue.

Sutures and Fontanels

In the adult skull, the edges of many of the bones are serrated so that the bones interlock closely with each other to form immovable joints. The serrated line of the joints is similar in appearance to a line of stitches and, for this reason, each joint is called a suture (see figure 3.6, *b* and *d*). The joints between the bones of the cranium

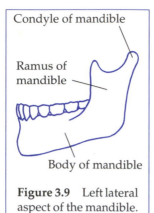

Condyle of mandible

Ramus of mandible

Body of mandible

Figure 3.9 Left lateral aspect of the mandible.

in an infant are also usually called sutures, even though the bones are joined by sheets of fibrous tissue and, consequently, do not interlock with each other (see figure 3.6, *a* and *c*). Fibrous tissue, described in more detail in chapter 4, is flexible, strong, and, in an infant, moderately elastic.

The flexible linkages allow the bones of the infant's cranium to override each other during the birth process. In doing so, the size of the cranium is effectively reduced, which facilitates an easier passage of the baby's head through the birth canal. After birth the normal orientation of the bones is quickly restored. At each of the angles (or corners) of the parietal bones, the fibrous tissue is in the form of a small sheet, called a *fontanel* (Williams et al. 1995). Since each parietal bone has four angles and the parietal bones join each other superiorly in the median plane, there are six fontanels. The anterior fontanel, which normally closes within the first 18 months, is at the junction of the parietal and frontal bones. At birth the frontal bone is in two halves, joined by the interfrontal or metopic suture; these halves normally fuse together within the first two years. The posterior fontanel is at the junction of the parietal and occipital bones and normally closes within the first two months. There are two sphenoid fontanels, one on each side of the skull at the junction of the parietal, frontal, sphenoid, and temporal bones. The sphenoid fontanels normally close within the first three months. There are also two mastoid fontanels, one on each side of the skull at the junction of the parietal, occipital, and temporal bones. The mastoid fontanels normally close within the first two years.

Sinuses

The frontal, ethmoid, sphenoid, maxillae, and temporal bones are partially hollow, resulting in cavities called sinuses. The sinuses communicate directly or indirectly with the nasal cavity by means of small channels (see figure 3.8*b*). Like the nasal cavity, the sinuses are filled with air. The frontal, ethmoid, sphenoid, and maxillary sinuses are linked directly to the nasal cavity and, as such, are referred to as paranasal sinuses. The sinuses in the temporal bones are linked indirectly to the nasal cavity via the middle ear—the temporal sinuses are linked to the middle ear and the middle ear is linked to the nasal cavity. During respiratory infections the sinuses may become inflamed, resulting in a painful condition called sinusitis. The sinuses make the skull marginally lighter and add resonance to the voice. However, they vary considerably in size among individuals and their precise function is not known (Williams et al. 1995).

> *The skull consists of 29 fairly flat or irregular bones that encase the brain, provide bases for the major sense organs, and form the upper and lower jaws. The main parts of the skull are the cranium (8 bones), face (13 bones), mandible, ear ossicles (3 in each ear) and the hyoid.*

The Vertebral Column

Prior to maturity, the vertebral column—also referred to as the backbone or spine—consists of 33 or 34 irregular bones called vertebrae. The vertebrae are divided into five fairly distinct groups: cervical, thoracic, lumbar, sacral, and coccygeal (figure 3.10, *a* and *b*). The neck consists of 7 cervical vertebrae. The thoracic or chest region consists of 12 thoracic vertebrae that provide articulation for the 12 pairs of ribs. The lower back consists of 5 lumbar vertebrae. The 5 sacral vertebrae form the posterior part of the pelvis; at maturity the sacral vertebrae fuse together to form the sacrum. The 4 or 5 coccygeal vertebrae are small and represent a vestigial tail. The coccygeal vertebrae normally fuse together at maturity to form the coccyx, or tailbone, which is approximately 3 cm long and is attached to the sacrum by ligaments.

Prior to maturity the vertebral column consists of 33 or 34 irregular bones called vertebrae, which fuse into 26 bones in the adult vertebral column. The function of the vertebral column is to provide a flexible supporting framework for the head, arms, and trunk.

When viewed from the side, the whole of the vertebral column of a newborn child is concave anteriorly (figure 3.10c). Between 3 and 6 months of age the child learns to hold her head upright and, as a result, the shape of the cervical region changes from concave anteriorly to convex anteriorly. Similarly, as the child learns to stand and walk—between 10 and 18 months of age—the shape of the lumbar region also changes from concave anteriorly to convex anteriorly. The cervical and lumbar curves are referred to as secondary curves since they develop as the child adopts an upright posture. The thoracic and sacrococcygeal curves are called primary curves since they are concave anteriorly throughout life. Figure 3.10a shows the shape of the adult vertebral column as viewed from the left lateral aspect.

Structure of a Vertebra

At birth, each vertebra, with the exception of the first two cervical vertebrae, consists of three bony elements united by cartilage (Williams et al. 1995) (figure 3.11a). The anterior element, the centrum or body, is basically a block of bone with slightly concave (waisted) sides and fairly flat kidney-shaped superior and inferior surfaces. The bodies of the vertebrae are mainly responsible for transmitting loads, especially the weight of the head, arms, and trunk. The posterior elements are curved struts that form the two halves of an arch, that is, the vertebral or neural arch. Each half of the arch consists of an anterior portion called the pedicle and a posterior portion called the lamina. The two laminae normally fuse together posteriorly during the first year (figure 3.11b). The pedicles normally fuse with the lateral superior posterior aspects of the body of the vertebra between the third and sixth years (figure 3.11c). The hole formed by the arch and the posterior aspect of the body, through which the spinal cord travels, is called the vertebral foramen. The vertebral arch mainly functions to protect the spinal cord.

After fusion of the laminae, seven processes arise from the arch. The spine of the vertebra extends backward from the point of fusion of the laminae. On each side of the arch, three processes arise from the junction of the pedicle and lamina. A transverse process extends laterally, a superior articular process extends upward, and an inferior articular process extends downward (figure 3.12). The spine and transverse processes are basically levers that provide areas of attachment for muscles, tendons, and ligaments. With respect to each pair of adjacent vertebrae, the superior articular processes of the lower vertebra articulate by means of facets with the inferior articular processes of the upper vertebra (figure 3.12a). These joints are called facet joints or apophyseal joints. In most upright postures the facet joints transmit some load. In general, the load transmitted by these

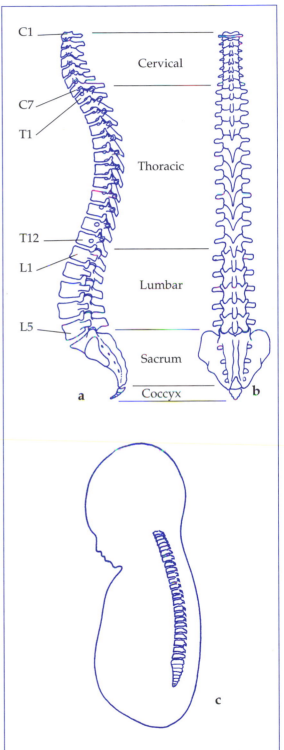

Figure 3.10   The vertebral column; (*a*) left lateral aspect (adult); (*b*) posterior aspect (adult); (*c*) left lateral aspect (infant).

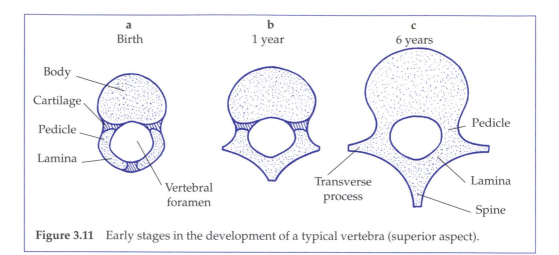

Figure 3.11   Early stages in the development of a typical vertebra (superior aspect).

joints decreases with flexion of the trunk (bending forward) and increases with extension of the trunk (bending backward). In addition to load transmission, the orientation of the superior and inferior articular processes somewhat determines the type and range of movement between adjacent vertebrae. The structure and function of facet joints is covered in chapter 6.

Each pair of adjacent vertebrae, except the first two cervical vertebrae, are joined by a tough rubbery disc of fibrocartilage called an intervertebral disc to form an intervertebral joint. Movement between adjacent vertebrae occurs due to deformation of the intervertebral discs (mainly in response to bending and torsion loads), and by sliding in the facet joints.

On each side of the vertebral arch is a depression in the superior aspect of the pedicle called the superior vertebral notch (see figure 3.12b). Since the pedicle joins the posterior superior aspect of the body, there is a much larger inferior vertebral notch beneath the pedicle. With respect to each pair of adjacent vertebrae, the inferior vertebral notch of the upper vertebra and the superior vertebral notch of the lower vertebra form a hole called the intervertebral foramen (figure 3.13a). A spinal (peripheral) nerve occupies the intervertebral foramen (figure 3.13b).

Distinguishing Features of Vertebrae

The vertebrae gradually increase in size from the second cervical vertebra down to the sacrum. This gradual increase reflects the gradual increase in weight the vertebrae have to support (see figure 3.10a). There are also changes in the size, shape, and orientation of the processes of the vertebrae. These changes are fairly gradual within each of the cervical, thoracic, and lumbar regions, but tend to be more marked at the junctions between the regions (see figure 3.10, a and b). The vertebrae in each region of the column have characteristics that distinguish them from vertebrae in other regions.

Cervical   All cervical vertebrae (C1 to C7) have a hole called a transverse foramen in each of their transverse processes. Only cervical vertebrae have this characteristic (figure 3.14). The first cervical vertebra is called the atlas. It has no body and basically consists of an anterior arch and a posterior arch, which together form a bony ring (figure 3.15a). The facets on the superior articular processes of the atlas articulate with the occipital condyles to form the atlanto-occipital joint linking the skull with the vertebral column. The spine of the atlas is rudimentary and the transverse processes of the atlas, as in all the other cervical vertebrae, are short.

The second cervical vertebra is called the axis. The axis has a body and a vertebral arch. Projecting upward from the superior aspect of the body of the axis is a process called the dens or odontoid process (figure 3.15, b and c). A facet on the anterior aspect

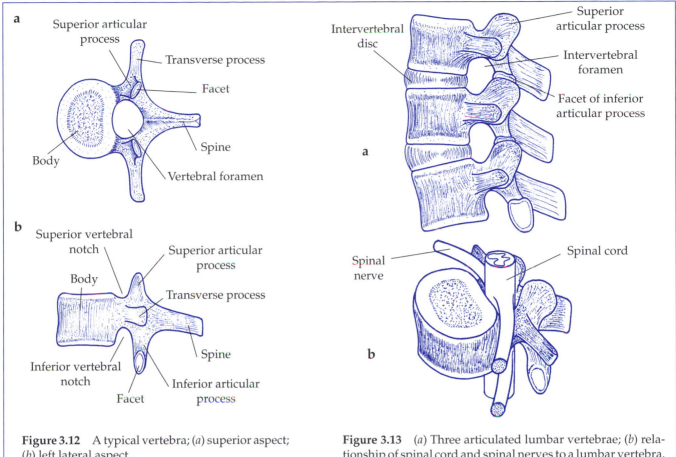

Figure 3.12 A typical vertebra; (*a*) superior aspect; (*b*) left lateral aspect.

Figure 3.13 (*a*) Three articulated lumbar vertebrae; (*b*) relationship of spinal cord and spinal nerves to a lumbar vertebra.

of the dens articulates with a facet on the posterior aspect of the anterior arch of the atlas. The dens is held against the anterior arch of the atlas by the transverse ligament of the atlas, which spans the posterior part of the anterior arch of the atlas and runs in a groove on the posterior aspect of the dens (figure 3.15, *a*, *c*, and *d*). To a certain extent the atlas rotates (in a transverse plane) around the dens; hence the name of the axis.

The dens represents the major portion of the body of the atlas that, during fetal growth, separates from the rest of the body of the atlas and fuses with the axis. There is no intervertebral disc between the atlas and axis. The spine of the axis is fairly short and the tip of the spine is bifid, that is, divided into two branches (see figure 3.15*b*).

The remaining five cervical vertebrae (C3, C4, C5, C6, and C7) are similar in that each consists of a kidney-shaped body and a vertebral arch (see figure 3.14). The spines of C3 to C6 are all bifid and gradually increase in length. The spine of C3 is slightly shorter than that of the axis, and the spine of C6 is slightly longer than that of the axis. The spine of C7 is much longer than that of C6 and can easily be felt as a prominence at the posterior inferior aspect of the neck. Due to its unusually long spine, C7 is sometimes referred to as the vertebra prominens.

The facets of the superior and inferior articular processes of the cervical vertebrae articulate in oblique planes that slope downward laterally and posteriorly. The orientation of the facet joints, the short transverse processes, the relatively thick intervertebral discs, and the relatively short spines of C3 to C6 all combine to give a fairly large range of movement in the cervical region as a whole compared to other regions of the column.

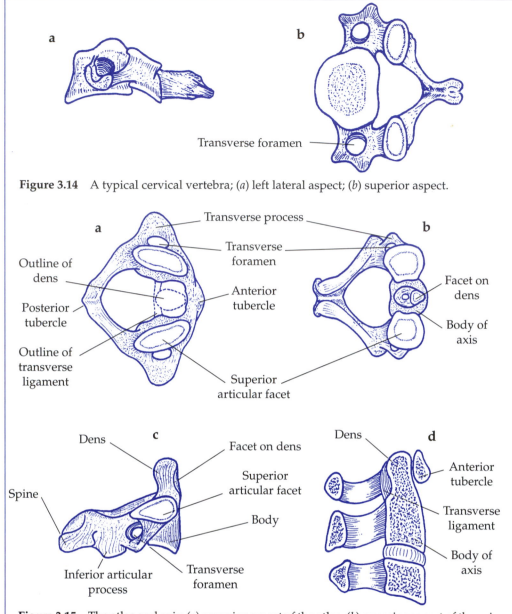

Figure 3.14 A typical cervical vertebra; (*a*) left lateral aspect; (*b*) superior aspect.

Figure 3.15 The atlas and axis; (*a*) superior aspect of the atlas; (*b*) superior aspect of the axis; (*c*) right lateral aspect of the axis; (*d*) median section through the atlas, axis, and third cervical vertebra.

Thoracic Thoracic vertebrae (T1 to T12) can be identified by the presence of facets on the bodies for articulation with the heads (posterior ends) of the ribs. The upper 10 thoracic vertebrae also articulate with the corresponding pairs of ribs by means of facets on the anterior lateral aspects of the transverse processes (figure 3.16). On each side of the body of T1 there is a superior whole facet and an inferior demifacet (half-facet); the heads of the uppermost pair of ribs articulate with the whole facets on the sides of T1 (figure 3.16*a*). On each side of the bodies of T2 to T8 there is a superior demifacet and an inferior demifacet. The heads of the second pair of ribs articulate, on the corresponding side, with the inferior demifacet of T1 and the superior demifacet of T2 (figure 3.16*a*). On each side of the body of T9 there is a superior demifacet and the heads of the third to ninth pairs of ribs articulate with the sides of the bodies of T2 to T9 in the same manner as the second pair of ribs. On each side of the bodies of T10 (figure 3.16*b*) to T12 there is a whole facet that articulates with the tenth to twelfth pairs of ribs.

The spines of the thoracic vertebrae are fairly long and tend to closely overlap each other, especially in the middle of the region (see figure 3.10*a*). The transverse processes of the thoracic vertebrae are also fairly long; they gradually decrease in length from T1 to T12. The superior and inferior articular facets articulate in a plane that slopes sharply downward posteriorly. The overlapping spines, relatively thin intervertebral discs, and splinting effect of the ribs result in a smaller overall range of movement in the thoracic region than in the cervical region.

Lumbar The lumbar vertebrae (L1 to L5) have fairly long transverse processes and large, flat spines, rectangular in shape (figure 3.17). The main distinguishing feature of the lumbar vertebrae is the orientation of the facets on the superior and inferior articular processes. The facets on the superior articular processes face medially and posteriorly, and the facets on the inferior articular processes face laterally and anteriorly (figure 3.17). The orientation of the facet joints severely limits rotation of the lumbar vertebrae about a vertical axis. However, the relatively thick intervertebral discs in the lumbar region ensure a much greater range of movement in other directions.

Sacrum The sacral vertebrae (S1 to S5) become progressively smaller from S1 through S5. The sacrum is formed by the fusion or partial fusion of the sacral vertebrae. When viewed from the front (or the back) the sacrum is more or less triangular with the apex pointing downward (figure 3.18*a*). The anterior edge of the upper surface of the first sacral vertebra projects forward and is called the sacral promontory. The anterior aspect of the sacrum is concave, largely due to the orientation of S3, S4, and S5 (figure 3.18*b*). However, in the anatomical position the large upper portion of the sacrum (S1 and S2) is angled downward and backward, which tends to accentuate the lumbar curve (see figure 3.10*a*).

The sacrum is sandwiched between the left and right hip bones and thus provides a firm base for the rest of the vertebral column. The anterior aspect of the sacrum is fairly smooth except for four transverse lines resulting from the fusion or partial fusion of the bodies of the sacral vertebrae (figure 3.18*a*). At the end of each of the transverse lines is a hole called the sacral foramen. The four pairs of sacral foramen are formed by the fusion of the ends of the transverse processes of the sacral vertebrae.

In comparison to the fairly smooth anterior surface, the posterior surface of the sacrum provides attachment for a large number of ligaments and aponeuroses and is quite rough. The posterior surface has five ridges (called sacral crests), that run vertically parallel to one another. The fusion of the upper four sacral spines forms the median sacral crest. The intermediate sacral crests and lateral crests (two each) are formed by the fusion of the articular processes (except for the superior processes of S1) and the ends of the transverse processes of the sacral vertebrae, respectively (figure 3.18*b*). Processes called sacral tubercles or transverse tubercles, which give the crests an undulating appearance, mark the sites of

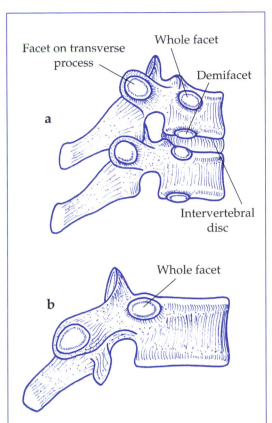

Figure 3.16 Thoracic vertebrae; (*a*) right lateral aspect of the first two thoracic vertebrae; (*b*) right lateral aspect of the tenth thoracic vertebra.

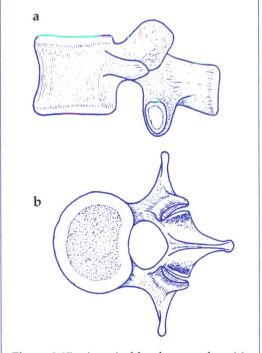

Figure 3.17 A typical lumbar vertebra; (*a*) left lateral aspect; (*b*) superior aspect.

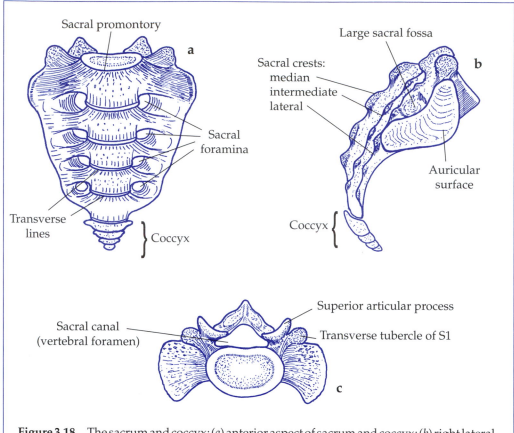

Figure 3.18   The sacrum and coccyx; (*a*) anterior aspect of sacrum and coccyx; (*b*) right lateral aspect of sacrum and coccyx; (*c*) superior aspect of the sacrum.

fusion. In addition to these tubercles there are usually two other tubercles, one on each side, that project posterior laterally from the body of S1 just lateral to the superior articular processes of S1; these tubercles are referred to as the transverse tubercles of S1 (figure 3.18*c*). On each superior lateral aspect of the sacrum is a fairly large C- or L-shaped articular surface called the auricular surface (auricle = ear-shaped) (figure 3.18*b*). The auricular surfaces are formed by the lateral expansions of the fused transverse processes of S1, S2, and S3. The auricular surfaces of the sacrum articulate with the hip bones (innominate bones) to form the sacroiliac joints (see chapter 6).

A fairly large oval or circular sacral fossa is adjacent to the posterior border of the angle of each auricular surface. There is usually a smaller sacral fossa at the distal end of each auricular surface (figure 3.18*b*). The sacrum and the left and right hip bones form a complete bony ring called the pelvis or pelvic girdle. Consequently, the sacrum is an important part of the vertebral column and the pelvis.

> *Most of the vertebrae consist of a body (responsible for transmitting loads), a vertebral arch (responsible for protecting the spinal cord), and seven processes arising from the arch. Three of the processes (spine and transverse processes) provide areas of attachment for muscles, tendons, and ligaments. The other processes, the superior and inferior articular processes that form facet joints, help to transmit loads in most postures and help to determine the type and range of movement between adjacent vertebrae.*

The Rib Cage

The rib cage is roughly the shape of an upright cone partially flattened from front to back (figure 3.19). The rib cage consists of 12 pairs of ribs and the sternum. Since the ribs form the major part of the rib cage (most of the wall of the upright cone), they have a distinct curved shape (figure 3.20). The heads (posterior ends) of the ribs articulate with the thoracic vertebrae as previously described. The anterior ends of the upper 10 pairs of ribs are attached to the sternum by pieces of cartilage called costal cartilages (costa = rib). The upper seven pairs of ribs are attached to the sternum by separate costal cartilages and are therefore sometimes referred to as true ribs. The costal cartilages of the eighth, ninth, and tenth pairs of ribs fuse with each other before fusing with the costal cartilages of the seventh ribs (see figure 3.19). Consequently, whereas the upper seven pairs of ribs have direct cartilaginous attachments to the sternum, the eighth, ninth, and tenth pairs of ribs have an indirect cartilaginous attachment to the sternum. The lower two pairs of ribs do not attach onto the sternum at all; the anterior ends of these ribs are free, and these ribs are referred to as floating ribs. Because none of the lower five pairs of ribs has a direct cartilaginous or other attachment to the sternum, these ribs are sometimes referred to as false ribs.

The Ribs

The upper 10 pairs of ribs are attached to the body and transverse processes of corresponding thoracic vertebrae. In each of these ribs the head is separated from the facet that articulates with the transverse process by a short neck (see figure 3.20). Laterally adjacent to the facet is a tubercle, which provides attachment for ligaments that support the vertebral column. Since the facet and tubercle are adjacent to each other, the facet is often called the tubercular facet. Lateral to the tubercle, at about the same distance from the tubercle as the head, the rib bends abruptly forward; this bend in the rib is called the angle of the rib, and it provides attachment for muscles of the back. The part of a rib between the angle and the anterior end is called the shaft.

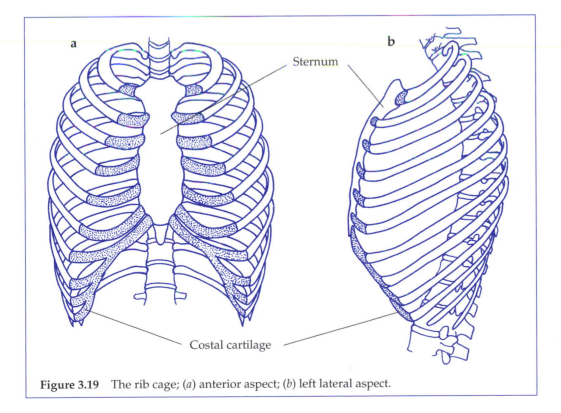

Sternum

Costal cartilage

Figure 3.19   The rib cage; (*a*) anterior aspect; (*b*) left lateral aspect.

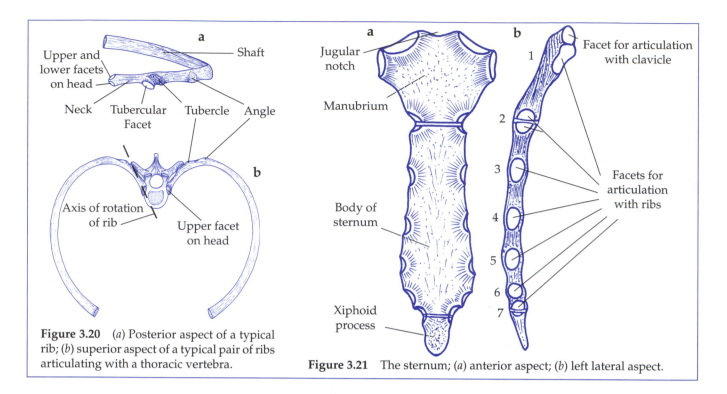

Figure 3.20 (*a*) Posterior aspect of a typical rib; (*b*) superior aspect of a typical pair of ribs articulating with a thoracic vertebra.

Figure 3.21 The sternum; (*a*) anterior aspect; (*b*) left lateral aspect.

The Sternum

The sternum is a fairly flat bone and consists of three parts: manubrium, body, and xiphoid process (figure 3.21). The manubrium occupies the upper quarter of the sternum. In the center of its superior border is a depression called the jugular notch or suprasternal notch (supra = above). On each side of this notch is a facet for articulation with the medial end of the corresponding clavicle (collarbone). On each of the lateral aspects of the manubrium is a facet for articulation with the costal cartilage of the corresponding first rib. At each end of the transverse junction between the manubrium and the body of the sternum is a facet for articulation with the costal cartilage of the corresponding second rib. The body of the sternum, which occupies over half the length of the sternum, also has facets for articulation with the costal cartilages of the third to seventh pairs of ribs. The costal cartilages of the seventh pair of ribs are attached at each end of the transverse junction between the body of the sternum and the xiphoid process. The xiphoid process provides attachment for some of the abdominal muscles; it consists of cartilage that usually becomes transformed into bone by the age of 40 (Tortora and Anagnostakos 1984).

Movement of the Ribs

The spaces between the ribs are called intercostal spaces; muscles largely occupy these spaces. These muscles in association with other muscles of the thorax move the ribs during breathing. During inspiration (breathing in) each rib swings upward and outward about an oblique axis that passes through the costovertebral joints of the rib—the joints between tubercular facet and transverse process and between head and vertebral body or bodies (see figure 3.20*b*). The upward and outward movement of the ribs is accompanied by a slight upward and forward swing of the body of the sternum about the manubrium. The combined movements of the ribs and sternum decrease the pressure inside the thorax, and air rushes into the lungs. During expiration (breathing out) the ribs and sternum are pulled back down by the elasticity of both the costal cartilages and the lungs themselves, and by the pull of various ligaments and muscles that are stretched during inspiration.

The rib cage consists of the sternum and 12 pairs of ribs. The rib cage as a whole is a fairly flexible structure providing protection for the heart and lungs, and is also very important in the ventilation of the lungs during breathing.

The Appendicular Skeleton

The appendicular skeleton consists of the bones of the upper and lower limbs. In an adult, 32 bones make up each upper limb and 31 bones make up each lower limb; the appendicular skeleton as a whole is comprised of 126 bones.

The Upper Limb

For descriptive purposes, I've divided each upper limb into five regions: shoulder (scapula and clavicle), upper arm (humerus), lower arm (radius and ulna), wrist (8 carpals), and hand (5 metacarpals and 14 phalanges) (figure 3.22).

The upper limbs consist of 64 bones, 32 bones in each limb, and can be divided into five regions: shoulder, upper arm, lower arm, wrist, and hand.

The Shoulder

The shoulder region consists of the scapula (shoulder blade) and clavicle (collarbone). Together with the manubrium, the scapulae and clavicles of both upper limbs form an incomplete ring of bone called the shoulder girdle (figure 3.23). The arms are suspended from the shoulder girdle.

The medial end of each clavicle articulates with the manubrium to form a sternoclavicular joint, and the lateral end of each clavicle articulates with the acromion process of the corresponding scapula to form an acromioclavicular joint (figures 3.23 and 3.24). The scapulae are not joined to the axial skeleton, but are held

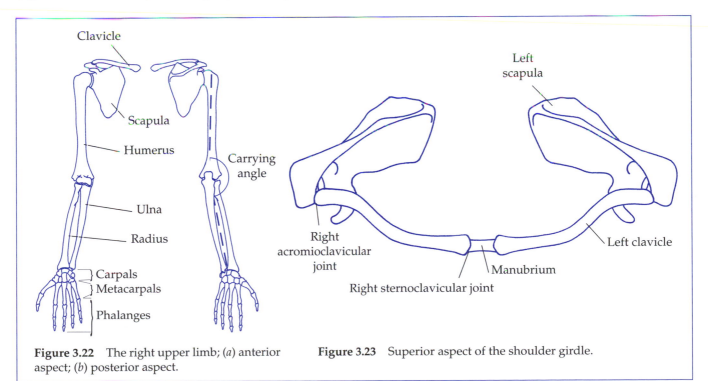

Figure 3.22   The right upper limb; (*a*) anterior aspect; (*b*) posterior aspect.

Figure 3.23   Superior aspect of the shoulder girdle.

in position at the lateral superior posterior aspects of the rib cage by muscles. Consequently, each scapula has a considerable range of movement. Most movements of the shoulder region involve movements at the sternoclavicular and acromioclavicular joints (see chapter 7).

From a superior aspect, each clavicle is S-shaped—concave anterior laterally and posterior medially (figures 3.23 and 3.24a). In a transverse plane the lateral one-third of each clavicle is fairly flat; however, the medial two-thirds becomes progressively thicker and more rounded toward the medial end (figure 3.24b). In addition to forming the only bony articulations of the upper limb with the axial skeleton, the clavicles act as horizontal struts that maintain the lateral position of the scapulae and, as such, give width to the shoulders.

Each scapula consists of a relatively large, flat, triangular portion called the blade, with three prominent features arising from the blade (figure 3.24, c, d, and e). The apex of the blade is called the inferior angle and points directly downward. The large anterior surface of the blade is called the subscapular fossa. A large process called the spine of the scapula arises from an oblique line running laterally and upward across the upper third of the posterior surface of the blade. Like the blade, the spine is fairly flat and triangular in shape; the posterior border of the spine forms a crest that can easily be felt beneath the skin. The superior surface of the spine and the posterior surface of the blade above the spine form a V-shaped trough called the supraspinous fossa. The inferior surface of the spine and the large posterior surface of the blade below the spine form a large area called the infraspinous fossa.

Projecting laterally and slightly upward from the lateral end of the spine is the acromion process forming the posterior part of a bony-fibrous arch above the shoulder joint. The acromion process can also be felt beneath the skin at the tip of the shoulder. At the superior lateral angle of the blade there is a fairly large oval-shaped shallow articular surface called the glenoid fossa. In the anatomical position the glenoid fossa faces laterally, slightly forward, and slightly upward. The glenoid fossa forms the shoulder joint (glenohumeral joint) with the head of the humerus.

Arising from the superior anterior part of the base of the glenoid fossa is a fingerlike projection called the coracoid process; this process curves laterally so that its tip is in front of the shoulder joint. The coracoid process, acromion process, crest of the spine and medial border of the blade all provide areas of attachment for muscles that are mainly concerned with the movement of the scapula and clavicle about the sternoclavicular and acromioclavicular joints, and the humerus about the shoulder joint. In contrast, the large anterior and posterior surfaces of the blade (subscapular, supraspinous, and infraspinous fossae) provide attachment for muscles that are mainly concerned with stabilizing the shoulder joint—keeping the glenoid fossa and head of the humerus in close contact.

The shoulder region consists of the scapula and the clavicle. The medial end of the clavicle articulates with the manubrium to form the sternoclavicular joint linking the upper limb and the axial skeleton. The lateral end of the clavicle articulates with the scapula to form the acromioclavicular joint. Together with the manubrium, the scapulae and clavicles of both upper limbs form an incomplete ring called the shoulder girdle.

The Upper Arm

There is only one bone in the upper arm, the *humerus*. The humerus is a typical long bone consisting of a relatively long shaft between two fairly bulbous ends (figure 3.25). The proximal end of the humerus has four main features: head, greater tuberosity, lesser tuberosity, and bicipital groove. The head is an almost perfect hemisphere, which, as mentioned above, articulates with the glenoid fossa to form the shoulder joint. The head faces medially, upward, and backward. The articular surface

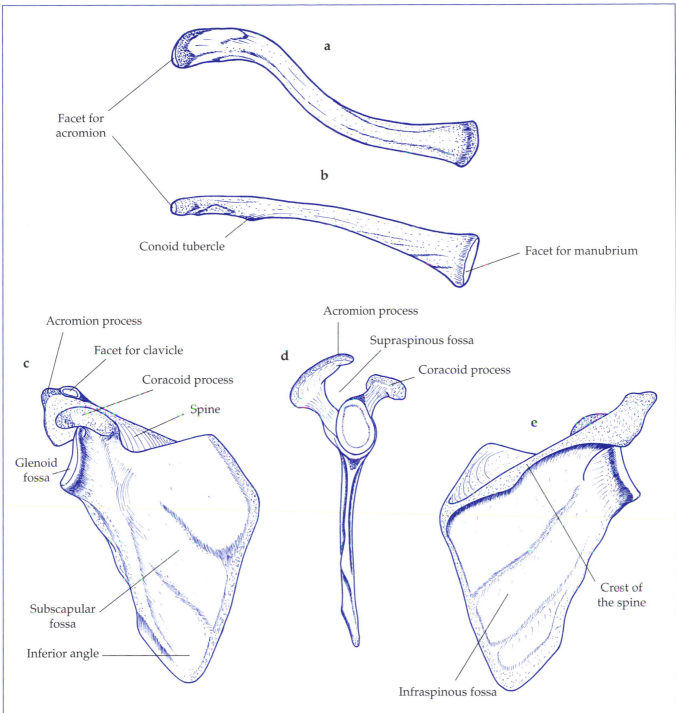

Figure 3.24   The clavicle and scapula; (*a*) superior aspect of right clavicle; (*b*) anterior aspect of right clavicle; (*c*) anterior aspect of right scapula; (*d*) lateral aspect of right scapula; (*e*) posterior aspect of right scapula.

of the head is much larger than that of the glenoid fossa. This difference in size between the articulating surfaces combined with the shallowness of the glenoid fossa permits a large range of movement in the shoulder joint. Adjacent to the head, occupying the whole of the lateral aspect of the proximal end of the humerus, is the greater tuberosity. Adjacent to the head on the anterior aspect is the much smaller lesser tuberosity, which projects directly forward. Running vertically downward

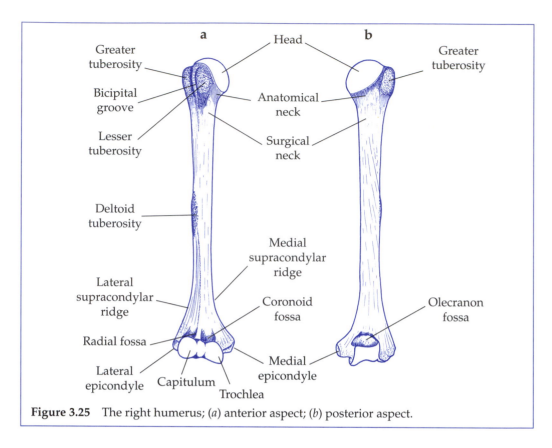

Figure 3.25 The right humerus; (*a*) anterior aspect; (*b*) posterior aspect.

between the two tuberosities is the bicipital groove (intertubercular groove). The head is separated from the two tuberosities by a rather ill-defined anatomical neck, and the proximal end as a whole is joined to the main shaft by a short, tapered surgical neck.

The upper two-thirds of the shaft of the humerus is more or less cylindrical. The lower one-third gradually becomes broader (in the coronal plane) toward the distal end. The surface of the shaft is fairly smooth apart from one roughened area in the middle of the anterior lateral aspect—the deltoid tuberosity—where the deltoid muscle attaches to the bone.

The distal end of the humerus has a cylindrical-shaped articular surface consisting of two condyles fused together side by side. The lateral condyle is called the capitulum and the larger pulley-shaped medial condyle is called the trochlea. On the anterior aspect just above the capitulum is a small depression called the radial fossa. There is a similar depression called the coronoid fossa just above the trochlea. On the posterior aspect there is a relatively large depression just above and continuous with the trochlea—the olecranon fossa. The medial epicondyle, easily felt beneath the skin on the medial aspect of the elbow, projects medially from the trochlea. The smaller lateral epicondyle projects laterally from the capitulum. Extending upward from the lateral epicondyle to the main part of the shaft is a distinct ridge called the lateral supracondylar ridge. A similar ridge, the medial supracondylar ridge, extends upward from the medial epicondyle.

The humerus is the one bone in the upper arm. The proximal end articulates with the scapula to form the shoulder joint (glenohumeral joint), and the distal end articulates with the radius and ulna to form the elbow joint. The humerus is a typical long bone consisting of a relatively long shaft between two expanded ends; the expanded ends increase the area of articulation in the shoulder and elbow joints and, thus, decrease the compression stress in these joints.

The Lower Arm

There are two long bones in the lower arm (forearm), the radius and the ulna. In the anatomical position, the radius is lateral to the ulna (see figure 3.22; figure 3.26). The anterior aspect of the proximal end of the ulna is dominated by a large pulley-shaped concave articular surface—the trochlea notch. This notch articulates with the trochlea of the humerus to form part of the elbow joint (figure 3.26). The proximal half of the trochlea notch forms the anterior part of the olecranon (or olecranon process). The tip of the elbow, easily felt beneath the skin, is the posterior superior point of the olecranon. When the elbow is fully extended, the proximal part of the rim of the trochlea notch occupies the olecranon fossa on the posterior aspect of the humerus (see figure 3.25b). The distal half of the trochlea notch forms the anterior superior part of the coronoid process projecting anteriorly from the shaft of the ulna. When the elbow is fully flexed, the coronoid process occupies the coronoid fossa on the anterior aspect of the humerus (see figure 3.25a). The anterior inferior aspect of the coronoid process, together with a small part of the shaft with which it is continuous, is usually roughened. This unnamed area is the area of attachment of one of the muscles (brachialis) that flex the elbow joint. Adjacent to and continuous with the inferior lateral edge of the trochlea notch is a smaller articular surface called the radial notch (figure 3.26b).

The shaft of the ulna tapers slightly from the proximal to the distal end. Whereas most of the shaft is fairly smooth, the lower two-thirds of the lateral aspect has a rather sharp ridge called the interosseous border of the ulna. The distal end of the ulna has a small drum-shaped head with a small projection on its posteromedial aspect called the styloid process of the ulna.

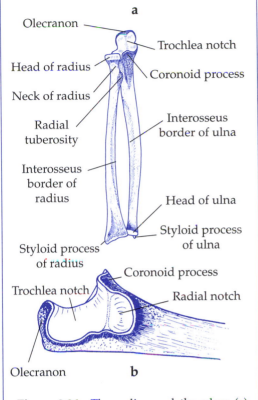

Figure 3.26   The radius and the ulna; (a) anterior aspect of right radius and ulna; (b) lateral aspect of proximal end of right ulna.

The proximal end of the radius consists of a drum-shaped head separated from the main part of the shaft by a short cylindrical neck (figure 3.26a). The circular side of the head and the superior surface of the head form a continuous articular surface. The side articulates with the radial notch on the ulna, and the superior surface articulates with the capitulum on the humerus. The elbow joint consists of the joints between the trochlea and trochlea notch, and the capitulum and head of the radius. In the elbow joint, the trochlea notch and superior surface of the head of the radius form a virtually continuous articular surface.

A roughened projection at the anterior medial part of the base of the neck is the radial tuberosity. The shaft of the radius is fairly smooth apart from a sharp ridge along the medial aspect called the interosseous border of the radius. In contrast to the ulna, the distal end of the radius is much thicker than the proximal end. The lateral part of the distal end of the radius forms a small projection called the styloid process of the radius. The inferior aspect of the distal end is dominated by a fairly large, more or less, quadrangular concave articular surface. Adjacent to and continuous with the medial edge of this surface is the ulnar notch, a small articular surface. This notch articulates with the side of the head of the ulna.

In the anatomical position, the radius and ulna lie side by side with their long axes more or less parallel to each other (see figure 3.22a). With the radius and ulna in this position, the lower arm is described as supinated. The radius is able to move with respect to the ulna by means of the proximal and distal joints between the two bones. As the head of the radius rotates within the radial notch, the distal end of the radius moves around the head of the ulna; the radius as a whole rotates about an axis passing through the head of the radius and the head of the ulna (figure 3.27, a and b). Consequently, medial rotation of the radius about the ulna from the anatomical

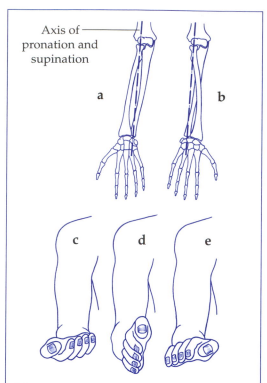

Axis of pronation and supination

a

b

c

d

e

Figure 3.27 Supination and pronation of the right forearm; (a) forearm supinated; (b) forearm pronated; (c) maximum supination; (d) neutral position; (e) maximum pronation.

position results in the radius crossing over the ulna (figure 3.27b). When the radius crosses over the ulna the lower arm is described as pronated.

The anatomical position is close to the position of extreme supination, and the position in which the radius is fully crossed over the ulna represents the position of extreme pronation. When the forearm moves from the position of extreme supination to that of extreme pronation the hand is rotated about its long axis through approximately 180° (figure 3.27, c and e). With the rest of the upper limb in the anatomical position, the position of the forearm in which the plane of the hand occupies a paramedian plane is usually referred to as the neutral position (figure 3.27d).

In the anatomical position the long axes of the upper and lower arms do not coincide and form an obtuse angle on the lateral aspect (see figure 3.22b). This angle is called the carrying angle and tends to be in the region of 165° in females and 175° in males. The carrying angle is largely due to the shape of the trochlea of the humerus. The medial end of the trochlea projects downward approximately 6 mm farther than the lateral end, which has the effect of tilting the ulna outwards (Williams et al. 1995). The functional significance of the carrying angle is not clear, but it is thought to increase the precision with which the hand can be controlled in movements involving elbow extension combined with pronation of the forearm (Williams et al. 1995). The significance of the difference in the size of the carrying angle between females and males is unknown, but it may be related to the relatively narrow shoulders, small waist, and broad hips in females compared to males.

The radius and ulna are the two long bones in the lower arm. The proximal ends of the radius and ulna articulate with the distal end of the humerus to form the elbow joint, and the distal ends articulate with the proximal row of carpals to form the wrist joint. The radius and ulna articulate with each other at their proximal and distal ends such that the radius is able to move relative to the ulna. Movement of the radius toward the anatomical position is called supination and movement of the radius away from the anatomical position is called pronation.

The Wrist and Hand

The wrist consists of eight small irregular bones called carpals. When articulated, the carpals form the carpus linking the distal ends of the radius and ulna to the proximal end of the hand (figure 3.28). The carpals are closely packed together; seven of them articulate with three or four other bones from amongst the other carpals, the radius, ulna, and metacarpals of the hand. The carpals, whose names tend to reflect their shapes, are arranged in a proximal and a distal row. From lateral to medial, the proximal row consists of the scaphoid (boat shaped), lunate (half-moon shaped), triquetrum (triangular), and pisiform (pea shaped). The proximal surfaces of the scaphoid, lunate, and triquetrum form a biconvex (surface rounded outwards) elliptical surface that articulates with the biconcave (surface rounded inwards) elliptical surface formed by the distal ends of the radius and ulna. The joints between these two elliptical surfaces constitute the wrist joint. The pisiform has one articulation with the anterior medial aspect of the triquetrum. From lateral to medial the distal row of carpals consists of the trapezium (four sided with two parallel sides), trapezoid (four sided), capitate (the central carpal), and hamate (with a distinct hooklike process anteriorly). The series of joints between the proximal and distal rows constitute the midcarpal joint.

> *The wrist consists of eight small irregular bones called carpals that articulate with each other to form the carpus. The carpals are arranged in a proximal and distal row. The proximal row articulates with the distal ends of the radius and ulna to form the wrist joint. The series of joints between the proximal and distal rows is called the midcarpal joint.*

The hand consists of 5 metacarpals and 14 phalanges (or digits) (see figure 3.28). In life, the metacarpals are joined together by soft tissues and form the palm of the hand on the anterior aspect. The metacarpals are miniature long bones and each metacarpal consists of a base (the proximal end), a shaft, and a head (the distal end). The proximal surfaces of the bases of the metacarpals articulate with the distal row of carpals to form the carpometacarpal joints. The combined ranges of movement in the wrist, midcarpal, and carpometacarpal joints facilitate a large range of movement for the hand as a whole. The bases of the medial four metacarpals articulate with each other side by side. Although the base of the first metacarpal (thumb) is close to that of the second metacarpal (index finger), the bases of these two metacarpals do not usually articulate with each other. The shafts of the metacarpals are fairly flat posteriorly and fairly rounded anteriorly. The head of each metacarpal has a convex condylar articular surface.

Each of the four fingers consists of three phalanges (proximal, middle, and distal), whereas the thumb has only two (proximal, distal). In each finger and the thumb the phalanges become progressively smaller from proximal through to distal. Like the

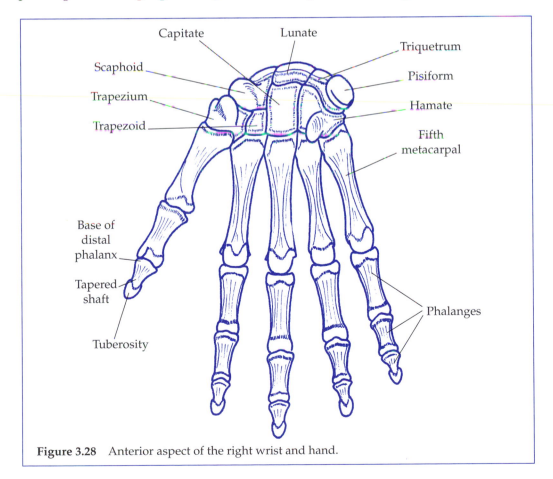

Figure 3.28   Anterior aspect of the right wrist and hand.

metacarpals, each phalanx consists of a base (proximal), a shaft, and a head (distal). The proximal end of the base of each proximal phalanx consists of a concave condyle articulating with the head of its corresponding metacarpal to form a metacarpophalangeal joint. Whereas the shafts of the metacarpals are more or less cylindrical, the shafts of the phalanges are almost semicircular in cross section; the posterior surface of each phalanx is fairly flat and the anterior surface is rounded. The joints between the phalanges, the interphalangeal joints (two joints in each finger and one joint in the thumb) are similar in terms of the shape of the articulating surfaces. The heads of the proximal and middle phalanges all have a pulley-shaped articular surface made up of a lateral condyle and a medial condyle, both of which are convex. Each of these pulley-shaped heads articulates with a biconcave condylar surface on the base of the corresponding middle or distal phalanx. The distal phalanges are quite small, especially those of the fingers. Each distal phalanx has a relatively broad base, a tapered shaft, and a rounded tuberosity at the head.

The hand consists of 5 metacarpals and 14 phalanges that are all miniature long bones. Each finger consists of three phalanges, whereas the thumb has only two phalanges. The joints between the phalanges, two joints in each finger and one joint in the thumb, are called interphalangeal joints.

The Lower Limb

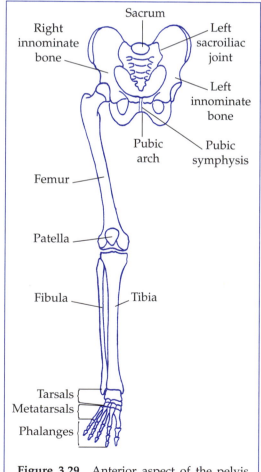

Figure 3.29 Anterior aspect of the pelvis and right lower limb.

In the discussion below I've divided each lower limb into four regions: hip (innominate), upper leg (femur and patella), lower leg (tibia and fibula), foot (7 tarsals, 5 metatarsals, and 14 phalanges; figure 3.29).

The lower limbs consist of 62 bones (31 bones in each limb).

The Hip

Together with the sacrum, the right and left innominate bones form a complete ring of bone called the pelvis or pelvic girdle (see figure 3.29). Consequently, the innominate bones attach the legs to the axial skeleton. Each innominate bone develops from three bones called the ilium, ischium, and pubis, which fuse together at maturity. The region where the three bones fuse together is dominated by a large semispherical concavity called the acetabulum (figure 3.30*a*). The acetabulum articulates with the head of the femur to form the hip joint. The acetabulum, which faces laterally forward, and downward, consists of an outer horseshoe-shaped articular surface called the acetabular rim and a deep central region called the acetabular fossa. The gap between the two ends of the acetabular rim is continuous with the acetabular fossa and is called the acetabular notch. The acetabular fossa and notch are deeper than the acetabular rim would be if it formed a complete articular cup. The acetabular fossa accommodates a ligament that joins the head of the femur to another ligament spanning the acetabular notch (see chapter 7). Consequently, as the head of the femur slides on the acetabular rim, the acetabular fossa prevents the ligament attached to the head of the femur from being crushed.

The ilium comprises the upper two-fifths of the acetabulum and the large more or less flat portion of the innominate bone above the acetabulum. The large flat upper part of the ilium is called the

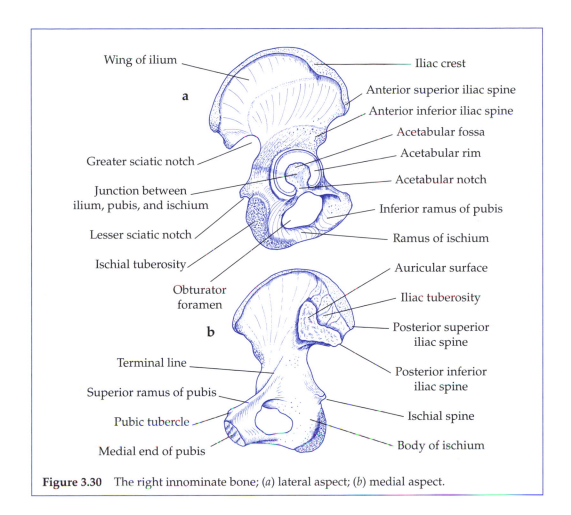

Figure 3.30 The right innominate bone; (*a*) lateral aspect; (*b*) medial aspect.

wing of the ilium. The superior border of the wing of the ilium is a broad crest called the iliac crest, felt beneath the skin just above the hip joint. The iliac crest, which has a shallow S shape when viewed from above, provides attachment for muscles comprising the anterior wall of the abdomen.

There is a projection at the anterior end of the iliac crest called the anterior superior iliac spine (ASIS). From the ASIS the anterior border of the ilium runs downward and backward to terminate in another projection called the anterior inferior iliac spine (AIIS). The AIIS lies just above the anterior superior part of the acetabulum and is separated from the ASIS by a notch. At the posterior end of the iliac crest is a projection called the posterior superior iliac spine (PSIS). From the PSIS the posterior border of the ilium runs downward and forward to terminate in a projection called the posterior inferior iliac spine (PIIS). A small notch separates the PSIS and PIIS. Whereas the ASIS, AIIS, and PSIS are usually easy to identify, the PIIS is not usually as well defined. Although normally referred to as spines, these four projections often more closely resemble tubercles or tuberosities.

On the posterior medial aspect of the wing of the ilium there is a large C- or L-shaped auricular surface that articulates with the auricular surface on the corresponding side of the sacrum to form the corresponding sacroiliac joint (see figure 3.30). Posterior to the angle of the auricular surface is the iliac tuberosity or sacral tuberosity of the ilium. The large lateral and medial surfaces of the wing of the ilium provide attachment for muscles that move the hip joint. The medial surface of the wing also supports the contents of the abdomen.

The ischium, which forms the posterior inferior portion of the innominate bone, consists of the body and the ramus. The body is a more or less vertical pillar that transmits the weight of the trunk, head, and arms to the support surface when the individual is sitting on a chair or stool. The superior part of the body forms the posterior inferior two-fifths of the acetabulum. Below the acetabulum the body of the ischium is characterized by the large ischial tuberosity on its lateral and inferior aspects (figure 3.30a). Arising from the posterior part of the body, just above the ischial tuberosity, is a process called the ischial spine projecting medially backward. The posterior border of the ilium and ischium between the PIIS and the ischial spine forms the greater sciatic notch. There is a smaller notch—the lesser sciatic notch— between the ischial spine and the ischial tuberosity. The ramus of the ischium is a broad, flat process that arises from the base of the body and projects medially forward and upward.

The pubis forms the anterior inferior portion of the innominate bone. It consists of the body, superior ramus, and inferior ramus. The body forms the anterior inferior one-fifth of the acetabulum. The superior ramus extends medially and also slightly forward and downward from the body to join the medial end of the inferior ramus. The junction between the two rami forms a fairly broad, flat region. The medial surface of this junction—the medial surface of the pubis—is elliptical in shape and lies in the median plane. The long axis of the ellipse is inclined at an angle of approximately 45° to the coronal plane. The medial surfaces of the right and left pubic bones are joined in the median plane by a disc of fibrocartilage. This joint is called the pubic symphysis (see figure 3.29).

The inferior ramus of the pubis projects downward and backward laterally to join the anterior end of the ischium ramus (see figure 3.30a). The inverted V-shaped notch formed by the inferior borders of the right and left inferior pubic rami is called the pubic arch (see figure 3.29). On the superior border of each superior ramus is a process called the pubic tubercle situated a short distance from the pubic symphysis. Running between the two pubic tubercles is a ridge, often poorly defined, called the pubic crest. The lateral end of the superior border of the superior ramus of each pubis is continuous with a distinct curved ridge on the medial aspect of the ilium terminating at the anterior inferior margin of the auricular surface of the ilium. This ridge is called the terminal line (see figure 3.30b). The pubis and ischium are both essentially V-shaped and joined at their free ends (see figure 3.30a). Consequently, when fused together the two bones create a large hole called the obturator foramen on account of the close proximity of the obturator nerve.

> The innominate bones link the lower limbs to the axial skeleton. Each innominate bone develops from three bones called the ilium, ischium, and pubis, which fuse together at maturity.

The Pelvis

Pelvis is a Latin word meaning basin (due to the large wings of the ilia, which give the impression of an incomplete bowl when the pelvis is viewed from an anterior superior aspect) (figure 3.31). From a mechanical point of view, the pelvis is made up of the upper pelvis and the lower pelvis. The upper pelvis consists of the two wings of the ilia and the upper third of the sacrum, which together form just over half of the upper part of the bowl. The anterior part of the bowl is missing and, thus, the upper pelvis is sometimes referred to as the false pelvis or the greater pelvis. The upper pelvis provides a base of support for the upper body.

The lower pelvis (also called the true or lesser pelvis), may be described very loosely as an incomplete cylinder; it consists of the ischia and pubic bones together with the inferior parts of the ilia and the inferior two-thirds of the sacrum. The lower pelvis transmits the weight of the upper body to the legs for standing and to the seat of a chair or stool for sitting. The margin between the upper and lower pelvis is called the inlet or pelvic brim. The inlet corresponds to a continuous ridge made up of the

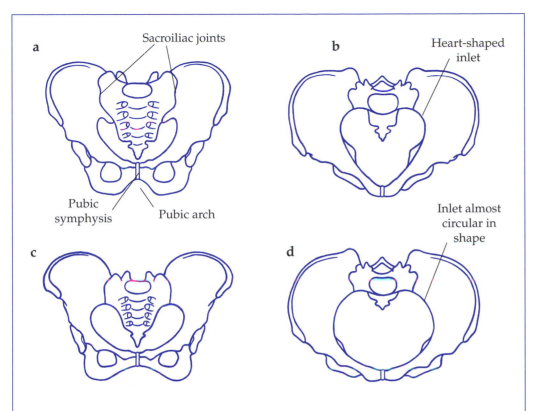

Figure 3.31 The pelvis; (a) anterior aspect of the male pelvis; (b) anterior superior aspect of the male pelvis; (c) anterior aspect of the female pelvis; (d) anterior superior aspect of the female pelvis.

pubic crest, the superior borders of the superior rami of the pubic bones, the terminal lines on the medial surfaces of the ilia, and a transverse ridge on the anterior aspect of the sacrum just below the uppermost transverse line.

Whereas the male and the female pelvises have the same basic structure, the shapes of the various parts, especially in the lower pelvis, differ considerably due to the childbearing functions of the female. During childbirth the child passes from the upper to the lower pelvis and then exits the abdomen via the inferior aspect of the lower pelvis. Consequently, the broader the lower pelvis, from side to side and from front to back, the easier it is for the child to pass through the pelvis. The shapes of the male and female pelvises differ in four main ways (see figure 3.31):

1. The inlet of the male pelvis is heart shaped, whereas that of the female pelvis is more circular.

2. The pubic bones are more in line with each other in the female than in the male. Consequently, the angle of the pubic arch is obtuse in the female and acute in the male.

3. The relative distance between the acetabulums is greater in the female than in the male. This results in a relatively greater girth around the hips in the female compared to the male.

4. In the male, the sacrum is curved such that the lower half of the sacrum and the coccyx bend forward. This curvature reduces the front to back dimension of the lower pelvis. In the female, the sacrum is relatively straight, which tends to maintain a fairly constant front to back dimension in the lower pelvis.

Together with the sacrum, the right and left innominate bones form the pelvic girdle or pelvis. The male pelvis and female pelvis have the same basic structure, but differ in shape due to the childbearing function of the female.

The Upper Leg

The upper leg or thigh contains a long bone called the femur and a relatively small bone called the patella (knee cap), which articulates with the lower end of the femur. The femur is the longest and strongest bone in the skeleton. The proximal end of the femur consists of a nearly spherical shaped head, which articulates with the acetabulum to form the hip joint (see figure 3.29; figure 3.32). The head is joined obliquely to the shaft by a thick neck that runs laterally downward, and backward from the head to the anterior medial region of the proximal end of the shaft (see also figure 3.1b). The neck resembles a truncated cone, partially flattened from front to back, its smaller surface joined to the head and its larger surface joined to the shaft. A large process called the greater trochanter dominates the superior posterior lateral region of the proximal end of the shaft. At the base of the neck on the posterior medial aspect of the shaft is another fairly large process called the lesser trochanter. The greater and lesser trochanters are linked on the posterior aspect by a distinct ridge called the intertrochanteric crest. The line of unison between the neck and the shaft on the anterior aspect is marked by another distinct ridge called the intertrochanteric line.

Like the humerus, the upper two-thirds of the shaft of the femur is cylindrical and the lower one-third gradually becomes broader (in the coronal plane) toward the

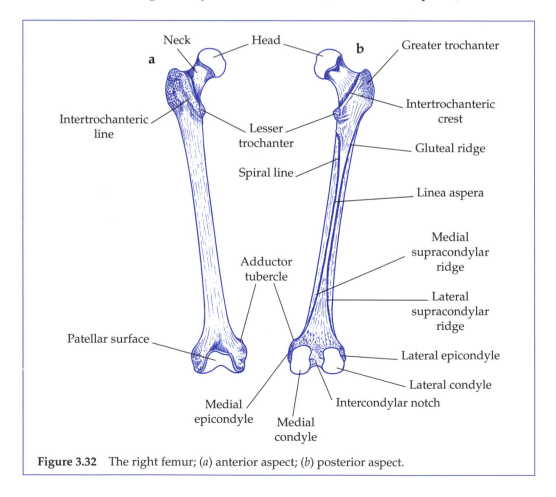

Figure 3.32   The right femur; (*a*) anterior aspect; (*b*) posterior aspect.

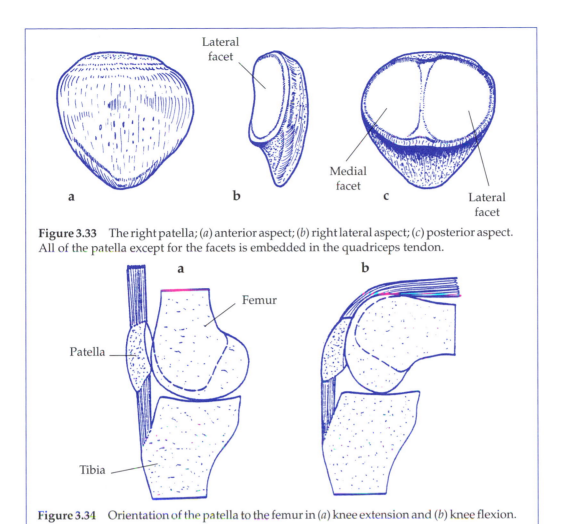

Figure 3.33   The right patella; (*a*) anterior aspect; (*b*) right lateral aspect; (*c*) posterior aspect. All of the patella except for the facets is embedded in the quadriceps tendon.

Figure 3.34   Orientation of the patella to the femur in (*a*) knee extension and (*b*) knee flexion.

distal end. In a paramedian plane, the anterior surface of the femur is slightly convex and the posterior surface is slightly concave (see figure 3.1*b*). The anterior and lateral surfaces are fairly smooth, whereas the posterior surface is dominated by two longitudinal ridges—one lateral and one medial—that run the length of the shaft. The ridges converge and join in the middle one-third of the shaft to form the linea aspera (rough line). Above and below the linea aspera the ridges diverge. Above the linea aspera the lateral ridge is called the gluteal ridge, which runs to the greater trochanter. The medial ridge above the linea aspera is the spiral line. The spiral line runs toward the lesser trochanter for half of its length, but then curves around the medial part of the shaft in a spiral manner to terminate at the lower end of the intertrochanteric line. Below the linea aspera the lateral and medial ridges are called the lateral supracondylar ridge and medial supracondylar ridge, respectively. At the bottom end of the medial supracondylar ridge is a projection called the adductor tubercle.

The distal end of the femur consists of two large convex condyles, the lateral and medial condyles, fused together side by side anteriorly. The condyles are separated posteriorly by a large notch called the intercondylar notch (or intercondylar fossa). Consequently, the articular surface of the condyles is V-shaped. The upper part of the common anterior portion of the articular surface is called the patellar surface. The patellar surface is pulley shaped—depressed in the middle in the sagittal plane—and articulates with the posterior surface of the patella to form the patellofemoral joint (see chapter 7). During extension and flexion of the knee joint, the patella slides up and down on the patellar surface and condyles of the femur.

The femur and the patella make up the upper leg. The femur is the longest and strongest bone in the skeleton. The proximal end of the femur articulates with the innominate bone to form the hip joint. The distal end articulates with the tibia to form the tibiofemoral joint and with the patella to form the patellofemoral joint.

The patella is a sesamoid (Greek for resembling a sesame seed)—a bone that is partially embedded in a tendon (figure 3.33). A sesamoid bone tends to increase the mechanical efficiency of the associated musculotendinous unit and prevent the tendon from rubbing on an adjacent bone. Consequently, the patella, embedded in the posterior part of the quadriceps tendon, increases the mechanical efficiency of the quadriceps muscle group and prevents the quadriceps tendon from rubbing against the patellar surface of the femur.

The anterior aspect of the patella is rounded superiorly and pointed inferiorly (figure 3.33, *a* and *b*). The whole of the anterior surface and the inferior quarter of the posterior surface are embedded in the quadriceps tendon. The upper three-quarters of the posterior surface articulates with the patellar surface of the femur when the knee joint is extended and with the condyles of the femur when the knee joint is flexed (figure 3.34). The articular surface of the patella is V-shaped with a sagittal ridge. The medial and lateral aspects of the ridge are referred to as the medial and lateral facets, respectively. The proximal end of the patella is sometimes referred to as the superior pole and the distal end as the inferior pole or apex of the patella.

The patella increases the mechanical efficiency of the quadriceps muscle group and prevents the quadriceps tendon rubbing against the femur.

The Lower Leg

The lower leg or shank contains two long bones, the tibia and the fibula, aligned with their shafts more or less parallel to each other (figure 3.35, *a* and *b*). The tibia is the larger of the two bones and is situated medial to the fibula. The proximal end of the tibia consists of two large condyles, the lateral and medial condyles, which are fused together side by side (figure 3.35c). The tibial condyles articulate with the femoral condyles to form the tibiofemoral joint (knee joint). The articular surfaces of the tibial condyles are oval in outline and almost flat. The lateral surface is usually slightly convex and slightly smaller than the medial surface. The medial surface may be slightly convex. The surfaces occupy the same plane more or less horizontally in the anatomical position. This orientation of the condylar surface gives rise to the term tibial table, sometimes used to describe the proximal end of the tibia. Between the two condylar surfaces at the center of the tibial table there are two small processes lying side by side; these processes are referred to as the lateral and medial tibial spines (or tibial eminences) or intercondylar tubercles. The area of the tibial table in front of the tibial spines and between the anterior aspects of the tibial condyles is referred to as the anterior intercondylar area. The corresponding area behind the tibial spines is referred to as the posterior intercondylar area.

The shaft of the tibia is fairly smooth apart from a distinct ridge on the upper one-third of the posterior aspect of the shaft. This ridge, the soleal line, runs obliquely downward and medially from the inferior posterior aspect of the lateral condyle. The middle two-thirds of the shaft is teardrop shaped in cross section; the posterior aspect is rounded, whereas the anterior aspect consists of two fairly flat areas, anterior lateral and anterior medial, which converge anteriorly to form a distinct ridge called the anterior crest. The anterior crest can easily be felt beneath the skin as a ridge running down the bone. The anterior medial surface of the tibia, covered only by skin, is usually referred to as the shin. Above and below the anterior crest the shaft broadens out toward the proximal and distal ends of the bone. Above the upper end

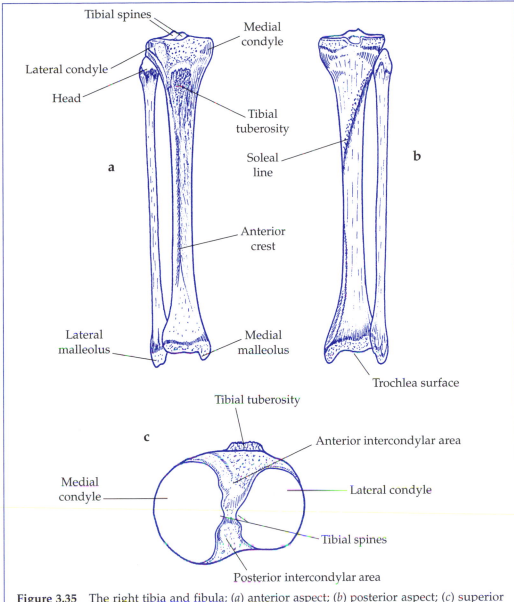

Figure 3.35   The right tibia and fibula; (*a*) anterior aspect; (*b*) posterior aspect; (*c*) superior aspect of tibial table.

of the anterior crest, on the anterior aspect of the shaft, is a fairly large process called the tibial tuberosity.

On the medial side of the distal end of the tibia there is a downward projection called the medial malleolus. The lateral aspect of the medial malleolus articulates with the medial aspect of the talus to form the medial part of the ankle joint (figure 3.36*a*). The remainder of the distal end of the tibia is dominated by a large biconcave condylar surface called the trochlea surface of the tibia (see figure 3.35*b*). The trochlea surface articulates with the superior aspect of the talus to form the main part of the ankle joint. The trochlea surface of the tibia and the articular surface of the medial malleolus are continuous with each other.

The tibia is almost completely responsible for transmitting loads from the upper leg to the foot and vice versa. In contrast, the fibula is a thin, relatively weak bone only marginally involved in load transmission between the upper leg and foot. The main functions of the fibula are to help form the ankle joint and to provide additional area for the attachment of muscles that move the ankle and foot. The proximal end of the fibula is called the head. The medial two-thirds of the superior aspect of the head articulates with the posterior inferior lateral aspect of the lateral tibial condyle to form the proximal tibiofibular joint.

The shaft of the fibula is characterized by four longitudinal ridges that give rise to four faces of varying width and length along the shaft. The distal end of the fibula is called the lateral malleolus. The medial aspect of the lateral malleolus articulates with the lateral aspect of the talus to form the lateral part of the ankle joint. The medial part of the shaft of the fibula immediately above the articular surface of the lateral malleolus articulates with the lateral part of the distal end of the tibia to form the distal tibiofibular joint.

> The tibia and the fibula are the two long bones in the lower leg. The tibia is much thicker than the fibula and is almost completely responsible for transmitting loads between the upper leg and foot. The fibula helps form the ankle joint and provides additional area of attachment for muscles of the lower leg. The tibia and fibula articulate with each other at their proximal and distal ends but have little or no movement relative to each other.

The Foot

The foot consists of 7 tarsals, 5 metatarsals, and 14 phalanges (figure 3.36). When articulated, the tarsals form the tarsus (figure 3.36*b*). The tarsus corresponds to the carpus in the upper limb, but the tarsals are all much larger than the carpals. Whereas the carpus is not usually considered to be part of the hand, the tarsus forms the posterior half of the foot. The foot articulates with the lower leg at the ankle joint—the joint between the tibia, fibula, and talus.

The talus, the second largest tarsal, has a convex pulley-shaped articular surface on its superior aspect called the trochlea surface of the talus (figure 3.36*c*); it articulates with the trochlea surface of the tibia. The trochlea surface of the talus is continuous with articular surfaces on its lateral and medial aspects, which articulate with the lateral malleolus and medial malleolus, respectively.

The inferior aspect of the talus articulates with the anterior half of the superior aspect of the calcaneus by means of two or, in some cases, three articular facets, which together constitute the subtalar joint (talocalcaneum joint) (see figure 3.36, *a* and *b*). The anterior aspect of the talus articulates with the posterior aspect of the navicular, on the medial aspect of the foot, to form the talonavicular joint.

The calcaneus—the largest tarsal—is often referred to as the heel bone. The posterior aspect of the calcaneus is characterized by a large tuberosity called the calcaneal tuberosity. The anterior aspect of the calcaneus articulates with the posterior aspect of the cuboid, on the lateral aspect of the foot, to form the calcaneocuboid joint. The calcaneocuboid and talonavicular joints are continuous with each other and constitute the midtarsal joint (see figure 3.36*c*). The anterior aspect of the navicular articulates with the posterior aspects of the three cuneiforms (medial, middle, and lateral), which lie side by side and articulate with each other. The posterior two-thirds of the lateral aspect of the lateral cuneiform articulates with the medial surface of the cuboid. The anterior aspects of the cuneiforms articulate with the bases of the first, second, and third metatarsals. The anterior aspect of the cuboid articulates with the bases of the fourth and fifth metatarsals. These joints between the four anterior tarsals and the metatarsals are referred to as the tarsometatarsal joints. The lateral four metatarsals are similar in length, but tend to increase in girth from the second through to the fifth. In comparison, the first metatarsal is shorter, but has a greater girth than the other four. The short length of the first metatarsal is thought to increase the efficiency of the arches of the feet (Williams et al. 1995).

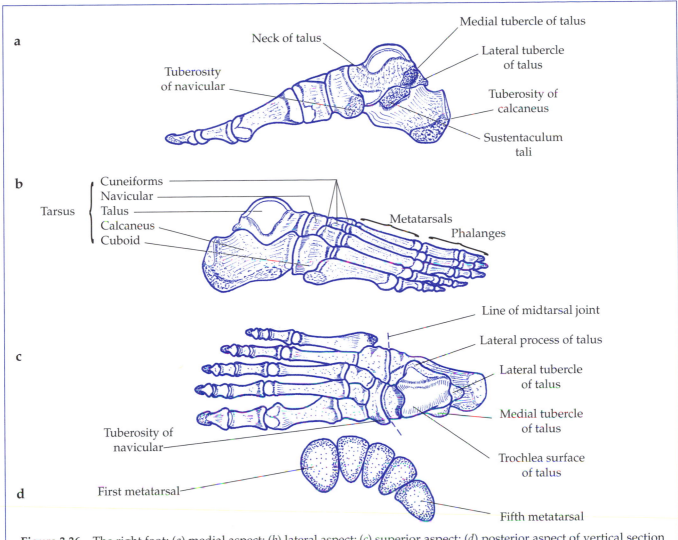

a
Neck of talus
Medial tubercle of talus
Lateral tubercle of talus
Tuberosity of navicular
Tuberosity of calcaneus
Sustentaculum tali

b
Tarsus
{ Cuneiforms
Navicular
Talus
Calcaneus
Cuboid
Metatarsals
Phalanges

c
Line of midtarsal joint
Lateral process of talus
Lateral tubercle of talus
Medial tubercle of talus
Trochlea surface of talus
Tuberosity of navicular

d
First metatarsal
Fifth metatarsal

Figure 3.36 The right foot; (*a*) medial aspect; (*b*) lateral aspect; (*c*) superior aspect; (*d*) posterior aspect of vertical section through the proximal ends of the metatarsals.

The foot consists of 7 tarsals, 5 metatarsals, and 14 phalanges. The tarsals are irregular bones articulating with each other to form the tarsus, which forms the posterior half of the foot. The lower leg articulates with the tarsus at the ankle joint.

The tarsals and metatarsals are arranged in the form of two longitudinal arches (medial and lateral) and a single transverse arch. The medial longitudinal arch is formed by the calcaneus, talus, navicular, the three cuneiforms, and the first, second, and third metatarsals (see figure 3.36, *a* and *c*). The lateral longitudinal arch, which is much flatter than the medial arch, is formed by the calcaneus, cuboid, and the fourth and fifth metatarsals (see figure 3.36, *b* and *c*). In combination, the longitudinal arches form a single arched structure between the posterior inferior aspect of the calcaneus and the heads of the metatarsals. The transverse arch runs across the foot from medial to lateral and is formed by the anterior five tarsals and the bases of the metatarsals. The shape of the arch is due to the cuboid, the middle and lateral cuneiforms, and the bases of the middle three metatarsals, which are wedge shaped in coronal section (figure 3.36*d*). Muscles largely maintain the arches, which function to cushion impact loads on the foot occurring in activities such as walking, running, jumping, and landing (see chapter 7).

The distribution of phalanges in the foot is similar to that in the hand—two in the great toe (big toe) and three in each of the other toes. As in the hand, the phalanges of the toes become progressively shorter from proximal to distal. In comparison with the corresponding phalanges of the thumb, the phalanges of the great toe are slightly longer and have a much greater girth. However, the phalanges of the other four toes are much shorter and, in general, smaller in girth than the corresponding phalanges in the hand. The interphalangeal and metatarsophalangeal joints are similar in structure to their counterparts in the hand.

> *The anterior aspect of the tarsus articulates with the proximal ends of the metatarsals to form the tarsometatarsal joints. The distal ends of the metatarsals articulate with the distal phalanges to form the metatarsophalangeal joints.*

Summary

This chapter described the bones of the skeleton with particular reference to the open-chain arrangement of the bones in the axial and appendicular parts of the skeleton. The skeleton has three main mechanical functions: to provide a supporting framework for all the other systems of the body, to protect organs such as the brain and spinal cord, and to provide a system of levers, operated by the skeletal muscles, to facilitate force transmission throughout the skeleton.

Bone is a connective tissue and all connective tissues are concerned to some extent with transmitting forces within and between body systems. The next chapter describes the functional anatomy of the various types of connective tissue.

Review Questions

1. Describe the three main mechanical functions of the skeleton.
2. Describe the three main reference planes and define the spatial terminology associated with the planes.
3. List the bones of the axial skeleton.
4. Describe the following features of the skull:
   - Sutures
   - Fontanels
   - Sinuses
5. Describe the primary and secondary curves of the vertebral column.
6. Describe the components of a typical vertebra.
7. Describe the distinguishing features of cervical vertebrae, thoracic vertebrae, and lumbar vertebrae.
8. Describe the difference between true ribs and false ribs.
9. List the bones of the upper limb.
10. Describe the shoulder girdle.
11. Describe the elbow joint.
12. Describe supination and pronation of the lower arm.
13. List the bones of the lower limb.
14. Describe the differences between the male pelvis and the female pelvis.
15. Describe the tibiofemoral and patellofemoral joints.
16. Describe the arches of the feet.

Chapter 4

Connective Tissues

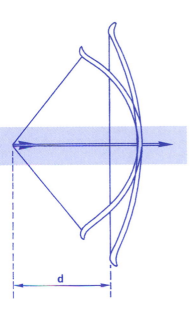

In muscle, nerve, and epithelial tissue the cells predominate and are directly related to the function of the tissues. In contrast, connective tissues have relatively few cells distributed within a large amount of noncellular matrix. The physical characteristics of the matrix directly determine the function of each type of connective tissue.

Connective tissues have two main functions: mechanical support and intercellular exchange. The connective tissues are continuous with each other throughout the body, but the composition of the matrix gradually changes from one part of an organ or system to another depending upon function. The purpose of this chapter is to describe the structure and functions of the various connective tissues.

Objectives

After reading this chapter you should be able to do the following:

1. Define or describe the key terms.
2. Describe the structure and function of ordinary connective tissues.
3. Describe the structure and functions of the three main types of cartilage.
4. Describe the growth and development of bone.
5. Describe the structure of mature bone.
6. Explain the difference between remodeling and modeling in bone.
7. Describe the effects of aging on bone.

Functions of Connective Tissues

Matrix: the noncellular component of connective tissue

Connective tissues, which include tendons and aponeuroses, have relatively few cells distributed within a large amount of noncellular material called **matrix** that is produced by the cells. A variety of connective tissues differ from each other, structurally as well as functionally, based on differences in the physical characteristics of the matrix. The matrix ranges from a semiliquid material in one type of connective tissue (areolar tissue) to a very hard solid material in another (bone). Connective tissues have two main functions: mechanical support and intercellular exchange.

> *Connective tissues consist of relatively few cells within a large amount of matrix produced by the cells. The various connective tissues differ from each other mainly in terms of the physical characteristics of the matrix.*

Mechanical Support

Most of the connective tissues help maintain or transmit forces to carry out a wide range of mechanical functions. These functions, largely concerned with providing strength or elasticity, include the following:

1. Binding together the cells of the body in the various tissues, organs, and systems
2. Supporting and holding in place the various organs
3. Providing stability and shock absorption in joints
4. Providing flexible links between bones in certain types of joints, and providing smooth articulating surfaces between bones in other types of joints
5. Transmitting muscle forces

Intercellular Exchange

In multicellular organisms the cells rely on the circulating body fluids such as blood to supply them with nutrients, oxygen, and other substances, and to carry away waste products such as carbon dioxide. This involves exchange of nutrients, gases, and other substances between the vessels of the circulating body fluids and cells adjacent to the vessels, and between cells adjacent to each other. Intercellular exchange ensures that all cells can be supplied with nutrients, gases, and other substances, and can excrete waste products, even if the cells do not receive a direct supply of the circulating body fluids.

Connective tissues have two main functions: mechanical support and intercellular exchange.

Distribution and Classification of Connective Tissues

Although all types of connective tissue are continuous with each other throughout the body, the composition of the matrix gradually changes from one part of an organ or system to another, depending on the function of the connective tissue at each particular location. For example, in a skeletal muscle the individual muscle cells are bound together by connective tissue whose function is to facilitate intercellular exchange and to bind the cells together. In contrast, the belly of the muscle is attached to bone at each end by means of tendons or aponeuroses whose sole function is to provide a strong link between the muscle and the bony attachments (figure 4.1).

Connective tissues are classified according to their function into ordinary and special connective tissues (Williams et al. 1995). Ordinary connective tissues, distributed widely throughout the body, have two main functions:

1. Binding cells together into tissues, organs, and systems
2. Providing mechanical links between bones at joints, and between muscles and bones

There are two special connective tissues: cartilage and bone. The main functions of cartilage are to transmit loads across joints efficiently and to allow movement between bones at certain joints. The mechanical functions of bone were described in chapter 3; this chapter develops the relationship between these functions and the structure of bone.

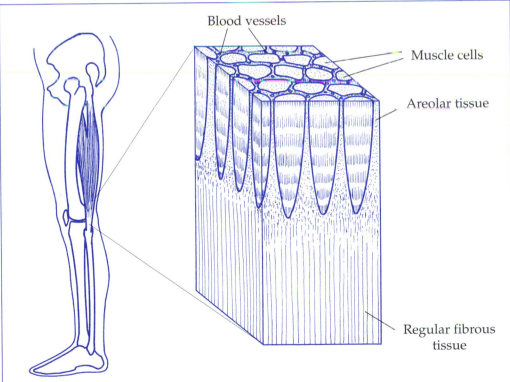

Figure 4.1   Location of different kinds of connective tissue in skeletal muscle. Areolar tissue binds the muscle cells together and facilitates intercellular exchange. Regular fibrous tissue provides a mechanical link between muscle and bone.

Ordinary Connective Tissues

Elastin: one component of the matrix of ordinary connective tissue; elastin fibers break after being stretched approximately 200% of their rest length

Collagen: one component of the matrix of ordinary connective tissue; collagen fibers break after being stretched approximately 10% of their rest length

Ground substance: the nonfibrous component of the matrix of ordinary connective tissue

The matrix of ordinary connective tissues consists of three basic components: **elastin** fibers, **collagen** fibers, and **ground substance**. The main difference in structure of the various types of ordinary connective tissue is in the proportion of these basic components in the matrix.

Elastin and Collagen Fibers

Both elastin and collagen fibers are proteins. A protein molecule consists of a long chain of amino acids. In elastin, the molecules are arranged randomly in terms of orientation and attachment to one another and individual shape (Alexander 1975) (figure 4.2*a*). When elastin is stretched, the molecules themselves do not stretch, but tend to straighten in the direction of stretching (figure 4.2*b*). The molecules resist straightening; that is, they experience tension stress (see chapter 1) and the greater the stretch, the greater the stress. When the stretching load is removed, the elastin molecules restore their original orientation and shape. Elastin is, therefore, elastic; hence its name.

An elastin fibril is formed by a number of elastin molecules and an elastin fiber consists of a number of fibrils grouped together. An elastin fiber is similar in shape, strength, and elasticity to a long, thin rubber band. Elastin fibers can be stretched by about 200% of their rest length before breaking (Nordin and Frankel 1989). They have a yellowish appearance and are often referred to as yellow elastic fibers or yellow fibers.

In contrast to elastin, the collagen molecules are arranged in a more regular manner; they tend to run in the same overall direction and for the most part are aligned parallel to each other (Alexander 1975) (figure 4.2*c*). Like elastin molecules, collagen molecules are attached to each other at various points. When stretched in the direction of their main orientation, collagen molecules quickly straighten so that the amount of extension is limited (figure 4.2*d*). Like elastin molecules, the collagen molecules experience tension stress when stretched; the greater the stretch, the greater the stress. They are also elastic and, as such, return to their resting orientation when the stretching load is removed. Each group of closely aligned parallel molecules constitutes an individual collagen fibril, and a collagen fiber consists of a number of fibrils grouped together. A collagen fiber is similar in shape, strength, and elasticity to a shoelace; it is virtually inextensible and, in relation to elastin, it is extremely strong. Collagen fibers break after being stretched by approximately 10% of their rest length (Nordin and Frankel 1989). Collagen fibers are white and are often referred to as white collagen fibers or white fibers.

The Ground Substance

Viscosity: the resistance of a fluid to flowing

The ground substance forms the nonfibrous part of the matrix. It is a viscous gel consisting mainly of large carbohydrate molecules (molecules consisting of carbon, hydrogen, and oxygen) and carbohydrate-protein molecular complexes (molecules consisting of carbon, hydrogen, oxygen, and nitrogen) suspended in a relatively large volume of water (Williams et al. 1995; Alexander 1975). The number and type of carbohydrate and carbohydrate-protein substances determine the actual volume of water. Many of these substances have an affinity for water and, as such, determine not only the volume of water in the ground substance, but also the **viscosity** of the ground substance. Viscosity refers to the resistance of a fluid to flowing; for example, oil is more viscous than water.

In contrast to the elastin and collagen fibers, whose sole function is to provide mechanical support, the ground substance is responsible not only for facilitating

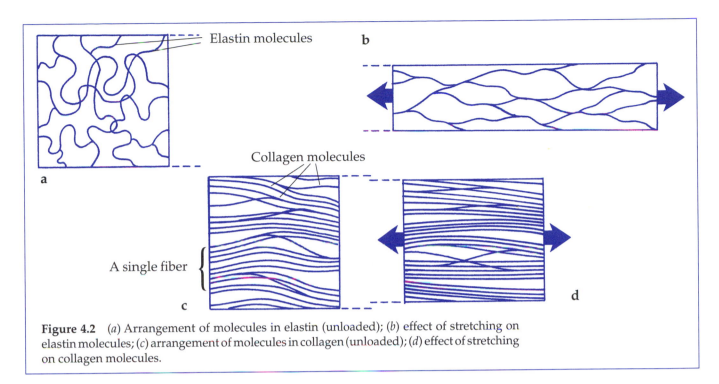

Figure 4.2 (*a*) Arrangement of molecules in elastin (unloaded); (*b*) effect of stretching on elastin molecules; (*c*) arrangement of molecules in collagen (unloaded); (*d*) effect of stretching on collagen molecules.

intercellular exchange, but also for providing some mechanical support. The glue-like viscosity of the ground substance enables it to bind cells together within the other main tissues (muscle, nerve, and epithelia). In epithelial tissue the ground substance is the main bonding material between the cells. In muscle and nerve tissue the cells are bound together by a combination of ground substance and fibers.

The ground substance in ordinary connective tissues is sometimes referred to as tissue fluid or extracellular fluid. In addition, it is also referred to as amorphous ground substance (amorphous = without definite structure) since it appears, even under a microscope, as a featureless fluid.

> *Throughout the body, ordinary connective tissues bind cells together into tissues, organs, and systems and provide mechanical links between bones at joints and between muscles and bones. The matrix of ordinary connective tissues consists of elastin and collagen fibers and ground substance. Elastin fibers provide elasticity and collagen fibers provide strength. The ground substance facilitates intercellular exchange and helps to bind cells within the other main tissues.*

Ordinary Connective Tissue Cells

The number and type of cells found in ordinary connective tissues depend on the type of connective tissue and the state of health of the individual (Williams et al. 1995). When present, the various types of cells are found suspended in the ground substance or, in some cases, attached to the collagen fibers. In general, there are six main types of cells found in ordinary connective tissues:

- **Fibroblasts**: Usually the most numerous type of cells, fibroblasts are often found attached to the collagen fibers and are responsible for producing the matrix (the ground substance and the elastin and collagen fibers).

- **Macrophages**: The macrophages are responsible for engulfing and digesting bacteria and other foreign bodies. They also dispose of dead cellular material that occurs as a result of injury, or as cells become old and die.

- **Plasma cells**: Plasma cells occur in large numbers in response to infection. They produce antibodies that inactivate and, with the macrophages, destroy harmful bacteria and other substances.
- **White blood cells**: The number and type of white blood cells increase in response to infection. They work with the plasma cells and macrophages to identify and destroy harmful bacteria and other substances.
- **Mast cells**: Mast cells, widespread throughout ordinary connective tissues, are responsible for producing heparin, which prevents the blood plasma from clotting inside blood vessels.
- **Fat cells**: Fat cells have a variety of functions and occur in great numbers in one particular type of ordinary connective tissue (adipose tissue).

The proportion of elastin fibers, collagen fibers, ground substance, and the number and type of cells within any particular ordinary connective tissue determines its function. Collagen fibers predominate where great strength is required, whereas elastin fibers predominate where considerable elasticity is needed. Similarly, the ground substance tends to predominate where intercellular exchange is of major importance. Under normal circumstances a wide variety of cells are present within ordinary connective tissues. In response to infection, there is an increase in the number of cells responsible for identifying and destroying harmful bacteria.

Irregular Ordinary Connective Tissues

Ordinary connective tissues are classified into irregular and regular tissues according to the arrangement of the fibrous content of the matrix. In irregular tissues the fibers tend to run in all directions throughout the tissue with no set pattern. In contrast, the fibers in regular tissues tend to be orientated in the same overall direction. There are four main types of irregular ordinary connective tissue: loose (areolar), adipose, irregular collagenous, and irregular elastic.

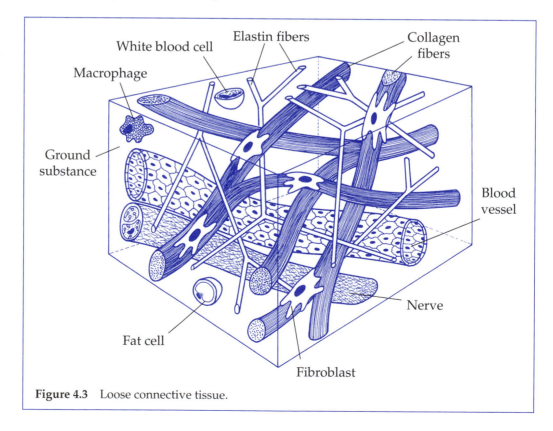

Figure 4.3 Loose connective tissue.

Loose Connective Tissue

Loose connective tissue is the most widely distributed of all the connective tissues. It is the basic cement substance or glue that binds cells within the other main tissues (muscle, nerve, and epithelia) and that binds these tissues into organs. It consists of a loose irregular network of elastin and collagen fibers suspended within a relatively large amount of ground substance (figure 4.3). The large amount of amorphous ground substance gives the impression of a lot of space between the fibers and cells of loose connective tissue. For this reason, loose connective tissue is also referred to as areolar tissue (areola = a small open area).

The loose network of elastin and collagen fibers, both of which branch freely, provides moderate elasticity and strength. Consequently, loose connective tissue is well suited to bind cells into tissues and tissues into organs, and to provide a supporting framework for nerves and the vessels of the circulating body fluids (blood and lymph). The viscosity of the ground substance is important for binding cells within the other main tissues, and the large amount of ground substance reflects the importance of loose connective tissue in the facilitation of intercellular exchange.

Adipose Connective Tissue

Adipose tissue has a loose network of elastin and collagen fibers similar to loose connective tissue. However, in contrast to loose connective tissue, there is little ground substance and a large number of closely packed fat cells. Each fat cell consists of a thin cell membrane surrounding a relatively large globule of fat (figure 4.4). Adipose tissue is widely distributed around the body in the following four main locations (McArdle, Katch, and Katch 1996).

1. In bone marrow.
2. In association with the various layers of loose connective tissue within certain organs, especially skeletal muscles.
3. As padding around certain organs and joints.
4. As a continuous layer beneath the skin. Skin is sometimes referred to as cutaneous tissue and the layer of fat as the subcutaneous fat layer.

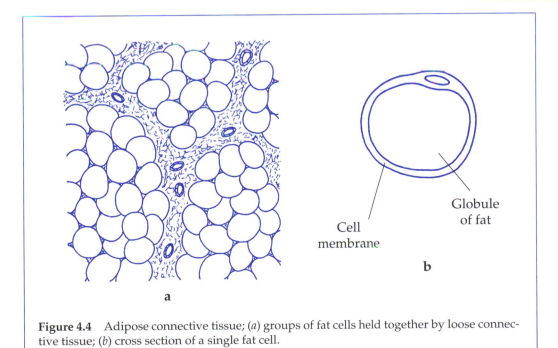

Figure 4.4   Adipose connective tissue; (*a*) groups of fat cells held together by loose connective tissue; (*b*) cross section of a single fat cell.

Adipose tissue is a poor conductor of heat and, consequently, the subcutaneous fat layer acts as an insulator, reducing the loss of body heat through the skin. Adipose tissue is moderately strong due to its collagen fiber content, and considerably elastic due to its elastin fiber content and the elasticity of the fat cells. Consequently, adipose tissue is well suited to provide mechanical support and protection (cushioning) in the form of padding around and between certain organs such as the heart, lungs, liver, spleen, kidneys, and intestines. Adipose tissue also acts as padding around joints such as the knee and over certain bones such as the heel bone. In addition to its heat insulation and mechanical functions, adipose tissue is the body's main food store. Adipose tissue provides approximately twice as much energy per gram as any other tissue in the body (McArdle, Katch, and Katch 1996).

Irregular Collagenous Connective Tissue

Irregular collagenous connective tissue consists of few elastin fibers and little ground substance. A dense, irregular network of collagen fiber bundles dominates the matrix (figure 4.5). The collagen bundles and their irregular arrangement enable the tissue to resist being stretched in any direction. However, though it is strong, the tissue has a certain amount of elasticity due to the wavy orientation of the collagen bundles. When stretched in a particular direction, the collagen bundles tend to straighten in the direction of stretching. Irregular collagenous connective tissue is most frequently found as a tough cover around certain organs, where it provides mechanical support and protection. For example, it is found as

1. a sheath around skeletal muscles (epimysium) and spinal nerves (epineurium);
2. a capsule or envelope around certain organs such as the kidneys, liver, and spleen that holds the organs in place;
3. the perichondrium of cartilage (discussed later in this chapter); and
4. the periosteum of bones (discussed later in this chapter).

Irregular Elastic Connective Tissue

Irregular elastic connective tissue is made up of very few collagen fibers and a moderate amount of ground substance. A dense, irregular network of elastin fibers dominates the matrix (figure 4.6). In comparison to irregular collagenous connective tissue, irregular elastic connective tissue is not as strong, but much more elastic. It is found where moderate amounts of strength and elasticity are required in more than one direction as, for example, in the walls of arteries and the larger arterioles, the trachea (windpipe), and bronchial tubes. There are few cells in irregular collagenous and irregular elastic connective tissues. The cells that are present are mainly fibroblasts.

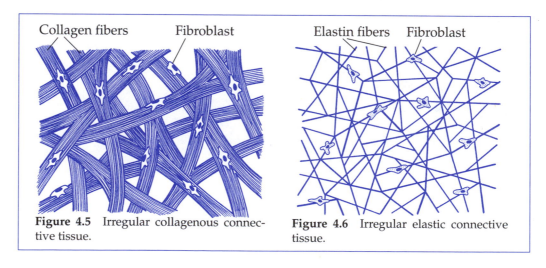

Figure 4.5 Irregular collagenous connective tissue.

Figure 4.6 Irregular elastic connective tissue.

Regular Ordinary Connective Tissues

There are two main types of regular ordinary connective tissue: regular collagenous and regular elastic.

Regular Collagenous Connective Tissue

Regular collagenous connective tissue consists almost entirely of collagen fiber bundles arranged parallel to each other. Usually, there are few elastic fibers and little ground substance. The only cells present are fibroblasts arranged in columns between the collagen bundles (figure 4.7). The collagen bundles are gathered together in the form of thick cords, bands, or sheets of various widths. In the unloaded state the collagen bundles have a slightly wavy orientation. When stretched, the bundles quickly straighten, and the tissue becomes taut. Regular collagenous connective tissue is extremely strong and virtually inextensible. It has three main forms:

1. Tendons and aponeuroses: mechanical links between skeletal muscle and bone

2. Ligaments and joint capsules: mechanical links between bones at joints

3. Retinacula: mechanical restraints on tendons that increase the mechanical efficiency of the muscle-tendon units or the stability of the associated joints

Tendons and Aponeuroses   Skeletal muscles are attached to other structures, usually bones, by regular collagenous connective tissue in the form of tendons and aponeuroses (figure 4.8).

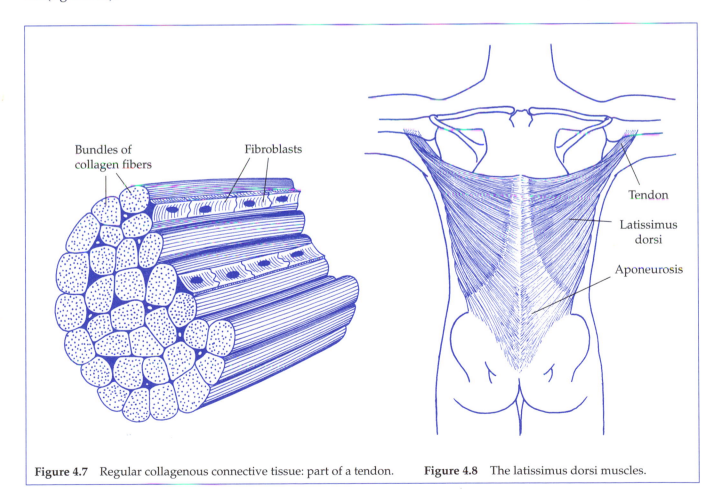

Figure 4.7   Regular collagenous connective tissue: part of a tendon.   **Figure 4.8**   The latissimus dorsi muscles.

Capsular ligament: a distinct thickening in part of the joint capsule that provides additional strength in one direction

Noncapsular ligament: a distinct band separate from the joint capsule or only partially attached to it

Extracapsular ligament: a noncapsular ligament outside the joint cavity

Intracapsular ligament: a noncapsular ligament inside the joint cavity

Ligaments and Joint Capsules   Skeletal muscles provide active links—contractile links—between bones. Ligaments and joint capsules provide passive links—noncontractile links—between bones. In association with the skeletal muscles, the ligaments and joint capsules bring about normal joint movement. Chapters 5, 6, and 7 describe the various joints of the body. However, at this point it is sufficient to appreciate that each synovial (freely moveable) joint—a joint involving sliding or rolling between the free surfaces of the ends of bones, as in the shoulder and hip—is enclosed within its own joint capsule (figures 4.9 and 4.10).

The joint capsule encloses a space, usually quite small, called the joint cavity. A joint capsule is composed of two or more layers of regular collagenous connective tissue forming a sleeve around the joint, rather like a piece of rubber tubing joining two glass rods together. Whereas the collagen bundles in each layer are parallel to each other, the bundles in adjacent layers run in different directions. This arrangement enables the capsule to strongly resist stretching in a number of different directions and, therefore, helps to maintain joint integrity (figure 4.11).

In all synovial joints the joint capsule is supported by a number of ligaments. These ligaments may be **capsular** or **noncapsular**. A capsular ligament is a distinct thickening in part of the joint capsule that provides additional strength in one direction. For example, the superior iliofemoral ligament, inferior iliofemoral ligament, and pubofemoral ligament are capsular ligaments that strengthen the anterior aspect of the capsule of the hip joint (see figure 4.10b). A noncapsular ligament is a distinct band separate from the joint capsule or only partially attached to it. Noncapsular ligaments may be **extracapsular** (outside the joint cavity) or **intracapsular** (inside the joint cavity). For example, the ligamentum teres of the hip joint is intracapsular, but the lateral ligament, medial ligament, and cruciate ligaments of the knee joint are all extracapsular (see figure 4.10a; figure 4.12). Noncapsular ligaments usually consist of a single layer of tissue, but broad ligaments may consist of two or more layers, similar to a joint capsule.

In addition to the noncapsular ligaments associated with synovial joints, there are other ligaments similar in structure to noncapsular ligaments that help stabilize other parts of the skeleton; for example, the ligaments between the clavicle and scapula, and between the clavicle and first rib (see figure 4.9b).

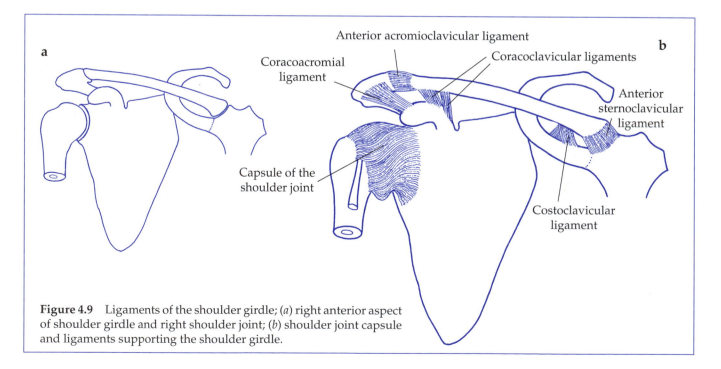

Figure 4.9   Ligaments of the shoulder girdle; (*a*) right anterior aspect of shoulder girdle and right shoulder joint; (*b*) shoulder joint capsule and ligaments supporting the shoulder girdle.

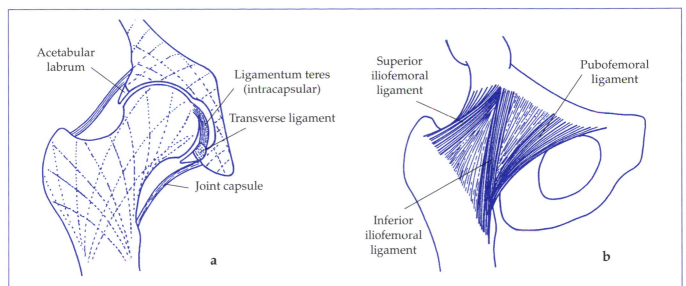

Figure 4.10 Ligaments of the hip joint; (*a*) coronal section through the right hip joint showing the joint capsule and ligamentum teres; (*b*) anterior aspect of the right hip joint showing the joint capsule and anterior capsular ligaments.

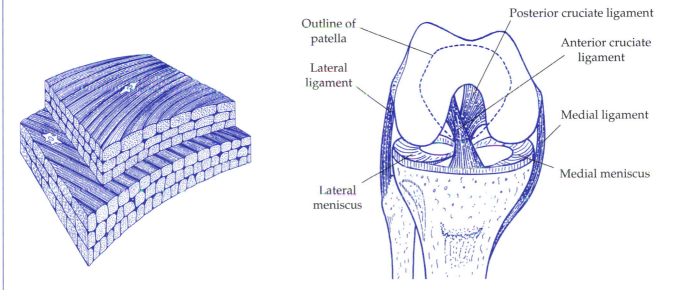

Figure 4.11 Three-dimensional section through a two-layered joint capsule.

Figure 4.12 Extracapsular ligaments of the knee joint. Anterior aspect of the right knee joint, flexed at 90°, with patella removed to show cruciate ligaments and femoral condyles slightly raised to show the menisci. The cruciate ligaments are located at the center of the joint, but lie outside the joint cavity.

Retinacula A retinaculum is a fairly broad single-layered sheet of regular collagenous connective tissue that holds in position the tendons of some muscles where they cross certain joints. There are basically two forms of retinacula:

1. In the form of a guy rope that restricts the side-to-side movement of a tendon. For example, there are two retinacula, one on each side of the knee joint, that restrain the patella and, therefore, the quadriceps tendon (figure 4.13). These retinacula help to maintain normal movement between the patella and the femur during flexion and extension of the knee.

2. In the form of a pulley that prevents one or more tendons from springing away from a joint when the muscles contract. For example, the retinacula at the

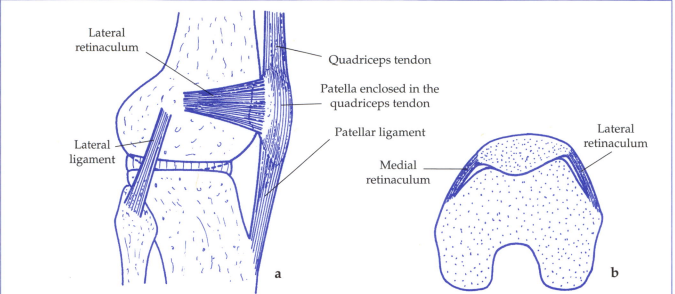

Figure 4.13   Retinacula of the patellofemoral joint; (*a*) lateral aspect of the right knee showing the lateral retinaculum and outline of the patella; (*b*) transverse section through the patellofemoral joint of the right leg showing the medial and lateral retinacula.

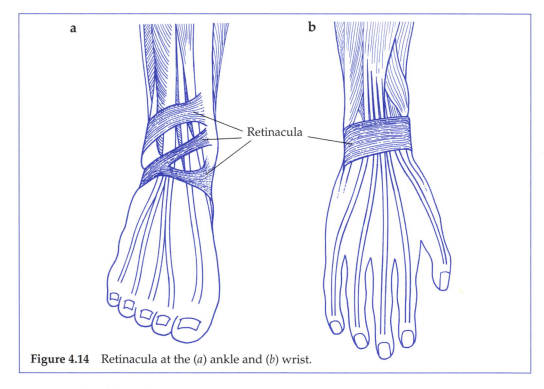

Figure 4.14   Retinacula at the (*a*) ankle and (*b*) wrist.

wrist and ankle hold down the tendons of the muscles that move these joints (figure 4.14). The effect of this form of retinaculum is to considerably increase the mechanical efficiency of the associated muscles.

Regular Elastic Connective Tissue

Regular elastic connective tissue consists largely of elastin fibers arranged parallel to each other. The proportion of collagen fibers and ground substance is usually fairly small. However, the proportion of collagen fibers and ground substance in regular elastic connective tissue is usually greater than the proportion of elastic fibers and ground substance in regular collagenous connective tissue (Akeson, Frank, Amiel, and Woo 1985). Regular elastic connective tissue is found where moderate amounts of strength and elasticity are required mainly in a single direction. As previously described, most ligaments consist of regular collagenous connective tissue, but a few consist of regular elastic connective tissue. Two of these so-called elastic ligaments (ligamentum nuchae, ligamentum flavum; see chapter 6) help to stabilize the vertebral column and to allow a certain amount of movement between the vertebrae.

Fibrous Tissue, Elastic Tissue, and Fascia

The matrixes of four of the six main types of ordinary connective tissue are dominated by collagen or elastin fibers: irregular collagenous, irregular elastic, regular collagenous, and regular elastic. Whereas all of these tissues could be described as fibrous, **fibrous tissue** normally refers only to regular or irregular collagenous tissue. **Elastic tissue** normally refers only to regular or irregular elastic tissue.

Fibrous tissue: regular or irregular collagenous tissue

Elastic tissue: regular or irregular elastic tissue

Fascia: any type of ordinary connective tissue in the form of a sheet

The term **fascia**, from the Latin for bandage, refers to any type of ordinary connective tissue in the form of a sheet. In this sense, all aponeuroses are fascia. However, fascia most often refers to superficial fascia and deep fascia. Superficial fascia refers to the continuous layer of loose connective tissue that connects the skin to underlying muscle or bone. This layer of loose connective tissue is closely associated with the subcutaneous layer of fat referred to earlier. Deep fascia describes the sheets of irregular collagenous tissue that form sheaths around muscles and groups of muscles, separating them into functional units. Figure 4.15 shows the six main types of ordinary connective tissue in relation to the dominant feature of their matrixes and the arrangement of their fibrous content.

The structural differences among types of ordinary connective tissue are in the proportions of elastin, collagen, and ground substance, and the arrangement—irregular or regular—of the fibrous content. There are four main types of irregular ordinary connective tissue: loose, adipose, irregular collagenous, and irregular elastic and two main types of regular ordinary connective tissue: regular collagenous and regular elastic.

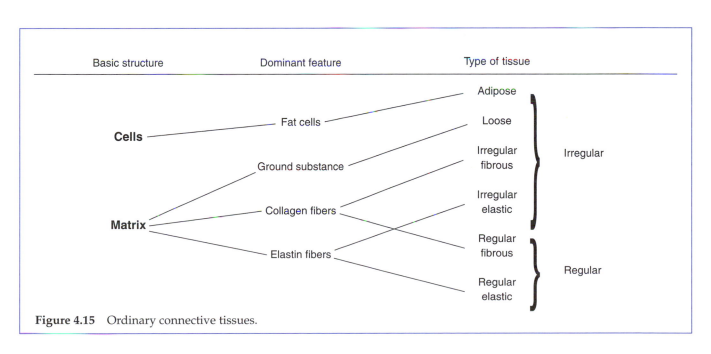

Figure 4.15 Ordinary connective tissues.

Cartilage

The matrix of cartilage is similar to that of fibrous and elastic ordinary connective tissues in that it consists mainly of collagen and elastin fibers embedded in ground substance. However, relative to ordinary connective tissues, the ground substance of cartilage is highly specialized (Caplan 1984). It consists of huge carbohydrate-protein molecular complexes, called proteoglycans, suspended in a large amount of water. The large amount of water is due to the proteoglycans, which have a high affinity for water; each proteoglycan complex is capable of attracting to itself a volume of water that is many times its own weight. Consequently, under normal circumstances, water is the chief constituent of cartilage. The proteoglycans and water produce a highly viscous gel usually referred to as proteoglycan gel. In combination with collagen and elastin, the proteoglycan gel forms a tough rubbery material, often called gristle, capable of strongly resisting tension, compression, and shear loads, and any combination of these loads, especially bending and torsion. In comparison, fibrous and elastic ordinary connective tissues are only designed to resist tensile loads.

Composite material: a material that is stronger than any of the separate substances from which it is made

Cartilage, like the other connective tissues, is a **composite material**—a material stronger than any of the separate substances from which it is made (Alexander 1968). Consequently, cartilage is stronger than either the fibers (collagen or elastin) or the proteoglycan gel. Wood and bone are other examples of natural composite materials. Fiberglass and the type of rubber from which tires are made are examples of manmade composite materials (Alexander 1968).

> *A cartilage matrix consists of collagen and elastin fibers embedded in proteoglycan gel. In combination with collagen and elastin, the proteoglycan gel forms a tough rubbery material capable of resisting all forms of loading. The main functions of cartilage are to facilitate load transmission and joint movement.*

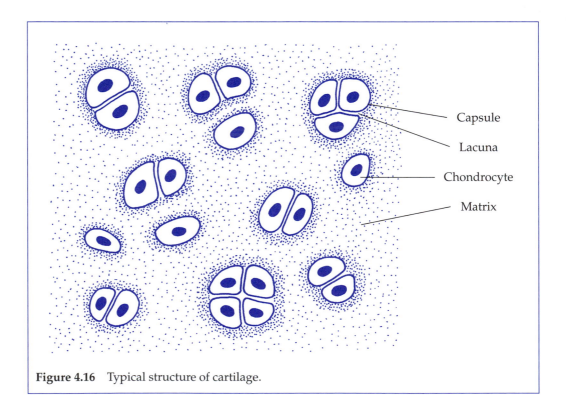

Figure 4.16   Typical structure of cartilage.

The only cells found in cartilage are cartilage cells that produce the cartilage matrix. The cells are called chondrocytes (mature cells) or chondroblasts (immature cells); they lie in fluid-filled spaces called lacunae distributed throughout the matrix (figure 4.16). The cells are arranged singly (parent cells) and in groups of two to five cells that originate from a single parent cell. As the cells become mature they separate from their parent groups and start to produce new groups. Whereas collagen and elastin fibers dominate the matrix of all three main types of cartilage, the region around each lacuna is usually free of fibers. This distinct region—the capsule of the lacuna—consists of proteoglycan gel, which is denser than in other parts of the matrix.

With the exception of the layer of cartilage covering the articulating surfaces of bones in synovial joints, the surface of other cartilages, such as the costal cartilages (chapter 3), is usually covered by a sheath of fibrous tissue called the perichondrium (peri = around, chondrium = cartilage) (Williams et al. 1995). Some of the fibroblasts of the perichondrium are transformed into chondroblasts, which eventually become chondrocytes. Nonweight-bearing perichondrium normally has a number of blood vessels running through it.

Mature cartilage contains no blood vessels (except in nonweight-bearing perichondrium) or nerves (Nordin and Frankel 1989). This reflects the mechanical functions of cartilage, in the sense that blood vessels and nerves would be destroyed by the deformation of cartilage in response to loading. In the absence of a direct blood supply, the cartilage cells depend on intercellular exchange via the proteoglycan gel for their nutrition and excretion. For this reason, and the fact that cartilage is often under considerable load, repair of cartilage is slow and may not take place at all (Caplan 1984).

When cartilage is loaded (subjected to tension, compression, shear, or any combination of these loads), water is forced out of the cartilage and the cartilage deforms. The rate and extent of deformation depends on the size and duration of the load. When the load is removed the proteoglycan structures restore the original level of water saturation and, consequently, the original size and shape of the cartilage, by absorbing water into the cartilage. The ability of a material to gradually deform in response to a load and, following unloading, to gradually restore its original size and shape is referred to as **viscoelasticity**. In comparison, **elasticity** refers to a material's ability to deform immediately in response to a load and, following unloading, to immediately restore its original size and shape, like a rubber band. Cartilage tends to behave viscoelastically in response to prolonged loading, and elastically in response to sudden impact loads (see chapter 10).

> *In mature cartilage there are no blood vessels (except in perichondrium) or nerves. The cartilage cells depend on intercellular exchange via the proteoglycan gel for nutrition and excretion. For this reason, and because cartilage is often under considerable load, repair of cartilage is slow and may not take place at all.*

The degree of viscoelasticity, elasticity, tensile strength, compressive strength, and shear strength of cartilage depends on the proportions of collagen fibers, elastin fibers, and proteoglycan gel in the matrix. In general, there are three main types of cartilage: hyaline cartilage, fibrocartilage, and elastic cartilage.

Hyaline Cartilage

Hyaline cartilage is the most abundant type of cartilage in the body (Tortora and Anagnostakos 1984). It has a pearly bluish-white tinge and under a low-power microscope the matrix appears amorphous and translucent (semitransparent) as in figure 4.16. Under a high-power microscope the matrix can be seen to consist of a

Viscoelasticity: the ability of a material to gradually deform in response to loading and, following unloading, to gradually restore its original size and shape

Elasticity: the ability of a material to deform immediately in response to loading and, following unloading, to immediately restore its original size and shape

dense network of very fine collagen fibrils and fibers embedded in proteoglycan gel (Williams et al. 1995). The size, shape, and arrangement of the cells and fibers in hyaline cartilage vary in different parts of the body depending on the function. Most of the skeleton is preformed in hyaline cartilage and prior to maturity the growth and development of many bones is largely determined by the hyaline cartilage content of the bones (discussed later in this chapter). In addition, hyaline cartilage comprises the following:

1. Articular cartilage—the smooth, tough, wear-resistant articulating surfaces of bones in synovial joints
2. The costal cartilages, which link the upper 10 pairs of ribs to the sternum and provide the rib cage with flexibility and elasticity
3. Supporting rings within the elastic walls of the trachea (windpipe) and the larger bronchial tubes
4. Part of the supporting framework of the larynx (voice box)
5. The external flexible part of the nose that forms the major part of the nostrils

Fibrocartilage

In fibrocartilage, or white fibrocartilage, the matrix is dominated by a dense regular network of bundles of collagen fibers arranged parallel to each other in several layers (figure 4.17). The bundles in adjacent layers run in different directions (like the layers in a joint capsule); this structure produces a strong material with a moderate amount of elasticity. Fibrocartilage is found in a number of locations and forms; for example:

1. It is present in complete or incomplete discs interposed between the articular surfaces of some synovial joints including the knee joint, the sternoclavicular joint, and the acromioclavicular joint. In these joints the discs improve the congruence (area over which the joint reaction forces are distributed) and stability of the joints (see chapter 5). In addition, the discs deform in response to loading and thereby provide shock absorption.

2. It exists in complete discs that join the bones in certain joints (symphysis joints; see chapter 5). These joints include the pubic symphysis and the joints

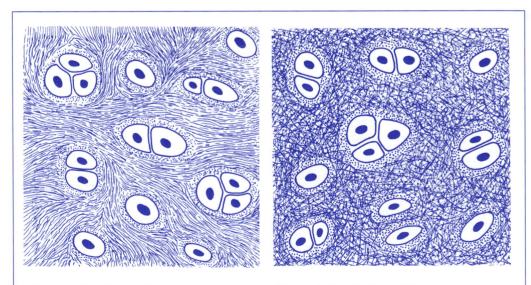

Figure 4.17   Fibrocartilage.   **Figure 4.18**   Elastic cartilage.

between the bodies of the vertebrae. In these joints, deformation of the disc in response to loading allows movement between the articulated bones and provides shock absorption.

3. It is present in the lip around the border of each glenoid fossa and acetabulum. The lips increase the areas of articulation and, therefore, the stability of the shoulder and hip joints.

4. It is in the lining of bony grooves, such as the bicipital groove (see figure 3.25), which are occupied by tendons. The grooves act as pulleys that normally increase the mechanical efficiency of the associated muscles.

Elastic Cartilage

In elastic cartilage, or yellow elastic cartilage, the matrix is dominated by a dense network of elastin fibers (figure 4.18). Elastic cartilage provides support with a moderate to high degree of elasticity. It is found mainly in the larynx, the external part of the ear (pinna), and the tube leading from the middle part of the ear to the throat (eustachian or auditory tube).

> *There are three main types of cartilage: hyaline cartilage, fibrocartilage, and elastic cartilage.*

Bone

The strongest and most rigid of all the connective tissues is bone. The matrix of bone consists of a dense, layered, regular network of collagen fibers embedded in a hard solid ground substance. The ground substance is called *bone salt* and consists of a combination of calcium phosphate and calcium carbonate with smaller amounts of magnesium, sodium, and chlorine (Alexander 1975). In mature bone, bone salt makes up about 70% of the total weight of bone, with collagen making up the remaining 30%. Bone salt is denser than collagen such that the bone salt and collagen both occupy about 50% of the total volume. The composite material made of bone salt and collagen produces a hard, tough, fairly rigid structure. Relative to cast iron, bone has the same tensile strength, is only one-third as heavy, and is much more elastic (Ascenzi and Bell 1971). The elasticity of bone, although slight relative to cartilage, is nevertheless important in enabling it to absorb a sudden impact without breaking. The ability of the ends of bones, together with the articular cartilage, to deform in response to loading is also important in maintaining normal transmission of loads across joints.

> *The matrix of bone consists of a dense, layered, regular network of collagen fibers embedded in a hard ground substance called bone salt. In combination with each other, the bone salt and collagen produce a hard, tough material with little elasticity.*

Bone Growth and Development

Around the third week of intrauterine (within the uterus) life, the embryo's skeleton starts to appear in the form of blocks and plates of tissue. The blocks and plates of most of the embryonic skeleton consist of hyaline cartilage. The top of the skull, the clavicles, and parts of the mandible, however, are formed in a highly vascular form of tissue called fibrous membrane. By the eighth or ninth week of intrauterine life the shapes of the embryonic bones are similar to their eventual adult shape (Williams et al. 1995).

Ossification

Intramembranous ossification: ossification of fibrous membranes

Intracartilagenous ossification: ossification of hyaline cartilage

Endochondral ossification: another name for intracartilagenous ossification

Periosteum: the layer of fibrous tissue that covers the nonarticular surfaces of a bone; the periosteum is responsible for growth in girth of a bone by appositional growth

Ossification or osteogenesis is the process by which the embryonic skeleton is transformed into bone (osteo = bone, genesis = creation). The ossification of fibrous membranes is called **intramembranous ossification** and the ossification of hyaline cartilage is called **intracartilagenous** or **endochondral ossification** (endo = within, chondral = cartilage). Both forms of ossification are similar and produce the same type of bone tissue. The process of endochondral ossification is described with reference to a typical long bone.

Growth in Girth

An embryonic long bone consists of a block of hyaline cartilage covered in a fibrous perichondrium (figure 4.19a). Between the fifth and twelfth weeks of intrauterine life, some fibroblasts in the perichondrium around the middle of the shaft of the cartilage model are transformed into osteoblasts. Osteoblasts are one of three types of bone cells and are responsible for the production of bone. The newly formed osteoblasts invade the hyaline cartilage immediately beneath the perichondrium and start to deposit calcium and other minerals in the matrix. Consequently, the hyaline cartilage is transformed into calcified cartilage. This process of mineralization is called calcification; calcified cartilage represents an intermediate stage in the process of ossification of cartilage into bone. Calcification continues until the calcified cartilage is transformed into bone. Consequently, a bony ring or collar is formed around the middle of the shaft of the otherwise cartilage model (figure 4.19b).

When the perichondrium starts to produce osteoblasts and, in turn, bone, it is called **periosteum**. The first site of bone formation, the middle of the shaft of the cartilage model, is called the primary center of ossification. The process of ossification proceeds from the bony collar in two directions: across the shaft from the outside toward the center, and toward the ends of the shaft. By the 36th week, around the time of birth, the bony collar has become a bony cylinder running the length of the shaft, but not progressing into the bulbous ends of the bone (figure 4.19c). The bony cylinder is thickest at its middle and thinnest at its ends. By this time the remaining hyaline cartilage in the middle of the shaft has been transformed into calcified cartilage. Soon afterward, a second type of bone cells called osteoclasts invades this central portion of calcified cartilage. Whereas osteoblasts produce new bone, osteoclasts remove bone and calcified cartilage. The osteoclasts start to remove the calcified cartilage in the middle of the shaft thereby creating a space called the medullary cavity (figure 4.19d). The medullary cavity gradually widens and extends toward both ends of the shaft. Simultaneously, the thickness of bone in the shaft gradually increases. The development of the skeleton prior to birth, especially the rate at which ossification occurs, is partly due to the loading exerted on the skeleton by the developing muscles, which is manifested in increased movement of the fetus.

Eventually all of the calcified cartilage is removed due to the combined effects of ossification across the shaft from the outside toward the center and osteoclastic activity from the center outward. By this time, the medullary cavity is occupied by yellow marrow consisting of loose connective tissue containing a large number of blood vessels, fat cells, and immature white blood cells (Williams et al. 1995). A layer of loose connective tissue containing many osteoblasts and a smaller number of osteoclasts lines the medullary cavity. This layer is called the endosteum.

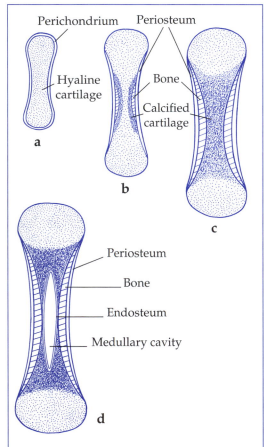

Figure 4.19 Endochondral ossification: early stages in the growth and development of a typical long bone.

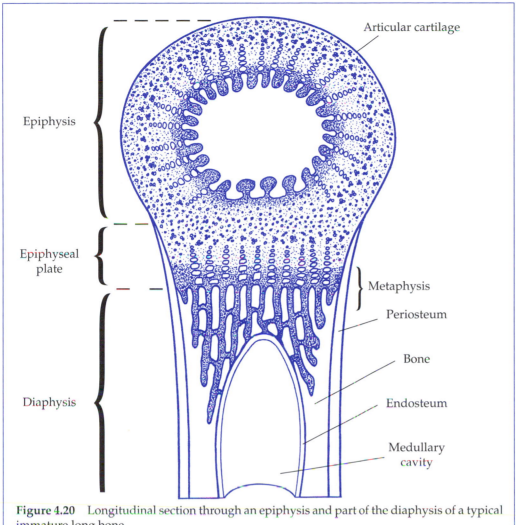

Figure 4.20 Longitudinal section through an epiphysis and part of the diaphysis of a typical immature long bone.

Long bones are designed, above all, to resist bending. For a given amount of bone tissue, a hollow shaft is stronger in bending than a solid one (Alexander 1968). This is the main reason the shafts of long bones are hollow. Growth in girth of the shaft of a long bone involves formation of new bone on the outside of the shaft by osteoblasts in the periosteum, and the removal of bone from the inside of the shaft by osteoclasts in the endosteum. The type of growth produced by the periosteum, which involves laying down new bone on the surface of older bone rather like the addition of rings in a tree, is called **appositional growth**. In mature bone the periosteum consists of irregular fibrous tissue. In addition to appositional growth, the periosteum has three other main functions:

1. To provide a protective cover around the shaft of the bone
2. To allow blood vessels to pass into the bone
3. To provide attachments for muscles, tendons, ligaments. and joint capsules

Growth in Length

Around the time of birth a secondary center of ossification occurs in the center of each end of a long bone. These new centers of ossification are responsible for the ossification of the ends of the bones; ossification proceeds from the center toward the

Appositional growth: the type of growth in which new tissue is laid down on the surface of existing tissue

Articular cartilage: the layer of hyaline cartilage that covers each articular surface of a bone (in a synovial joint)

Epiphysis: end of a bone separated from the diaphysis by an epiphyseal plate prior to maturity

Diaphysis: the shaft of a bone

Epiphyseal plate: the region of a bone between the epiphysis and the diaphysis that is responsible for growth in length of the bone by interstitial growth

periphery. After the secondary centers of ossification have been established, the only hyaline cartilage remaining from the original cartilage model is that covering the bulbous ends of the bone and the plates of cartilage separating the ends of the bone from the shaft (figure 4.20). These two regions of hyaline cartilage are continuous with each other and remain so until maturity.

Part of the cartilage covering each end of a bone forms an articular surface and, as such, is referred to as **articular cartilage**. Each end of a bone is called an **epiphysis** and the shaft is called the **diaphysis**. The plates of cartilage that separate the epiphyses and diaphysis are called **epiphyseal plates** (figure 4.20). The epiphyseal plates are responsible for growth in length of the bone. During normal growth they remain active until the bone has achieved its mature length.

An epiphyseal plate consists of four layers (Tortora and Anagnostakos 1984) (figure 4.21). The layer adjacent to the epiphysis is called the reserve or germinal layer. In this layer, which anchors the epiphyseal plate to the bone of the epiphysis, the chondrocytes are distributed throughout the matrix usually as single cells or in pairs. The second layer is called the proliferation layer. As its name suggests, it is responsible for chondrogenesis (production of new cartilage). The chondrocytes in this layer undergo fairly rapid cell division, and in turn, the cells produce new matrix that results in an increase in the amount of cartilage. Growth in length of the shaft of

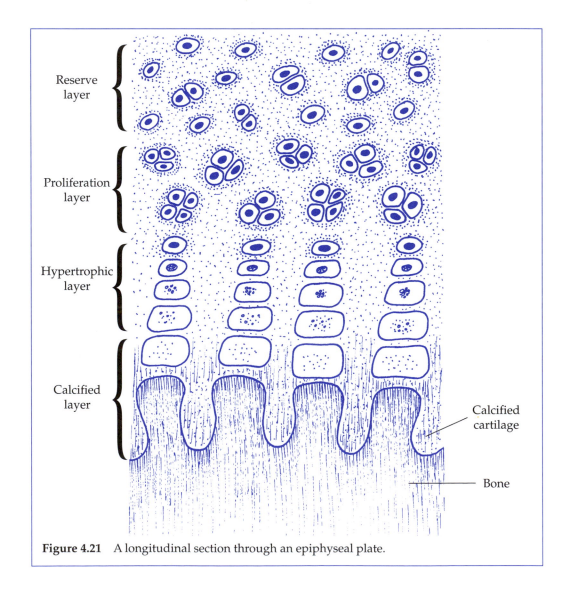

Figure 4.21　A longitudinal section through an epiphyseal plate.

a bone is due to chondrogenesis in the proliferation layers of the epiphyseal plates. This type of growth, in which additional new tissue is produced from within the mass of existing tissue, is called **interstitial growth.**

The third layer of the epiphyseal plate is called the hypertrophic layer. In this layer, the chondrocytes are arranged in columns and gradually increase in size, with the larger and more mature cells farthest from the epiphysis. The fourth layer of the epiphyseal plate is called the calcified layer. In this layer the hypertrophied chondrocytes and surrounding matrix are replaced by calcified cartilage. The calcified cartilage interdigitates with the underlying bone forming a relatively strong bond (see figures 4.20 and 4.21) that is able to resist shear loading. As new cartilage is formed in the proliferation layer, the calcified cartilage in contact with the underlying bone is itself gradually transformed into bone. The net result of these processes is that the epiphyseal plates, which remain about the same thickness, gradually move farther from the middle of the shaft as the shaft increases in length.

The **metaphysis** is the region where the epiphysis joins the diaphysis; in a growing bone this corresponds to the calcified layer of the epiphyseal plate together with the interdigitating bone (see figure 4.20). The interface between the hypertrophic and calcified layers is sometimes referred to as the tidemark.

> *The adult bony skeleton develops from an embryonic skeleton that forms during the second month of intrauterine life and consists mainly of hyaline cartilage and fibrous membrane. The process by which cartilage and membrane are transformed into bone is called ossification. Growth in girth of bones occurs by appositional growth; growth in length of bones occurs by interstitial growth.*

When a long bone has achieved its mature length, longitudinal growth in the epiphyseal plates ceases. Shortly afterward, the epiphyseal plates are replaced by bone so that the epiphyses are fused with the shaft. In most long bones one end usually fuses with the shaft before the other end. In the long bones of the arms and legs, fusion of both ends normally takes place between 14 and 20 years of age (Williams et al. 1995). In some other bones, such as the innominate bones (which consist of three bones—ilium, pubis, and ischium—prior to maturity), fusion takes place usually between 20 and 25 years of age. It follows that the epiphyseal plates of the various bones are vulnerable to injury for a relatively long period. Injury to an epiphyseal plate may, in severe cases, result in one of two types of bone deformity (Pappas 1983):

1. A complete cessation of growth and premature fusion resulting in, for example, a limb length discrepancy
2. An asymmetric cessation of growth across an epiphyseal plate resulting in an angular deformity and joint incongruity

The degree of bone deformity resulting from an epiphyseal plate injury depends on the following factors:

1. An individual's physical maturity; the more mature the individual, the lower the likelihood of serious deformity
2. The severity of the injury
3. Which epiphyseal plate is injured

The epiphyseal plates at each end of a long bone usually contribute different amounts to the length of the shaft. For example, the proximal and distal epiphyseal plates of the humerus contribute approximately 80% and 20%, respectively, to the total length of the bone. In contrast, the proximal and distal epiphyseal plates of the femur contribute approximately 30% and 70%, respectively, to the total length of the

Interstitial growth: the type of growth in which new tissue is produced from within the mass of existing tissue

Metaphysis: the region of a bone where the epiphysis joins the diaphysis; in a growing bone this corresponds to the calcified layer of the epiphyseal plate together with the interdigitating bone

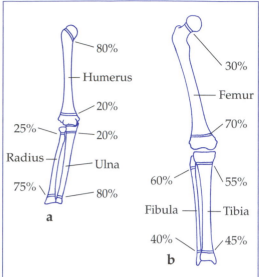

Figure 4.22 Contributions of the proximal and distal epiphyseal plates to growth in length of the long bones of the (*a*) upper and (*b*) lower limbs.

bone (Pappas 1983) (figure 4.22). Injury to the epiphyseal plate that makes the largest contribution to the total length of a bone is likely to have a greater effect on bone growth than injury to the other epiphyseal plate (Siffert 1987).

The vulnerability of epiphyseal plates to injury is largely due to the plates being the weakest parts of the immature skeleton. For example, ligaments and joint capsules are two to five times stronger than epiphyseal plates (Larson and McMahan 1966). When a ligament supporting a particular joint is inserted into the epiphysis (rather than the diaphysis), a load applied to the joint that tends to stretch the ligament is, in a child, more likely to result in a fracture through the epiphyseal plate than in a tear in the ligament. In an adult, the same type of loading would tend to cause a ligament tear since the epiphysis and diaphysis are fused (Pappas 1983) (figure 4.23).

When a bone has achieved its predetermined length the epiphyseal plates are replaced by bone so that the epiphyses are fused with the shaft. In general, fusion takes place between 14 and 20 years of age. Injury to epiphyseal plates during the growth period may result in a variety of skeletal abnormalities including discrepancies in limb length and angular deformities in joints.

Growth of the Epiphyses

Just as the epiphyseal plates are responsible for growth in length of a bone, the hyaline cartilage that covers the end of a bone is responsible for growth of the epiphysis. This

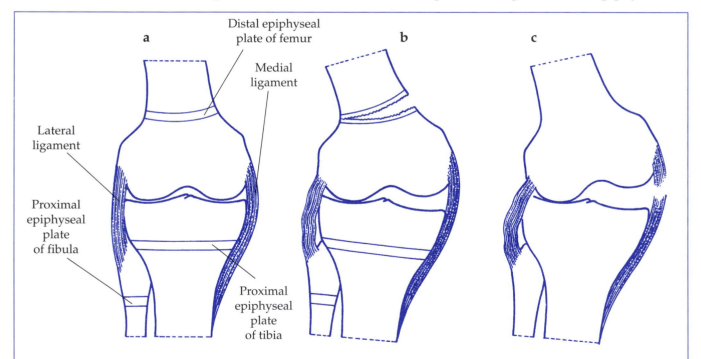

Figure 4.23 Effect of degree of skeletal maturity on type of injury; (*a*) anterior aspect of right knee joint showing normal alignment of femur, tibia, and fibula; (*b* and *c*) abduction of the lower leg which may result from a blow to the outside of the knee while the foot is in contact with the ground. In a child this is more likely to result in a fracture through the distal epiphyseal plate of the femur than tearing of the medial ligament (*b*). After maturity, excessive abduction of the knee will almost certainly result in partial or complete tearing of the medial ligament (*c*).

cartilage consists of an articular region and a nonarticular region. Like an epiphyseal plate, articular cartilage (and its adjacent nonarticular regions; see figure 4.20) consists of four layers. The only real difference in structure between articular cartilage and an epiphyseal plate is in the arrangement of the fibers in the reserve layer.

In an epiphyseal plate the collagen fibers cross each other obliquely, forming a strong bond between the epiphyseal bone and the reserve layer of the plate. In articular cartilage the reserve layer is the outer layer. Whereas the majority of the layer is similar in structure to the reserve layer of an epiphyseal plate, the outer surface of articular cartilage is cell free and consists of densely packed collagen fibers and fibrils arranged parallel to the articular surface. This arrangement produces a tough wear-resistant surface.

The type of growth produced by articular cartilage is the same as that produced by an epiphyseal plate—interstitial growth. During the growth period, the rate of ossification of an epiphysis is greater than the rate of growth of the epiphysis. Consequently, the thickness of the articular cartilage becomes relatively thinner with age (figure 4.24). At maturity, the thickness of articular cartilage is approximately 1 to 7 mm and this tends to decrease with age due to mechanical wear. Whereas bone growth is largely determined by genetic factors, the mechanical stress experienced by articular cartilage and epiphyseal plates, as a result of movement and the maintenance of an upright posture, also has a major effect on bone growth (see chapter 11).

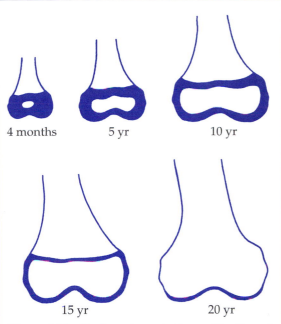

Figure 4.24 Various stages in the ossification of the distal epiphysis of the femur.

In a number of bones, such as some of the carpals and tarsals, ossification is completed from a single (primary) center of ossification. In other bones, such as metacarpals and metatarsals, there is a primary center of ossification, and only one secondary center of ossification; there is only one epiphysis. In all of the large long bones—the bones of the arms and legs—secondary centers of ossification occur in both ends of the bone around the time of birth.

> *The articular cartilage and associated nonarticular regions of an epiphysis are responsible for growth of the epiphysis by interstitial growth. During growth and development, the relative thickness of articular cartilage gradually decreases. At maturity it is approximately 1 to 7 mm thick.*

Growth of Apophyses

Secondary centers of ossification occur not only in the epiphyses of long bones, but also in some of the rudimentary tuberosities of some bones, including the femur, innominate bones, and calcaneus bones (figure 4.25). These secondary centers of ossification occur in regions of bone called **apophyses** around 10 to 14 months after birth. Each apophysis grows and ossifies in much the same way as an epiphysis. Apophyses provide areas of attachment for the tendons of powerful muscles such as the quadriceps (tibial tuberosity), hamstrings (ischial tuberosity) and calf muscles (calcaneal tuberosity) (figure 4.26). This form of attachment is different from that of most tendons, which attach directly onto the periosteum.

Prior to maturity each apophysis is separated from the rest of the bone by an **apophyseal plate**, which is very similar in structure and function to an epiphyseal plate. Each apophyseal plate is responsible for growth of the bone adjacent to the nonapophyseal side of the plate. Growth of the apophysis itself

Apophysis: a tuberosity separated from the rest of a bone prior to maturity by an apophyseal plate.

Apophyseal plate: a cartilaginous region of a bone that separates an apophysis from the rest of the bone prior to maturity; an apophyseal plate is responsible for growth of the bone adjacent to the nonapophyseal side of the plate.

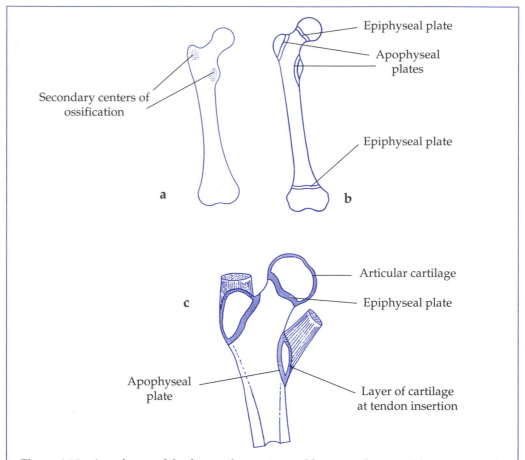

Figure 4.25   Apophyses of the femur: the greater and lesser trochanters; (*a*) occurrence of secondary centers of ossification in bony outgrowths; (*b*) apophyseal and epiphyseal plates of the femur; (*c*) growth areas of the head of the femur and the greater and lesser trochanters.

is due to a layer of cartilage (mixture of hyaline cartilage and fibrocartilage) outside of the apophysis into which the fibers of the tendon insert (see figure 4.25*c*). At maturity, the apophyses fuse with the rest of the bone.

> *Apophyses provide areas of attachment for powerful muscles. Prior to maturity each apophysis is separated from the rest of the bone by an apophyseal plate that is similar in structure and function to an epiphyseal plate. At maturity the apophyses fuse with the rest of the bone.*

Epiphyses, especially those that form weight-bearing joints, are most frequently subjected to compression loading and, therefore, are often referred to as pressure epiphyses. In contrast, apophyses are most frequently subjected to tension loading and are often referred to as traction epiphyses. Whereas apophyseal plates do not affect growth in bone length, they do affect the alignment and strength of the tendons attached to them. Consequently, injury to apophyseal plates may affect the mechanical characteristics of associated muscles, which, in turn, may affect normal joint function.

Studies of sport-related injuries in children show that the proportion of injuries involving growth plates (epiphyseal and apophyseal) is between 6% and 18% of the total number of injuries (Speer and Braun 1985; Krueger-Franke, Siebert, and Pfoerringer 1992; Gross, Flynn, and Sonzogni 1994). About 5% of these growth-plate injuries result in some type of bone deformity (Larson 1973). On the basis of these

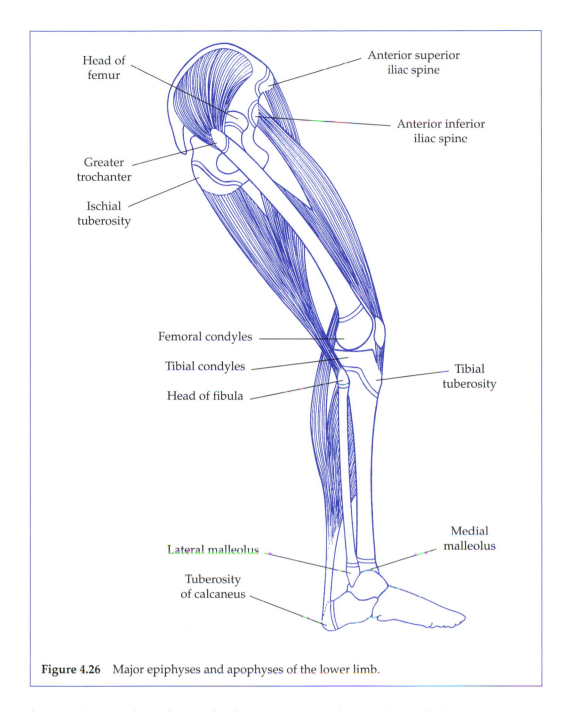

Figure 4.26   Major epiphyses and apophyses of the lower limb.

figures, the number of growth-plate injuries resulting in bone deformity is in the region of 3 to 9 per thousand. However, this estimate is likely to be conservative since many injuries that occur during free play and sports are not reported or are incorrectly diagnosed (Combs 1994).

Structure of Mature Bone

During the period of growth and development of a long bone, bone tissue is deposited in a manner that maximizes the strength of the bone as a whole. The development of a hollow shaft is just one example of this process. Different regions of the bone are subjected to different types and magnitudes of loading. For example, the epiphyses are mainly subjected to compression loading, whereas the shaft is

Compact bone: bone in which the osteones are closely packed with little or no space between them

Cancellous bone: bone in which the osteones are loosely packed with spaces between them filled with red marrow; the osteones are usually arranged in the form of trabeculae

Osteone: a column of bone consisting of three to nine concentric layers of bone surrounding a haversian canal

mainly subjected to bending and torsion loads. For this reason, not only is the shaft hollow, but the density and thickness of bone in the shaft is greater than that in the epiphyses since the shaft is subjected to the greatest bending and torsion loads. Dense bone is called **compact bone**. The thickness of compact bone in the shaft gradually decreases from the middle of the shaft toward the epiphyses (figure 4.27).

In each epiphysis the calcified layer of the articular cartilage gradually merges with the underlying bone. This transitional region is referred to as subchondral bone; it encloses a mass of low-density bone that makes up the remainder of the epiphysis. The low-density bone is in the form of a honeycomb or latticework consisting of thin curved bars of bone attached to each other by interconnecting bars of bone. The bars of bone are called trabeculae and the latticework formed by the trabeculae is called trabecular bone, spongy bone or, most often, **cancellous bone** due to the large number of spaces between the trabeculae. The majority of the trabeculae cross each other at right angles; this arrangement maximizes the strength of the trabecular bone. The spaces between the trabeculae are filled with red marrow—loose connective tissue containing a large number of blood vessels, some white blood cells and fat cells, and a large number of cells called erythroblasts responsible for producing red blood cells. The spaces in cancellous bone are continuous with the medullary cavity and, therefore, the red marrow is continuous with the yellow marrow.

In most joints, especially weight-bearing joints, the loading on the epiphyses is different in different positions of the joint. The change in loading may be in terms of the type, magnitude, or direction of the load, or a combination of these characteristics. The trabeculae are arranged to minimize the stress experienced by the epiphyses in all positions of the joint during habitual movements.

The only real structural difference between compact bone and cancellous bone is in the density of the bone. Compact bone is much more dense and, therefore, much less elastic than cancellous bone. The elasticity of cancellous bone is very important in ensuring congruity in joints during load transmission, thereby minimizing stress within the epiphyses and on the articular cartilages (Ascenzi and Bell 1971; Radin 1984).

Compact and Cancellous Bone

Compact bone consists of complete and incomplete columns of bone closely packed together (figure 4.28). Each column of bone is called an **osteone** or a haversian system. In long bones the osteones run parallel to the long axis of the bone. Each osteone consists of three to nine concentric rings (or layers) of bone surrounding a central open channel. The concentric rings of bone are called lamellae and the central channel is called a haversian canal. Each lamella basically consists of a single layer of closely packed collagen fibers arranged parallel to each other, with bone salt embedded between the fibers. Whereas the collagen fibers in each lamella are parallel to each other, the orientation of fibers in adjacent lamellae is different (figure 4.29). This arrangement, similar to the layers of collagen fibers in a joint capsule, enables the bone to strongly resist deformation in any direction. In addition to the lamellae in osteones, the outer surface of compact bone consists of a number of circumferential or surface lamellae that encircle the bone and tend to bind the osteones together (figure 4.28*b*).

Between the lamellae are numerous osteocytes. Osteocytes are basically osteoblasts that have become entrapped in the bone. They are mainly responsible for repairing bone damage, and are

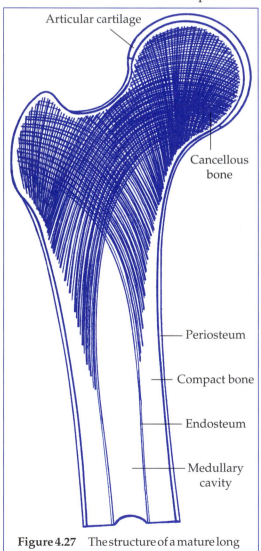

Articular cartilage

Cancellous bone

Periosteum

Compact bone

Endosteum

Medullary cavity

Figure 4.27   The structure of a mature long bone: a longitudinal section through the proximal third of the femur.

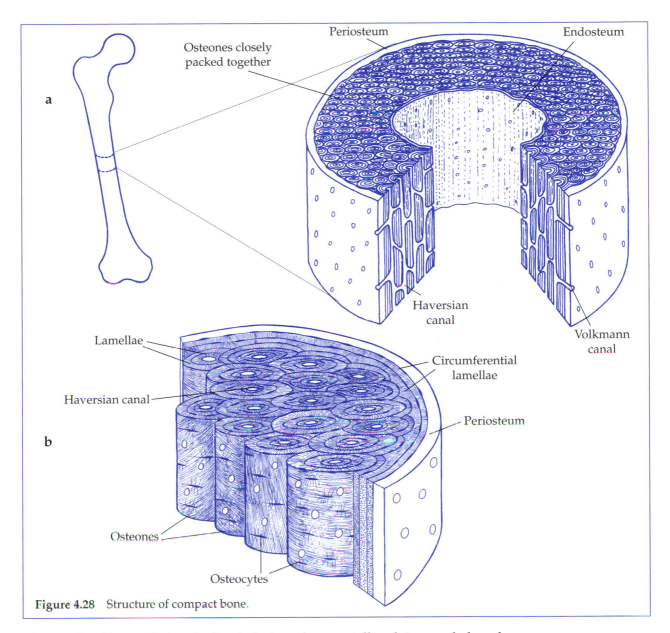

Figure 4.28   Structure of compact bone.

also involved in regulating the level of minerals, especially calcium and phosphorus, in the blood (Bailey et al. 1986). Each osteocyte lies in a lacuna, a small space, and the lacunae are linked together by tiny channels called canaliculi. The canaliculi run between the lamellae and through the lamellae from one side to the other. This system of canaliculi enables the osteocytes to communicate with each other both physically, by means of projections from the cell bodies into the canaliculi, and chemically, by means of secretions from the cells (see figure 4.29; figure 4.30). Communication between the osteocytes is important for coordinating the growth, development and repair of lamellae, and for facilitating intercellular exchange among osteocytes and between the osteocytes and blood vessels.

The haversian canal at the center of each osteone contains blood vessels and nerves supported by loose connective tissue. The canaliculi are linked to the haversian canals, thereby facilitating intercellular exchange between blood vessels and osteocytes. In addition to being linked together by canaliculi, the haversian canals of adjacent osteones are also linked together by channels called volkmann canals, which are similar in size to haversian canals.

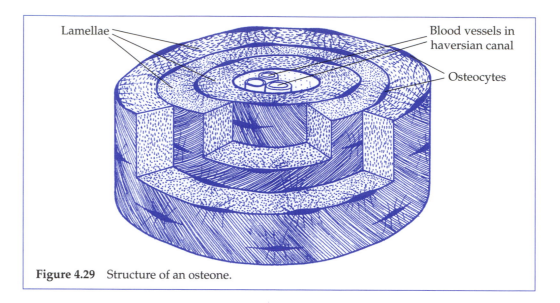

Figure 4.29   Structure of an osteone.

Volkmann canals, like haversian canals, contain blood vessels and nerves supported by loose connective tissue. The volkmann canals form a system of channels, which run from the outside of the bone to the medullary cavity, by linking together the periosteum, haversian canals, and endosteum (see figure 4.28*a*). The system of haversian and volkmann canals enables blood vessels and nerves to pass along, around, and across the bone.

In compact bone the osteones are very closely packed with little or no space between them. In cancellous bone the osteones are loosely packed with spaces in between them filled with red marrow; the osteones are usually arranged in small groups that form trabeculae. Many of the trabeculae do not contain haversian canals and consist of several lamellae in the form of a narrow strip. Consequently, cancellous bone consists of a mixture of osteonal and nonosteonal bone.

> *In a mature long bone the diaphysis consists of a cylinder of compact bone thickest in the middle and tapered toward each end. Each epiphysis consists of a relatively thin outer layer of subchondral bone enclosing a mass of cancellous bone. The cancellous bone extends into the diaphysis and tapers toward the middle. The articular surfaces are covered by articular cartilage.*

Modeling and Remodeling in Bone

Skeletal genotype: the process of genetically programmed change in the external form (size and shape) and internal architecture of the bones

Modeling: changes in the expression of the skeletal genotype that occur as a result of environmental influences

Remodeling: coordination of osteoblastic and osteoclastic activity resulting in changes in external form and internal architecture of the bones, including repair of bones

The processes of growth, development, and maintenance of the bones of the skeleton are carried out by the interaction of three subprocesses: the expression of the **skeletal genotype, modeling,** and **remodeling**. The expression of the skeletal genotype refers to the process of genetically programmed change in the external form (size and shape) and internal architecture of the bones. Modeling refers to the changes in the expression of the skeletal genotype that occur as a result of environmental factors such as nutrition and, in particular, the mechanical strains imposed by normal habitual activity. Remodeling refers to the coordination of osteoblastic and osteoclastic activity responsible for the actual changes in external form and internal architecture of the bones, including repair of bones (figure 4.31). Since bone is continuously being absorbed from some places (by osteoclasts) and deposited in others (by osteoblasts), the process of remodeling is sometimes referred to as turnover. Prior to maturity, all bones are in a continual state of change in external form and internal architecture. After skeletal maturity is achieved (approximately 20 to 25 years of age), modeling of external form

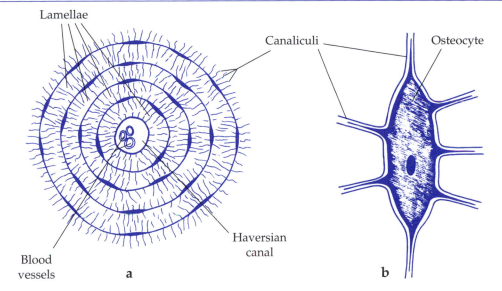

Figure 4.30 Osteocytes and canaliculi; (a) cross section of an osteone; (b) osteocyte lying in a lacuna showing projections from the cell body into the canaliculi.

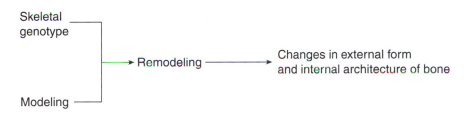

Figure 4.31 Relationship between skeletal genotype, modeling, and remodeling in bone growth, development, and maintenance.

decreases to negligible proportions, but modeling of internal architecture continues throughout life (Frost 1979; Bailey 1995).

> *Growth, development, and maintenance of the bones of the skeleton is determined by the interaction of three processes: the expression of the skeletal genotype, modeling, and remodeling.*

Porosity, Osteopenia, and Osteoporosis

Due to the various channels and spaces within compact and cancellous bone, any particular region of a bone consists of certain amounts of bone tissue and nonbone tissue. The term **porosity** describes the proportion of nonbone tissue. At skeletal maturity the porosity of compact and cancellous bone is approximately 2% and 50%, respectively; the density (amount of bone tissue per unit volume) of compact bone is approximately double that of cancellous bone (Radin 1984). The density of bone tissue depends on the degree of mineralization. During ossification, the degree of mineralization of bone tissue gradually increases and reaches a maximum level at skeletal maturity (Bailey et al. 1986). However, the amount of bone within the skeleton may continue to increase for 5 to 10 years after skeletal maturity, especially in physically active individuals (Stillman et al. 1986; Talmage and Anderson 1984). Consequently, bone mass peaks in males and females between 25 and 30 years of age. In terms of turnover, this means that from skeletal maturity to the age at which peak bone mass occurs, more new bone is formed than old and damaged bone is absorbed.

Porosity: the proportion of nonbone tissue in a bone or region of a bone

Following peak bone mass there is usually a stable period in which the amount of bone in the skeleton remains about the same; there is a balance between bone absorption and bone formation. This stable period is followed by a gradual decrease in bone mass for the rest of the life of the individual; the rate of bone absorption exceeds the rate of bone formation. Bone mass is the product of bone volume and bone density. The loss in bone mass that occurs with age following peak bone mass is the result of decreases in bone volume and bone density. **Osteopenia** refers to a level of bone density below the normal level for the age and sex of the individual (Bailey 1995).

Bone mass starts to decrease earlier and at a greater rate in females than in males. In males, bone loss normally starts to occur between 45 and 50 years of age and proceeds at a rate of 0.4% to 0.75% per year (Bailey et al. 1986; Smith 1982). In females, bone loss has three phases. The first phase starts around 30 to 35 years of age and proceeds at a rate of 0.75% to 1% per year until the menopause. From menopause until about five years after menopause the rate of bone loss increases to between 2% and 3% per year. During the final phase, the rate of bone loss is approximately 1% per year. On this basis, females may lose 40% of their peak bone mass by the age of 80 years. In contrast, males may lose 20% of their peak bone mass by the age of 80 years (figure 4.32).

Even though body weight tends to decrease with age, the rate of bone loss is usually much greater than the rate at which body weight decreases. Consequently, the effect of bone loss is that the bones, especially weight-bearing bones, become progressively weaker relative to the weight of the rest of the body. In addition to a gradual decrease in strength, the bones also gradually lose their elasticity and, as a result, become more and more brittle. In some individuals, especially females, a loss of bone mass and elasticity is eventually reached when some bones are no longer able to withstand the loads imposed by normal habitual activity. Consequently, these bones become very susceptible to fracture. This condition, the most common bone disorder in elderly people, is called **osteoporosis** (Bailey et al. 1986). Osteoporosis may cause severe disfigurement, especially of the trunk, due to fractured or crushed

Osteopenia: bone density below the normal level for the age and sex of the individual

Osteoporosis: loss of bone mass and elasticity to the extent that bones are no longer able to withstand the loads imposed by normal, habitual activity, resulting in a high susceptibility to fracture

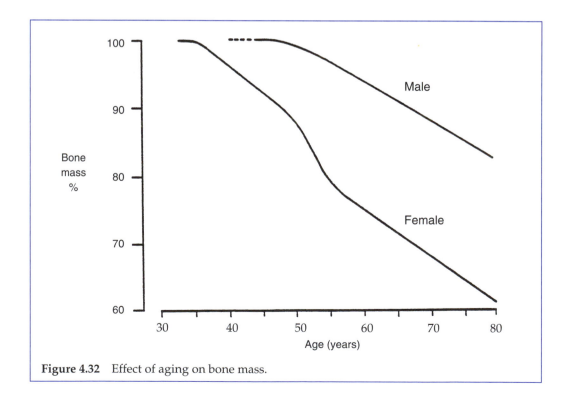

Figure 4.32   Effect of aging on bone mass.

vertebrae. Many deaths in the elderly are due to complications arising from bone fractures that occur as a result of osteoporosis (Kaplan 1983).

Bone loss tends to occur earlier and to proceed at a faster rate in cancellous bone than in compact bone (Bailey et al. 1986). Consequently, regions of bones with a high proportion of cancellous bone, such as the bodies of the vertebrae, the head and neck of the femur, and the distal end of the radius, are particularly vulnerable to osteoporosis and, therefore, to fracture in elderly people. This vulnerability is reflected in studies indicating a rapid increase in the incidence of bone fractures with age, especially in females. For example, the results of one study showed that the incidence of fracture to the distal end of the radius was seven times higher in 54-year-old women than in 40-year-old women (Bauer 1960). In another study the incidence of fracture of the neck of the femur was found to be 50 times higher in 70-year-old women than in 40-year-old women (Chalmers and Ho 1970). With regard to compact bone, bone loss occurs mainly on the endosteal surface so that bone width remains relatively unchanged into old age (Smith 1982).

Whereas the cause of osteoporosis is not yet clear, there is general agreement that four variables are mainly responsible: genetic factors, endocrine status, nutritional factors, and physical activity (Bailey et al. 1986; MacKinnon 1988). The relative contribution of these variables has not yet been established, but physical activity level seems to be the most important. In the absence of weight-bearing activity, no amount of endocrine or nutritional intervention will prevent rapid bone loss; there must be mechanical stress (Bailey et al. 1986). Research suggests that regular physical activity throughout life, within the moderate overload range (see chapter 11), can help to prevent osteoporosis in three ways (Bailey et al. 1986; Stillman et al. 1986; Talmage and Anderson 1984; Smith and Gilligan 1987):

1. Peak bone mass is directly related to the level of physical activity prior to peak bone mass; the higher the peak bone mass, the lower the risk of osteoporosis.
2. An above-average level of physical activity after peak bone mass will delay the onset of bone loss.
3. An above-average level of physical activity after peak bone mass will reduce the rate of bone loss.

From about 30 years of age in females and about 45 years of age in males there is a gradual decrease in bone mass and bone elasticity. Many individuals, especially females, develop osteoporosis, which may cause severe disfigurement and in some cases death due to complications arising from osteoporotic bone fractures. Whereas the cause of osteoporosis is not yet clear, research suggests that one of the main causes is lack of mechanical stress. Regular physical activity throughout life appears to be the best way of preventing osteoporosis.

Summary

This chapter described the structure and functions of the various connective tissues. The dominant structural feature of all connective tissues is a large mass of noncellular matrix; the physical characteristics of the matrix of each connective tissue determine its function. Connective tissues provide mechanical support at all levels of cellular organization and facilitate intercellular exchange. The mechanical function of connective tissues is clearly evident in the next chapter, which describes the articular system.

Review Questions

1. Describe the two main functions of connective tissues.
2. Describe the difference between the following:
   - Regular and irregular ordinary connective tissues
   - Loose connective tissue and adipose connective tissue
   - Capsular and noncapsular ligaments
   - Intracapsular and extracapsular ligaments
   - Ligaments and retinacula
   - Fibrous tissue and elastic tissue
3. Describe the different types of cells found in ordinary connective tissues.
4. Describe the two main functions of cartilage.
5. Explain the difference between elasticity and viscoelasticity.
6. Describe the difference in structure and function between regular fibrous tissue and fibrocartilage.
7. Explain why repair in cartilage is usually slow and may not take place at all.
8. Differentiate between or among the following:
   - Primary and secondary centers of ossification
   - Appositional and interstitial growth
   - Epiphyseal and apophyseal plates
   - Osteoblasts, osteoclasts, and osteocytes
   - Compact and cancellous bone
   - Osteopenia and osteoporosis

Chapter 5

The Articular System

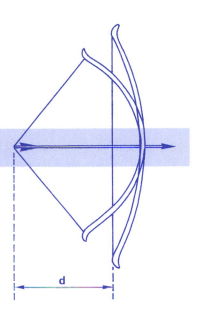

The human body is capable of a broad range of movements facilitated by the combined effects of the open-chain arrangement of the bones, the number of joints linking the bones, the different types of joints, and the range of movement in the joints. Most joints allow a certain amount of movement and all joints transmit forces. Joints differ in terms of the type and range of movement and the mechanism of force transmission; these differences are reflected in each joint's structure. This chapter describes the structure and function of the various types of joints.

Objectives

After reading this chapter you should be able to do the following:

1. Define or describe the key terms.
2. Distinguish between temporary and permanent joints.
3. Describe the structural classification of joints.
4. Describe the structure and specific functions of the different forms of fibrous and cartilaginous joints.
5. Describe the structure of a synovial joint.
6. Describe the stability-flexibility classification of joints.
7. Explain the relationship of the structural classification to the stability-flexibility classification of joints.
8. Describe the different forms of synovial joints.
9. Describe the functions of joint capsules and ligaments.
10. Differentiate between flexibility, laxity, stability, and congruence.

Joint: a region where two or more bones are connected

Articular system: all of the joints of the body

A **joint**, also referred to as an articulation or an arthrosis, is defined as a region where two or more bones are connected. The adult skeleton normally has 206 bones linked by approximately 320 joints. The **articular system** refers to all of the joints of the body. The function of joints is largely mechanical—to facilitate relative motion between bones and transmission of force from one bone to another.

The joints of the adult articular system may be regarded as permanent joints because they are present throughout the life of the individual, though the structure and function of some of these joints may change with increasing age. In addition to the permanent joints, the immature articular system also has a large number of temporary joints, which are concerned with bone growth, that gradually become less distinct and are virtually obliterated at maturity. Consequently, in relation to the immature articular system, the term joint refers not only to the regions where the separate bones are connected, but also to the regions that unite the bony parts of each immature bone.

> All of the joints in the adult articular system—permanent joints—are present in the immature articular system. However, the immature articular system also has a number of temporary joints, concerned with bone growth, which gradually become less distinct and are virtually obliterated at maturity. The adult joints facilitate relative motion between bones and transmission of force from one bone to another.

Fibrous joint: a joint in which the opposed surfaces of the bones are united by fibrous tissue

Cartilaginous joint: a joint in which the opposed surfaces of the bones are united by cartilage

Synovial joint: a joint in which the opposed surfaces are not attached to each other, but are held in contact with each other by ligaments and a joint capsule

Structural Classification of Joints

In terms of structure, there are basically two types of joints:

1. Joints in which the articular (opposed) surfaces of the bones are united by either fibrous tissue or cartilage are called **fibrous joints** and **cartilaginous joints**, respectively (figure 5.1a).

2. Joints in which the articular surfaces are not attached to each other but are held in contact with each other by a sleeve of fibrous tissue supported by ligaments (figure 5.1b) are referred to as **synovial joints**, and the fibrous sleeve is the joint capsule (see chapter 4).

Fibrous Joints

Fibrous joints are often referred to as **syndesmoses** (syn = with, desmo = ligament). The degree of movement in a syndesmosis is largely determined by the amount of fibrous tissue between the articular surfaces. In general, the smaller the amount of fibrous tissue, the more limited the movement. There are two types of syndesmoses: membranous and sutural.

Membranous Syndesmoses

In a membranous syndesmosis the articular surfaces are united by a sheet of fibrous tissue called an interosseous membrane (inter = between, osseous = bone). The interosseous membrane acts rather like webbing. The interosseous borders of the radius and ulna are connected by an interosseous membrane (figure 5.2*a*). The majority of the fibers in the membrane run obliquely downward and medially from the interosseous border of the radius to that of the ulna. The remaining fibers run obliquely downward and laterally from the interosseous border of the ulna to that of the radius. The interosseous membrane functions to stabilize the radius and ulna in all positions in the range from full supination to full pronation, and to provide areas of attachment for muscles on the anterior and posterior aspects of the

Syndesmosis: another name for a fibrous joint

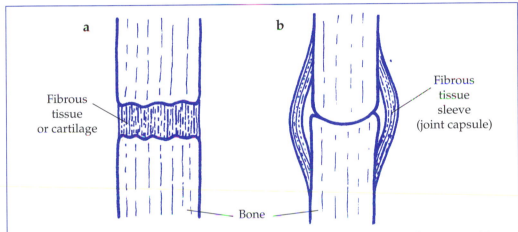

Figure 5.1   There are two basic types of joint structures; (*a*) articular surfaces united by fibrous tissue or cartilage; (*b*) articular surfaces not attached but held in contact by a fibrous tissue sleeve (joint capsule).

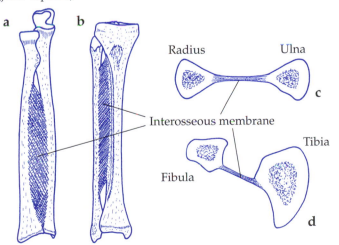

Figure 5.2   Anterior and cross-sectional views of the membranous syndesmoses between the radius and ulna, (*a*) and (*c*), and between the tibia and fibula, (*b*) and (*d*).

lower arm (figure 5.2c). There is an interosseous membrane connecting the medial border of the shaft of the fibula and the lateral border of the shaft of the tibia (figure 5.2b). Similar in structure to the interosseous membrane in the lower arm, it stabilizes the tibia and fibula in all positions of the ankle joint, and provides areas of attachment for muscles on the anterior and posterior aspects of the lower leg (figure 5.2d).

Whereas the interosseous membranes in the lower arms and lower legs are permanent features, there is a particular group of syndesmoses, the sutures and fontanels of the skull, that originate as membranous syndesmoses and then change to sutural syndesmoses (see also chapter 3).

Sutural Syndesmoses

In a sutural syndesmosis the articular surfaces are united by a thin layer of fibrous tissue, rather like the thin layer of concrete between bricks in a wall (figure 5.3). By late childhood all of the sutures and fontanels of the skull are converted from membranous to sutural syndesmoses. The thin layer of fibrous tissue, together with the interlocking of the articular surfaces, tends to severely limit movement in these joints. With increasing age the fibrous tissue in sutures is gradually replaced by bone so that each suture is converted to a synostosis (syn = with, osteo = bone).

Cartilaginous Joints

The degree of movement in cartilaginous joints is determined by the type and thickness of the cartilage. There are two kinds of cartilaginous joints: **synchondroses** and **symphyses**.

Synchondroses

Synchondrosis: a cartilaginous joint in which the opposed surfaces are united by hyaline cartilage

Symphysis: a cartilaginous joint in which the opposed surfaces are united by a combination of hyaline cartilage and fibrocartilage; a layer of hyaline cartilage covers each of the articular surfaces, and sandwiched between the layers of hyaline cartilage is a relatively thick pad of fibrocartilage

In a synchondrosis (syn = with, chondro = cartilage) the articular surfaces are united by hyaline cartilage. There are two types of synchondroses: temporary and permanent. The temporary synchondroses, sometimes referred to as physeal joints, include the following:

1. Joints between bones that eventually fuse together to form larger bones in the adult skeleton. For example, each innominate bone is formed by the fusion of the corresponding ilium, ischium, and pubis.
2. Joints formed by epiphyseal plates.
3. Joints formed by apophyseal plates.
4. The joints between the first ribs and the manubrium.

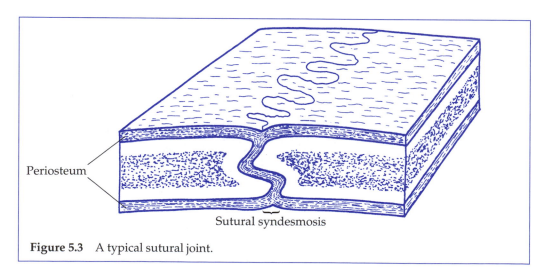

Periosteum

Sutural syndesmosis

Figure 5.3   A typical sutural joint.

All temporary synchondroses are converted to synostoses at some stage in the life of the individual. There are a few synchondroses that remain moderately flexible throughout life. These joints are the permanent synchondroses, and they include the joints between the anterior ends of the second to the tenth ribs and the sternum (the costal cartilages).

Symphyses

In a symphysis (sym = with, physeal = growth plate) the articular surfaces are united by a combination of hyaline cartilage and fibrocartilage. A layer of hyaline cartilage covers each articular surface, and sandwiched between the layers of hyaline cartilage is a relatively thick pad of fibrocartilage (figure 5.4). The fibrocartilage pad is often referred to as a disc even though it is usually kidney shaped or oval. The joint is normally supported by a number of ligaments that cross the outside of the joint and attach onto the periphery of the fibrocartilage pad. It is important to appreciate that the bone, hyaline cartilage, and fibrocartilage elements in a symphysis joint are intimately connected; there are no sharp divisions between the three tissues, but rather a gradual change from one tissue type to another. In effect, the joint consists of a single piece of material whose consistency varies across the joint.

Fibrocartilage readily deforms in response to bending and torsion loads, and the degree of movement in a symphysis joint is largely determined by the thickness of the fibrocartilage pad; the thicker the fibrocartilage, the greater the capacity for movement. The joints between the bodies of the vertebrae, the pubic symphysis, and the manubriosternal joint are all symphysis joints. Whereas most of the symphysis joints remain moderately flexible throughout life, some of them, such as the manubriosternal joint, may be converted to synostoses (Tortora and Anagnostakos 1984).

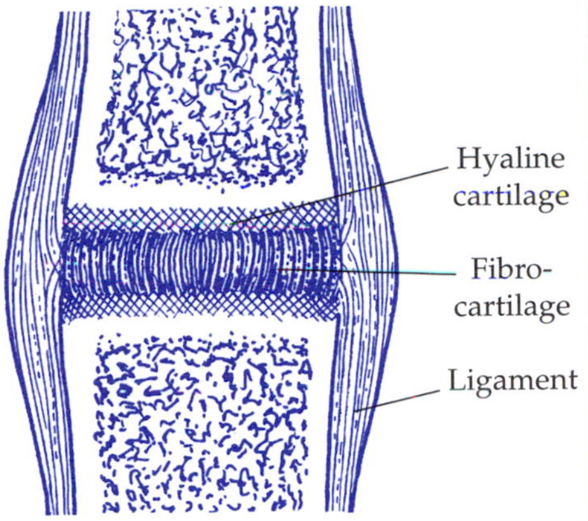

Figure 5.4   A typical symphysis joint.

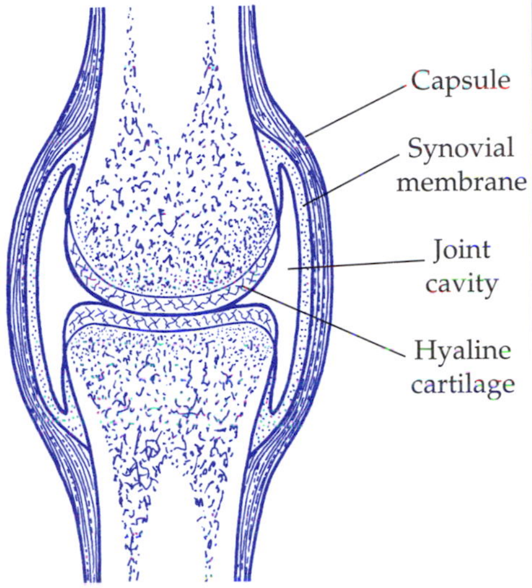

Figure 5.5   A typical synovial joint.

Synovial Joints

In the adult skeleton, approximately 80% of the joints are synovial (syn = with, vial = cavity). In general, synovial joints have greater range of movement than fibrous or cartilaginous joints. The body's capacity to adopt a broad range of postures is due largely to the range of movement in synovial joints. Virtually all of the joints in the upper and lower limbs are synovial.

As you recall from chapter 4, in a synovial joint each articular surface is covered with a layer of articular (hyaline) cartilage. The surfaces are not attached to each other but, under normal circumstances, are held in contact with each other, in all positions of the joint, by a joint capsule and various ligaments (figure 5.5). During movement of a synovial joint the articular surfaces slide and roll on each other. The capsule encloses a joint cavity that, due to the close contact between the articular surfaces, is normally very small. In figure 5.5, the joint cavity is shown much larger than normal to differentiate the various features of the joint. The inner wall of the capsule and the nonarticular bony surfaces inside the joint are covered with synovial membrane. The synovial membrane consists of areolar tissue (see chapter 4) with specialized cells that secrete synovial fluid into the joint cavity. Synovial fluid is viscous and resembles the consistency of raw egg whites. It has two important functions.

1. A mechanical function: The fluid lubricates the articular surfaces so they slide over each other easily, thereby preventing excessive wear.
2. A physiological function: The fluid seeps into the articular cartilage and nourishes the cartilage cells.

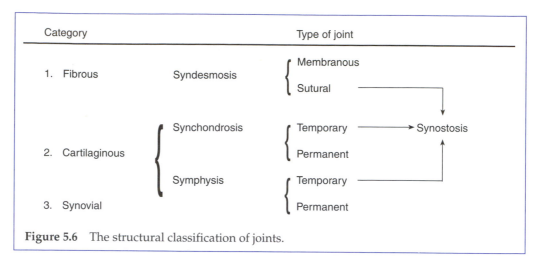

Figure 5.6 The structural classification of joints.

Figure 5.6 summarizes the structural classification of joints.

In terms of structure, there are basically two types of joints:

- *Fibrous or cartilaginous joints, where the articular surfaces are united by either fibrous tissue or cartilage*
- *Synovial joints in which the articular surfaces are not attached to each other, but are held in contact with each other by a joint capsule and ligaments*

Congruence, Articular Discs, and Menisci

The articular surfaces in most synovial joints are reciprocally shaped, which normally results in a large contact area (relative to the area of the articular surfaces) between the opposed articular surfaces in all positions of the joint. For any particular joint position and joint reaction force, the larger the contact area between the articular surfaces—the larger the area over which the joint reaction force is transmitted—the lower the compression stress on the articular surfaces, and vice versa. Some synovial joints, such as the tibiofemoral joint, do not have reciprocally shaped articular surfaces so that the actual area of contact between the articular surfaces in any particular joint position is likely to be relatively small. However, in such joints the effective area of contact between the articular surfaces is normally as large as in joints with reciprocally shaped articular surfaces because of one or more pieces of fibrocartilage that form wedges between the unopposed parts of the articular surfaces and distribute the joint reaction force over a large area of the articular surfaces. The fibrocartilage wedges are not attached to the articular surfaces, but are normally in contact with the articular surfaces. They are held in position by attachment to the inner wall of the joint capsule or by attachment to bone adjacent to the articular surface.

In the acromioclavicular joint, and sometimes the ulna-carpal joint, there is a single fibrocartilage wedge in the form of a ring that tapers from the outside toward the center (figure 5.7a). In the tibiofemoral joint there are normally two C-shaped fibrocartilage wedges; each wedge is called a **meniscus** (half-moon, crescent shape; figure 5.7b). In some joints, such as the sternoclavicular and ulna-carpal joints, there is usually a complete disc of fibrocartilage that effectively divides the joint into two joints (figure 5.7, c and d). A complete disc of fibrocartilage is referred to as an **articular disc**. The **congruence** of a joint refers to the area over which the joint reaction force is transmitted; in any particular joint position, the greater the area, the greater the congruence of the joint, and vice versa. Normally, congruence is maximized in order to minimize compression stress on the articular surfaces. One of the main functions of articular discs and menisci is to improve congruence in joints. Improved congruence will tend to

Meniscus: a C-shaped wedge of fibrocartilage that helps to increase congruence in the tibiofemoral joint; there are normally two menisci in each tibiofemoral joint

Articular disc: a piece of fibrocartilage that helps to increase congruence in some synovial joints; the fibrocartilage may be in the form of a ring that tapers from the outside toward the center

Congruence: the area over which the joint reaction force is transmitted in a synovial joint; in any particular joint position, the greater the contact area between the articular surfaces, the greater the congruence, and vice versa.

1. reduce compression stress on the opposed articular surfaces,
2. help to maintain normal joint movements and effective distribution of synovial fluid over the articular surfaces, and
3. improve shock absorption.

The articular surfaces in most synovial joints are reciprocally shaped, which allows a high level of congruence in all positions of the joint. Some synovial joints do not have reciprocally shaped articular surfaces, and congruence in these joints is improved by articular discs.

Joint Movements

Movement in joints is usually a combination of linear (figure 5.8, *a* and *b*) and angular (figure 5.8, *c* and *d*) movement. With regard to cartilaginous and synovial joints, the dominant type of movement is angular. Linear movement may occur, but in normal joint movements the degree of linear movement is usually quite small. For example, the pubic symphysis and the intervertebral symphyses may experience a certain degree of linear movement similar to that shown in figure 5.8*b*. When this happens the fibrocartilage pad in the joint is subjected to a shear load likely to tear the pad if the degree of strain is beyond a very small amount. A limited amount of linear movement may occur in some synovial joints such as the intercarpal and intertarsal joints, and the joints between the superior and inferior articular facets of the vertebrae. However, in most synovial joints, linear movement is normally slight (Basmajian 1970). Linear movement beyond a very small amount almost certainly results in partial separation of the articular surfaces and damage to the capsule and ligaments of the joint.

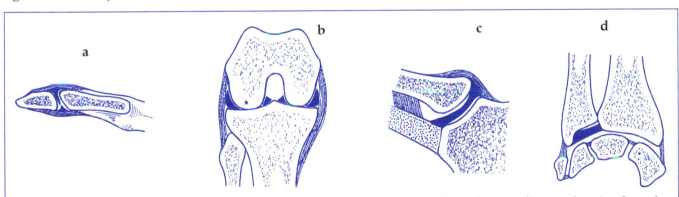

Figure 5.7   Articular discs and menisci; (*a*) coronal section through right acromioclavicular joint; (*b*) coronal section through right knee joint with knee flexed at approximately 90°; (*c*) coronal section through right sternoclavicular joint; (*d*) coronal section through left wrist joint.

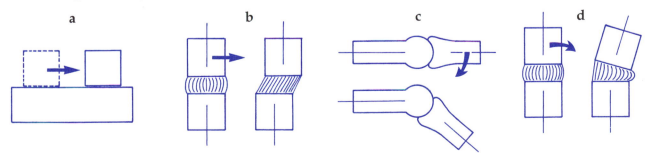

Figure 5.8   Types of movement in joints; (*a*) linear motion in a synovial joint; (*b*) linear motion in a fibrous joint; (*c*) angular motion in a synovial joint; (*d*) angular motion in a fibrous joint.

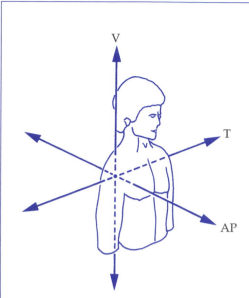

Figure 5.9 Reference axes with respect to the shoulder joint: V—vertical axis; T—transverse axis; AP—anteroposterior axis.

Degrees of freedom: the linear and angular directions of movement considered normal for a joint with respect to reference anteroposterior, transverse, and vertical axes providing six possible degrees of freedom—three linear directions (along the axes) and three angular directions (around the axes)

Reference Axes and Degrees of Freedom

In describing a particular joint's movement, it is useful to refer to three mutually perpendicular reference axes. The three reference axes denote anteroposterior (front to back), transverse (side to side), and vertical directions with respect to the anatomical position. Figure 5.9 shows the position of the reference axes in relation to the shoulder joint. With respect to the reference axes, there are six possible directions, called **degrees of freedom**, in which the shoulder joint, or any other joint, might be able to move, depending upon its structure. The six directions consist of three linear directions (along the axes) and three angular directions (around the axes). A joint with six degrees of freedom could move in any direction by a combination of linear and angular movements. Some cartilaginous joints have six degrees of freedom, albeit within a small range of movement. In contrast, the larger synovial joints tend to have no linear degrees of freedom, but they usually have one to three angular degrees of freedom with a relatively large range of movement.

Most movements in everyday life such as walking, bending, reaching, and writing involve simultaneous or sequential movement in two or more joints. In such multijoint movements, the degrees of freedom in the whole of the segmental chain responsible for the movement are the sum of the degrees of freedom of the individual joints in the chain. Consequently, there is an almost infinite number of combinations of joint movements that could be employed in all multijoint movements. Furthermore, temporary or permanent impairment in one joint in a segmental chain can usually be compensated by a change in the movement of other joints in the chain (see chapter 12).

Angular Movements

Angular movements in joints refer to rotations around the three reference axes. With the anatomical position as a reference position, special terms describe the various angular movements.

In most joints the terms abduction and adduction refer to rotations around the anteroposterior axis. In the shoulder, wrist, and hip joints, abduction and adduction refer to movement of the arm, hand, and leg away from and toward the median plane, respectively (figures 5.10 and 5.11a). In the hand and foot, abduction occurs when the fingers and toes are spread, and adduction occurs when the fingers and toes are returned to the reference position (figure 5.11b).

In most joints the terms flexion and extension refer to rotation around the transverse axis. In the shoulder, wrist, and hip joints, flexion refers to movement of the arm, hand, and leg forward, and extension refers to movement of the arm, hand, and leg backward (figure 5.11c; figure 5.12, a and b). In the elbow, knee, metacarpophalangeal, and interphalangeal (hand and foot) joints, flexion occurs when the joints bend and extension occurs when the joints straighten (figure 5.12, c and d). In the trunk—the vertebral column as a whole—flexion refers to bending the trunk forward and extension refers to the reverse movement (figure 5.13, a and b). Lateral flexion of the trunk occurs when the trunk bends to the side about an anteroposterior axis (figure 5.13c).

Circumduction describes a movement in which the distal end of a bone or limb moves in a circle while the proximal end (at which joint movement takes place) stays in the same place. Consequently, the movement of the bone or limb describes a cone shape. All joints capable of flexion and extension and abduction and adduction, such as the shoulder, wrist, metacarpophalangeal, and hip joints, are capable of circumduction.

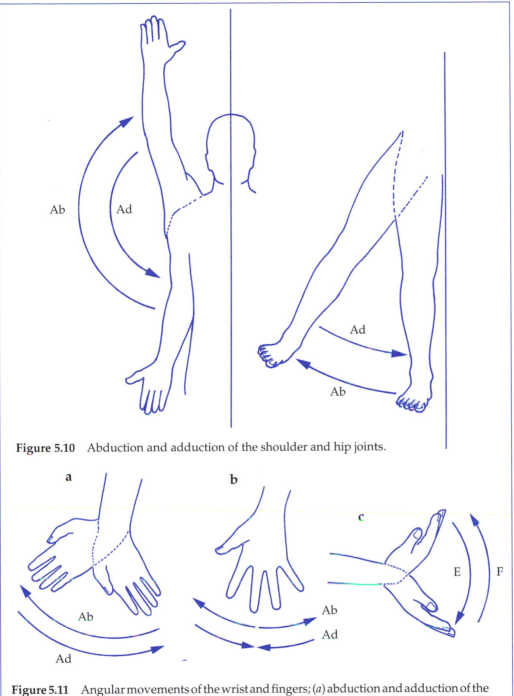

Figure 5.10   Abduction and adduction of the shoulder and hip joints.

Figure 5.11   Angular movements of the wrist and fingers; (*a*) abduction and adduction of the wrist; (*b*) abduction and adduction of the fingers; (*c*) flexion and extension of the wrist.

Some joints, such as the shoulder, hip, and the vertebral column as a whole, can rotate about a vertical axis with respect to the anatomical position. This form of rotation is usually described as axial rotation, rotation around an axis passing along, parallel to, or close to parallel to the shaft of the moving bone. Axial rotation of the shoulder occurs when the humerus rotates about an axis parallel to its long axis. For example, internal rotation (medial rotation) and external rotation (lateral rotation) of the shoulder occurs when cleaning a table top with a duster; in this situation the positions of the shoulder and elbow joints tend to move slightly, but the side-to-side sweeping movement of the hand is largely brought about by internal and external

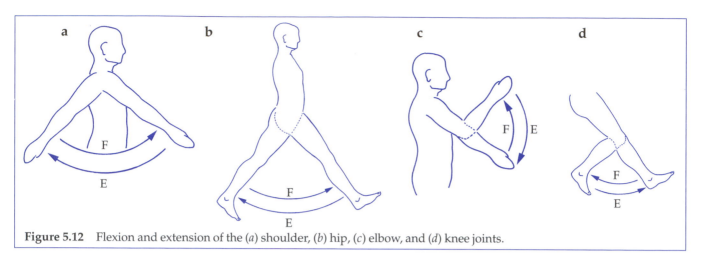

Figure 5.12   Flexion and extension of the (*a*) shoulder, (*b*) hip, (*c*) elbow, and (*d*) knee joints.

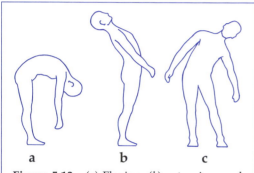

Figure 5.13   (*a*) Flexion, (*b*) extension, and (*c*) lateral flexion of the trunk.

rotation in the shoulder. Axial rotation of the humerus should not be confused with pronation of the forearm, which occurs as a result of movement in the proximal and distal radio-ulnar joints. Rotation of the trunk about a vertical axis is usually described as axial rotation (or twist) to the left or right.

In addition to abduction, adduction, flexion, extension, and rotation, there are a number of other terms used to describe specific movements in certain joints. These movements include supination and pronation of the forearm that occur, for example, when using a screwdriver to put in or take out a screw.

With respect to the anatomical position, abduction and adduction describe angular movement of a joint about the anteroposterior axis. Flexion and extension describe angular movement of a joint about the transverse axis, and axial rotation describes joint rotation about an axis passing along, parallel to, or close to parallel to the shaft of the moving bone.

The Stability-Flexibility Classification of Joints

Joint stability: the strength of the bond between the bones in a joint; the stronger the bond, the more stable the joint

Joint flexibility: the degree of movement in a joint

Synarthroses: sutural syndesmoses and temporary synchondroses; synarthroses are also referred to as immovable or fixed joints

Amphiarthroses: symphyses, membranous syndesmoses, and permanent synchondroses; amphiarthroses are also referred to as slightly moveable joints

Diarthroses: synovial joints; diarthroses are also referred to as freely moveable joints

Joints are most frequently classified on the basis of structure, as described earlier. However, there is another fairly widely used classification based on the degree of **stability** and **flexibility** of the joints. Joint stability refers to the strength of the bond between the bones; the stronger the bond, the more stable the joint. Joint flexibility refers to the degree of movement in the joint. There are three categories: **synarthroses**, **amphiarthroses**, and **diarthroses** (Tortora and Anagnostakos 1984).

A synarthrosis is a stable joint in which the degree of flexibility is nil or virtually nil. The sutures of the adult skull, the temporary synchondroses, and all synostoses fit into this category. In the structural classification the prefix syn means "with." In the stability-flexibility classification the prefix syn means "together"; this refers to the direct attachment between the articular surfaces.

A diarthrosis is a relatively unstable joint with a relatively high degree of flexibility. All synovial joints fit into this category. The prefix dia means "apart;" this refers to the fact that the articular surfaces are not attached to each other, even though they are in contact.

An amphiarthrosis is a joint whose stability and flexibility characteristics are between the extremes represented by synarthroses and diarthroses. Amphiarthroses (amphi = both) are sometimes referred to as slightly moveable joints, whereas

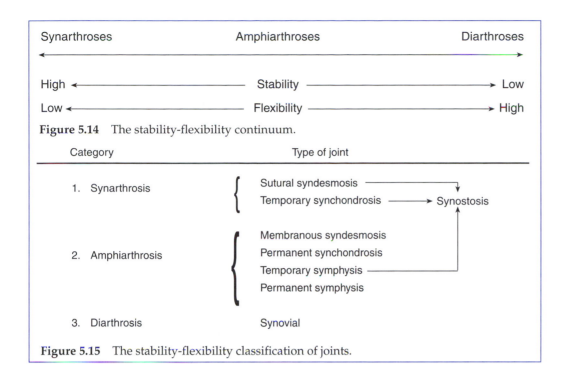

Figure 5.14   The stability-flexibility continuum.

Figure 5.15   The stability-flexibility classification of joints.

synarthroses and diarthroses are referred to as immovable and freely moveable joints, respectively. Membranous syndesmoses and symphyses fit into the category of amphiarthroses.

The stability-flexibility classification may be thought of as a continuum, with synarthroses and diarthroses at opposite ends of the continuum (figure 5.14). Each particular joint occupies a certain point on the continuum. The positions of these points may change as the structure and, therefore, the stability and flexibility of the joints change. Figure 5.15 summarizes the stability-flexibility classification of joints.

Whereas most joints fit exclusively into one of the main categories of both the structural and the stability-flexibility classifications, some joints have characteristics of more than one category (Williams et al. 1995). The following are examples of such joints:

1. A joint cavity is one of the main characteristics of synovial joints. However, partial cavities may develop in some nonsynovial joints such as the manubriosternal and pubic symphysis joints, and the joints between the ribs and the sternum.

2. In some synovial joints the articular surfaces are separated or partially separated by articular discs or menisci.

3. The anterior portion of each sacroiliac joint is basically synovial, whereas the posterior portion is a membranous syndesmosis (see chapter 6).

Consequently, in terms of structure and function, the range of joints is very broad. This range reflects the capacity of the skeletal system for structural adaptation—the capacity to modify the structure of joints in relation to their particular functional requirements.

> *Joints are classified on the basis of stability and flexibility into synarthroses, amphiarthroses, and diarthroses. Synarthroses such as sutures are stable with little or no flexibility. Diarthroses (including all synovial joints) are relatively unstable with a relatively high degree of flexibility. Amphiarthroses are joints whose stability and flexibility characteristics are between the extremes represented by synarthroses and diarthroses.*

Synovial Joint Classification

Synovial joints are classified according to the type of movement that occurs in the joints. There are two kinds of synovial joints:

1. Joints with basically linear, limited movement. In these joints the articular surfaces, which are fairly flat, slide on each other. Consequently, these joints are called gliding or plane joints. Sliding occurs in all synovial joints to a certain extent, but in gliding joints it is the main type of movement. Gliding joints include the intercarpal and intertarsal joints and the joints between the superior and inferior facets of the vertebrae.

2. Joints with basically angular movements. Movement in these joints is normally a combination of rolling and sliding between the articular surfaces. There are three groups: uniaxial, biaxial, and multiaxial. Each group is divided according to the shape of the articular surfaces.

Uniaxial

In uniaxial joints, movement takes place mainly about a single axis. There are two types of uniaxial joints—hinge joints and pivot joints. In a hinge joint, a convex, pulley-shaped (bicondylar) articular surface articulates with a reciprocally shaped concave surface. The elbow (humero-ulnar), interphalangeal, and ankle joints are the best examples of hinge joints (figure 5.16). The notch of the pulley prevents (or severely limits) side-to-side movement. The tibiofemoral joint is usually classified as a hinge joint even though the articular surfaces of the femoral and tibial condyles are not very congruent. However, in the normal tibiofemoral joint the congruence between the articular surfaces is considerably increased by the presence of menisci. If, for the purpose of comparison, the menisci are considered to be part of the tibia, the shapes of the articular surfaces of the knee joint are similar to those in the interphalangeal joints.

In a pivot joint, a cylindrical articular surface rotates about its long axis within a ring formed by bone and fibrous tissue. The joint between the dens of the axis and the fibro-osseous ring formed by the anterior arch and transverse ligament of the atlas is a pivot joint (refer to figure 3.15, *a* and *d*). The proximal radio-ulnar joint is also a pivot joint (figure 5.17). The head of the radius is held against the radial notch by a ligament called the annular ligament (annulus = ring). During supination and pronation of the forearm, the head of the radius rotates within the ring formed by the annular ligament and radial notch.

Biaxial

In biaxial joints, movement takes place mainly about two axes at right angles to each other, usually the anteroposterior (abduction/adduction) and transverse (flexion/extension) axes. There are three types of biaxial joints: condyloid, ellipsoid, and saddle. In a condyloid joint a convex condylar surface articulates with a concave condylar surface. The metacarpophalangeal joints are condyloid joints (see figure 5.16, *c* and *d*). In an ellipsoid joint such as the radiocarpal joint, an elliptical convex surface articulates with an elliptical concave surface. The articular surface on the distal end of the radius is elliptical, concave, and shallow. This surface articulates with the proximal articular surfaces of the scaphoid and

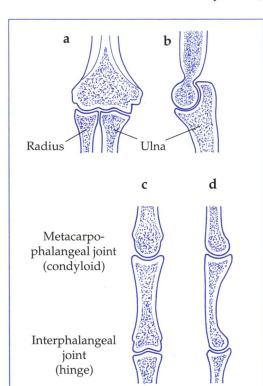

Figure 5.16   Hinge and condyloid joints; (*a*) coronal section through right elbow joint in extension; (*b*) sagittal section through elbow joint (humero-ulnar) in extension; (*c*) and (*d*) coronal and sagittal sections through metacarpophalangeal and interphalangeal joints.

lunate that together form a convex elliptical articular surface. Movements at the metacarpophalangeal joints and radiocarpal joints are normally combinations of flexion, extension, abduction, and adduction. In a saddle (or sellar) joint the articular surfaces are saddle shaped (figure 5.18*a*). Each articular surface is convex in one direction and concave in a direction at right angles to the convex direction. Movement takes place mainly in two planes at right angles to each other. The carpometacarpal joint of the thumb (figure 5.18*b*) and the calcaneocuboid joints are saddle joints.

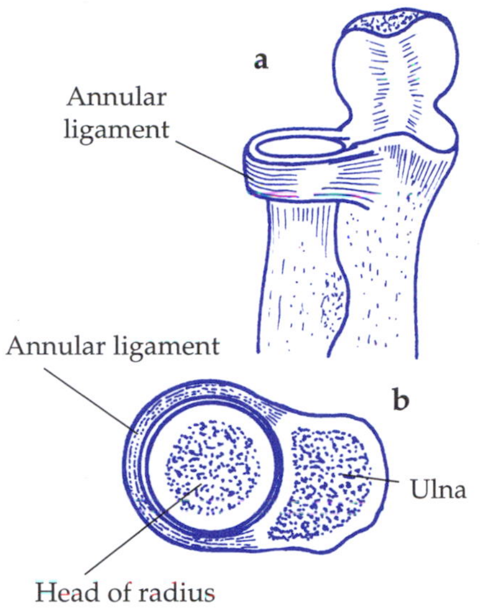

Multiaxial

Some joints, such as the shoulder and hip, can rotate about all three reference axes. By combining rotations about the three reference axes, these joints can rotate about any axis in between the three. Consequently, these joints are referred to as multiaxial joints. In this type of joint, a very rounded articular surface, like part of a ball, articulates with a cuplike concavity. Due to the shapes of the articular surfaces, these joints are usually referred to as ball and socket joints. The best examples of ball and socket joints are the shoulder and hip (refer to figures 4.9*a* and 4.10*a*). Figure 5.19 summarizes the classification of synovial joints.

Figure 5.17   A typical pivot joint: the proximal radio-ulnar joint; (*a*) anterior aspect of proximal radio-ulnar joint; (*b*) transverse section through the proximal radio-ulnar joint.

Synovial joints are classified according to the type of movements that occur in the joints. Joints in which the main type of movement is linear are called gliding or plane joints. Joints in which the main type of movement is angular are classified into uniaxial (hinge and pivot), biaxial (condyloid, ellipsoid, and saddle), and multiaxial (ball and socket) joints.

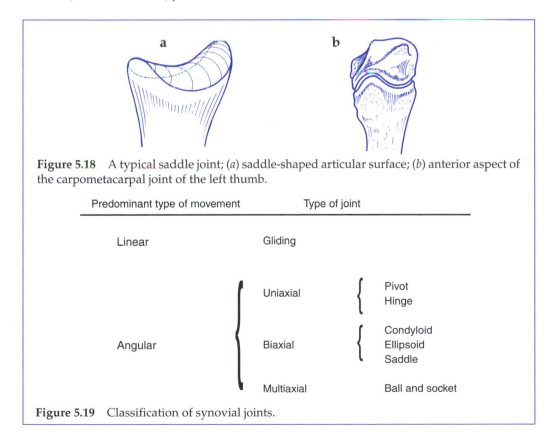

Figure 5.18   A typical saddle joint; (*a*) saddle-shaped articular surface; (*b*) anterior aspect of the carpometacarpal joint of the left thumb.

Predominant type of movement	Type of joint	
Linear	Gliding	
Angular	Uniaxial	Pivot Hinge
	Biaxial	Condyloid Ellipsoid Saddle
	Multiaxial	Ball and socket

Figure 5.19   Classification of synovial joints.

Flexibility, Stability, and Laxity in Synovial Joints

In previous sections of this chapter the terms flexibility and stability have been used in a general sense. At this point it is necessary to define these terms more specifically in relation to synovial joints.

Flexibility

In a synovial joint, flexibility refers to the range of movement in those directions (degrees of freedom) considered normal for the joint. For example, the knee joint is designed primarily to rotate about a transverse axis—to flex and extend. Consequently, flexion and extension are considered normal movements at the knee. However, rotations about an anteroposterior axis—abduction and adduction—are considered abnormal movements at the knee. The flexibility of a joint is determined by four factors:

1. Shape of the articular surfaces
2. Tension in the joint capsule and ligaments at the ends of the various ranges of motion
3. Soft tissue bulk, mainly skeletal muscle, surrounding the bones forming the joint
4. Extensibility of the skeletal muscles controlling the movement of the joint

Shape of Articular Surfaces

In some joints, flexibility is limited by the impingement or interlocking of the nonarticular surfaces of the bones at the end of certain ranges of motion. For example, elbow extension is restricted by the interlocking of the olecranon process of the ulna with the olecranon fossa of the humerus (figure 5.20a). Similarly, elbow flexion is restricted by the interlocking of the coronoid process of the ulna with the coronoid fossa of the humerus (figure 5.20b). Lateral and medial displacement of the radius and ulna relative to the humerus in the elbow joint is severely restricted by the interlocking of the articular surfaces (figure 5.20c).

Tension in Joint Capsule and Ligaments at Limit of Range of Movement

The function of ligaments is described in detail later in the chapter. At this point it is sufficient to appreciate that in a normal joint, the joint capsule and some of the ligaments that support the joint become taut at the end of each range of movement, and thereby restrict further movement (see figure 5.20).

Soft Tissue Bulk

Soft tissue bulk restricts flexibility in some joints. For example, elbow and knee flexion may be restricted in a heavily muscled individual due to the impingement of the adjacent body segments.

Extensibility of Muscles

The extensibility of a muscle is the maximum length the muscle-tendon unit can attain without injury. Muscle extensibility largely depends on the length range in which the muscle normally functions—the difference between its length when shortened and its length when extended. Generally speaking, the shorter the range of length, the lower the extensibility. In the absence of regular flexibility training, games and sports involving highly repetitive and exclusive movement patterns are likely to result in

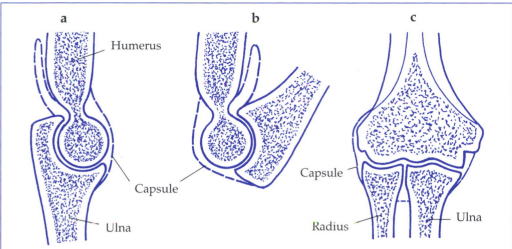

Figure 5.20 Effect of shape of articular surfaces and capsule on flexion-extension range of motion and abduction-adduction range of motion in the elbow joint; (*a*) sagittal section through elbow joint in extension; (*b*) sagittal section through elbow joint in flexion; (*c*) coronal section through elbow joint in extension.

reduced extensibility in some muscles. In most individuals, muscle extensibility is probably the main factor limiting joint flexibility (Bach et al. 1985; Steiner 1987; Nicholas and Marino 1987; Herbert 1988).

The flexibility of a synovial joint is determined by

- *the shape of the articular surfaces,*
- *tension in the joint capsule and ligaments at the ends of the various ranges of motion,*
- *soft tissue bulk, mainly skeletal muscle, surrounding the bones forming the joint, and*
- *extensibility of the skeletal muscles controlling the movement of the joint.*

Stability and Laxity

In a synovial joint, stability refers to the degree of congruence between the articular surfaces. During joint movement different parts of the articular surfaces come into contact with each other. However, in all positions of a joint, the better the congruence, the more stable the joint and the lower the risk of abnormal joint movements.

The term **laxity** refers to the degree of instability in a joint—the range of movement in those directions considered abnormal for the joint. The relationship of laxity to congruence and stability is shown in figure 5.21.

Laxity: the range of movement in those directions considered abnormal for a joint

Joint congruence is determined by three factors:

1. Shape of the articular surfaces
2. Suction between the articular surfaces
3. Supporting structures: skeletal muscles, ligaments, and the joint capsule

Shape of the Articular Surfaces

Most synovial joints have fairly congruent articular surfaces in all joint positions. In those joints where the articular surfaces are not very congruent, articular discs or menisci are usually present to improve congruence. Perfect congruence between articular surfaces would maximize joint stability and minimize joint laxity. However,

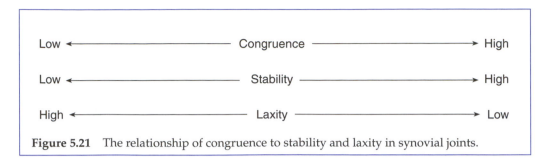

Figure 5.21   The relationship of congruence to stability and laxity in synovial joints.

perfect congruence would impair nutrition of the articular cartilage. Consequently, most joints are slightly incongruent and usually have a slight degree of laxity.

Suction Between Articular Surfaces

In a normal joint the articular surfaces are usually in close contact, with a thin film of synovial fluid between the surfaces. The film of fluid allows sliding between the surfaces but keeps the surfaces in contact with each other due to suction. In a similar manner, it is often difficult to separate two sheets of glass held in contact with each other by a thin film of water, even though the two sheets will slide on each other.

Supporting Structures

The supporting structures largely responsible for maintaining close contact between the articular surfaces in all positions of a joint are the skeletal muscles that control the movement of the joint (chapters 2 and 8) and the ligaments of the joint.

Functions of Joint Capsule and Ligaments

The joint capsule has two main functions:

1. To assist joint stabilization by helping prevent movement beyond normal range(s) and excessive laxity
2. To provide a base for the synovial membrane

At the end of each normal range of movement, part of the joint capsule will usually become taut to prevent movement beyond the normal range. For example, in full extension of the elbow, the anterior aspect of the capsule will be taut and the posterior aspect will be slack (see figure 5.20*a*). Similarly, in full flexion of the elbow, the posterior aspect of the capsule will be taut and the anterior will be slack (see figure 5.20*b*). With regard to abnormal ranges of movement, the regions of the joint capsule in the planes of abnormal ranges of movement are usually at their natural length—neither taut nor slack. However, these regions quickly become taut in response to abnormal movements, thereby helping to restrict the ranges of abnormal movements. For example, abduction and adduction are abnormal movements at the elbow and the joint capsule normally helps to prevent these movements (see figure 5.20*c*).

In association with skeletal muscles, ligaments bring about normal movement in joints. In a normal joint, ligaments help to maintain maximum stability in all positions of the joint, and in so doing, they effectively guide the movements of the joint. During normal movements the tension in ligaments is low to moderate—ligaments only become taut to prevent or restrict abnormal movements. An abnormal movement may be defined as any movement resulting in a decrease in normal congruence, such as distraction (separation) of the articular surfaces. Partial distraction, when the area of contact between the articular surfaces is reduced, is called **subluxation** (sub = toward, luxation = dislocation).

Subluxations are usually transient; normal congruence is usually restored as soon

Subluxation: a transient decrease in the normal area of contact between the articular surfaces in a synovial joint as the load causing the subluxation is removed. Subluxations often result in joint sprains—partial tearing of ligaments and the joint capsule, together with effusion (swelling) in the joint. Complete distraction, when the articular surfaces are completely separated, is called **luxation.** Like subluxations, luxations are usually transient. A complete distraction that persists after the load causing it is removed is a **dislocation** (figure 5.22). A dislocation usually requires manipulation by a physician or paramedic to restore the normal relationship between the articular surfaces (Grana, Holder, and Schelberg-Karnes 1987). Dislocation usually results in considerable damage to ligaments and the joint capsule.

Considerable force is required to cause severe subluxations and luxations. Ligaments and joint capsules are likely to be subjected to very high forces in two particular situations:

1. Unexpected situations in which the degree of muscular control of joint movement is less than adequate, for example, twisting an ankle by stepping on an uneven surface. In this situation, with the body moving forward over the foot, the ankle is likely to be rapidly and forcibly twisted.
2. High speed collisions, such as a tackle in football or soccer.

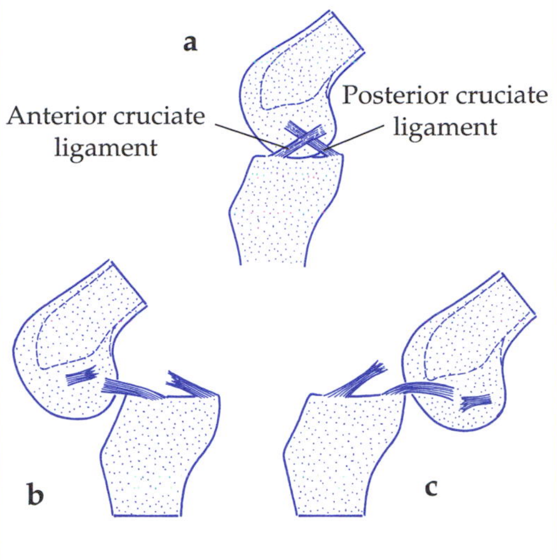

There are basically two kinds of abnormal movements in joints: hyperflexibility and excessive laxity. These will be described with reference to the knee joint.

Hyperflexibility

Hyperflexibility may be defined as movement beyond the normal range of movement in a direction considered normal for the joint. Extension is a normal movement in the knee joint. At full knee extension, the four main ligaments that support the knee will become taut to prevent further movement. If the knee is forcibly extended beyond this position, that is, hyperextended, the ligaments and joint capsule will be damaged (figure 5.23). Hyperextension of the knee is an example of hyperflexibility. In most individuals full knee extension normally corresponds to a position in which the leg is straight—the upper and lower leg are in line. Consequently, in most individuals hyperextension of the knee occurs when the knee is extended beyond the straight position. However, some individuals are able to voluntarily extend their knees slightly beyond the straight position. In these cases, extension of the knee beyond the straight position to the point of full extension would be considered normal for the individual provided that normal congruence was maintained.

Excessive Laxity

Excessive laxity is defined as movement beyond the normal degree of laxity in directions considered abnormal for the joint. For example, abduction and adduction are abnormal movements in the knee. In a normal knee joint these movements are restricted to a minimal level by the medial and lateral ligaments, respectively (figure 5.24). Abduction of the knee beyond a minimal level will damage the medial ligament and joint capsule. Similarly, adduction of the knee beyond a minimal level will damage the lateral ligament and joint capsule.

Movements of the femoral condyles backward and forward (along the anteroposterior axis) relative to the tibial condyles are abnormal movements. In a normal knee joint these movements are restricted to a minimal level by the anterior and posterior cruciate ligaments, respectively (see figure 5.22). Forward movement of the femoral

Figure 5.22 Dislocation of the knee joint; (*a*) normal orientation of the cruciate ligaments in the partially flexed knee joint; (*b*) anterior dislocation of the femur on the tibia showing torn posterior cruciate ligament; (*c*) posterior dislocation of the femur on the tibia showing torn anterior cruciate ligament.

Luxation: a transient separation of the articular surfaces in a synovial joint

Dislocation: a permanent (in the absence of treatment) separation of the articular surfaces in a synovial joint

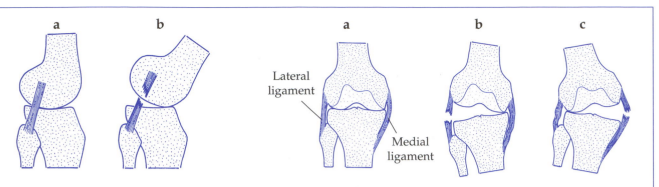

Figure 5.23 Hyperextension of the knee joint; (*a*) lateral aspect of extended right knee showing normal orientation of lateral ligament; (*b*) hyperextension of the knee showing torn lateral ligament.

Figure 5.24 Damage to the lateral and medial ligaments of the knee joint; (*a*) anterior aspect of extended right knee showing normal orientation of the lateral and medial ligaments; (*b*) adduction of the knee joint showing torn lateral ligament; (*c*) abduction of the knee joint showing torn medial ligament.

condyles beyond a minimal level will damage the posterior cruciate ligament and the joint capsule. Similarly, backward movement of the femoral condyles beyond a minimal level will damage the anterior cruciate ligament and the joint capsule.

Any degree of hyperflexibility or excessive laxity in a joint not only tends to damage ligaments and joint capsules, but also tends to result in damage to articular cartilage. The latter is due to localized overloading of articular cartilage resulting from reduced congruence caused by the abnormal movements.

> *A joint capsule provides a base for the synovial membrane and assists joint stabilization. In association with muscles, ligaments bring about normal joint movements and help maintain joint stability. During normal movements ligaments only become taut to prevent or restrict abnormal movements—hyperflexibility and excessive laxity.*

Movement Between Articular Surfaces

There are three basic types of movement that occur between articular surfaces: spinning, sliding, and rolling (figure 5.25). In most joints, normal movement involves a combination of these three types of movement, either simultaneously or sequentially. Sliding, and to a lesser extent spinning, subjects the articular cartilage to a combination of compression and shear loading. Rolling subjects the articular cartilage to compression loading. These loads tend to cause mechanical wear of the articular cartilage, breaking up the cartilage internally and wearing away the cartilage at the surface. Under normal circumstances, when the articular surfaces are congruent and properly lubricated, there is a balance between the amount of wear and the production of new cartilage. However, during subluxations, those parts of the articular cartilage that remain in contact are subjected to abnormally high loads. Over a period of time, the wear caused by the abnormal loads may outpace the production of new cartilage and result in permanent damage to the cartilage (see chapter 10).

> *In most synovial joints, movement between the articular surfaces involves spinning, rolling, and sliding. In a normal joint, the combination of spinning, sliding, and rolling is such that wear of the articular cartilage is minimized. Excessive laxity in a joint tends to overload certain parts of the articular cartilage resulting in excessive wear.*

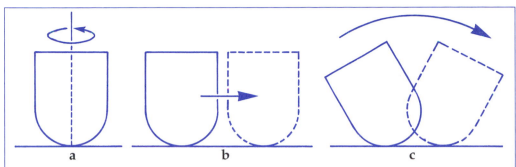

Figure 5.25   Types of movement between articular surfaces in synovial joints; (*a*) spinning; (*b*) sliding; (*c*) rolling.

Flexibility Training

In normal healthy individuals, ligaments and joint capsules, like muscle-tendon units, adapt to the length range in which they normally function. Consequently, if a joint is not regularly moved through the full range of movement, the ligaments and joint capsule shorten, preventing full range of normal movement. In this context, full range of normal movement is the range of movement between positions of the joint where subluxation of the articular surfaces occurs.

Flexibility training, in the form of a carefully prescribed program of exercises designed to lengthen ligaments and joint capsules and to increase the extensibility of muscle-tendon units, may restore full range(s) of movement. As previously mentioned, lack of muscle extensibility rather than shortened ligaments and joint capsules is probably the main factor limiting flexibility in most individuals. During flexibility training, take great care to ensure that ligaments and joint capsules are not stretched to the point where joints become hyperflexible. If this happens, the overstretched ligaments and joint capsules will not be able to function properly; the joints will be less stable in all joint positions, and abnormal movements will become more likely.

Summary

This chapter discussed the structure and functions of the various joints. Joints facilitate relative motion and transmission of force between bones. There is considerable variation in the type and range of movement and the mechanism of force transmission in different joints. The next two chapters consider the joints of the axial skeleton (chapter 6) and the appendicular skeleton (chapter 7).

Review Questions

1. Differentiate between permanent and temporary joints.
2. Describe the two main approaches to classifying joints.
3. Differentiate between the following:
   - A syndesmosis and a symphysis
   - A synarthrosis and a diarthrosis
4. Describe the basic structure of a synovial joint.
5. Describe the factors affecting congruence in synovial joints

6. With respect to synovial joints, differentiate between the following:
   - Flexibility and laxity
   - Subluxation and dislocation
7. Describe the factors affecting the flexibility of a synovial joint.
8. Describe the function of ligaments in relation to synovial joints.
9. Describe the types of movement occurring between the articular surfaces in a synovial joint.

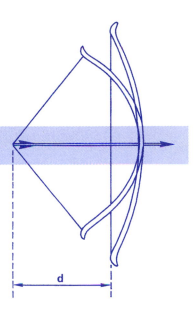

Chapter 6

Joints of the Axial Skeleton

A number of small uniaxial synovial joints are found in the adult axial skeleton, but the larger joints of the axial skeleton and the joints of the pelvis are mainly cartilaginous. In contrast, the large majority of joints in the adult appendicular skeleton are synovial. The difference in the types of joints that make up the axial and appendicular parts of the skeleton reflect differences in function. The synovial joints of the appendicular skeleton facilitate a large range of movement with limited shock absorption, whereas the cartilaginous joints of the axial skeleton and pelvis provide a high level of shock absorption with limited range of movement. This chapter describes the structure and function of the joints of the axial skeleton and pelvis.

Objectives

After reading this chapter you should be able to do the following:

1. Define or describe the key terms.
2. Describe the structure and function of intervertebral joints and facet joints.
3. Explain the relationship between intervertebral joints and facet joints in terms of transmission of load between vertebrae.
4. Describe the location and function of the ligaments that support the intervertebral and facet joints.
5. Describe the range of movement in different regions of the vertebral column.
6. Describe the effects of aging and excessive overload on degeneration and damage in intervertebral discs and vertebrae.
7. Describe normal and abnormal curvatures of the vertebral column in the median and frontal planes.
8. Describe the structure and function of the sacroiliac joints and pubic symphysis joint.
9. Explain the relationship between the sacroiliac and pubic symphysis joints in terms of transmission of load across the pelvis.

The Joints Between the Vertebrae

Each adjacent pair of vertebrae, apart from the atlas and axis, articulate with each other by the following means:

Intervertebral joint: symphysis joint linking the bodies of adjacent vertebrae in a motion segment

Facet joint: synovial gliding joint between superior and inferior articular processes in a motion segment

1. A symphysis joint between the vertebral bodies. These joints are usually referred to as **intervertebral joints** (figure 6.1).
2. Two synovial gliding joints between the inferior articular processes of the upper vertebra and the superior articular processes of the lower vertebra. These joints are usually referred to as **facet joints** or apophyseal joints (figure 6.1).
3. A number of longitudinal and intersegmental ligaments. These ligaments are described later in the chapter. At this point it is sufficient to appreciate that some of the intersegmental ligaments constitute membranous syndesmoses (see also chapter 5).

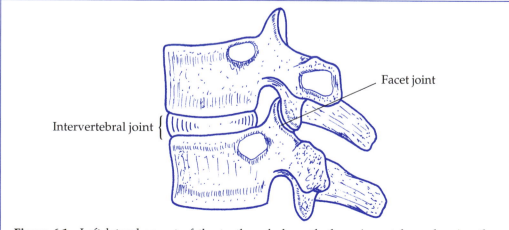

Figure 6.1 Left lateral aspect of the tenth and eleventh thoracic vertebrae showing the intervertebral joint and left facet joint.

There is no intervertebral joint (symphysis joint) between the atlas and axis, but there is a synovial pivot joint between the dens of the axis and the anterior arch of the atlas (see figure 3.15). In other respects, the links between the atlas and axis are similar to those between other vertebrae.

Movement between each pair of adjacent vertebrae involves simultaneous movement in all of the joints (cartilaginous, synovial, and fibrous) that link them together. It is not surprising, then, that the various joints function as a single unit to transmit loads between the vertebrae in the most mechanically efficient manner (Dunlop, Adams, and Hutton 1984). The functional unit consisting of two adjacent vertebrae and the joints and supporting soft tissues linking them together is often referred to as a **motion segment** (Brinckmann 1985). The vertebral column consists of a chain of motion segments.

Just as the various joints within a single motion segment function as one unit, so the chain of motion segments functions as a unit to transmit loads along the vertebral column. It follows that damage to any particular joint in a motion segment will likely affect the function of adjacent motion segments. For example, damage to an intervertebral joint may result in an individual adopting compensatory movements in adjacent motion segments to reduce the load on the injured joint. Even a slight amount of compensatory movement in a joint may alter the normal pattern of loading in the joint (Riegger-Krugh and Keysor 1996).

Motion segment: functional unit consisting of two adjacent vertebrae and the joints and supporting soft tissues linking them together

Intervertebral disc: the specialized fibrocartilage link within an intervertebral joint

Intervertebral Joints

The intervertebral joints and vertebral bodies form the main weight-bearing part of the vertebral column. The intervertebral joints are specialized symphysis joints. A thin layer of hyaline cartilage covers each articular surface of the vertebral bodies. Sandwiched between the layers of hyaline cartilage is an **intervertebral disc**, a fairly thick pad of highly specialized fibrocartilage. The vertebral bodies, hyaline cartilage, and intervertebral disc are intimately connected to each other (see figure 5.4).

> There is no intervertebral joint between the atlas and axis, but otherwise each adjacent pair of vertebrae articulate by means of the following:
>
> - An intervertebral joint
> - Two facet joints (apophyseal joints)
> - Longitudinal and intersegmental ligaments

In the transverse plane the shape of the discs is the same as that of the vertebral bodies—kidney shaped. In the frontal plane the discs are rectangular. With regard to the median plane, the discs in the cervical and lumbar region tend to be thicker in front than behind, so that the discs contribute to the anterior convexity of these regions (Williams et al. 1995). The discs in the thoracic region tend to be of uniform thickness in the median plane. The anterior concavity of the thoracic region is largely due to the shape of the vertebral bodies, which are slightly shorter at the front compared to the back (Williams et al. 1995) (see figure 3.10a). The thickness of the discs, like the size of the vertebral bodies, increases from the superior end of the column downward. On average, the thickness of the discs in the cervical, thoracic, and lumbar regions in an adult is approximately 3 mm, 5 mm, and 9 mm, respectively (Kapandji 1974). In a normal adult the discs contribute approximately 20% to the total length of the vertebral column (Williams et al. 1995).

> The intervertebral joints and vertebral bodies form the main weight-bearing part of the vertebral column. The size of the vertebral bodies and thickness of the intervertebral discs increase downward.

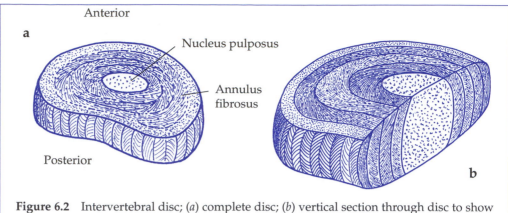

Figure 6.2 Intervertebral disc; (*a*) complete disc; (*b*) vertical section through disc to show relationship between nucleus pulposus and annulus fibrosus.

Each intervertebral disc consists of an outer ring of fibrous material called the annulus fibrosus (fibrous ring), which surrounds an inner gelatinous mass called the nucleus pulposus (pulpy center) (figure 6.2). The annulus fibrosus consists of a relatively thin outer ring of regular fibrous tissue and a broader inner ring of fibrocartilage. The fibrocartilage portion is in the form of a number of concentric layers. The collagen fibers in each layer run in the same direction, but the fibers in adjacent layers run in slightly different directions. In association with the nucleus pulposus, this layered arrangement of the fibrocartilage ring enables the disc to strongly resist all types of loading.

Osmotic swelling pressure (OSP): the water-absorbing force exerted by an intervertebral disc

The nucleus pulposus is structurally similar to proteoglycan gel (see chapter 4) and, as such, has a high affinity for water; the nucleus pulposus continually exerts an **osmotic swelling pressure** (OSP) to absorb water (Lindh 1989). The annulus fibrosus also exerts a certain amount of OSP, but less than that of the nucleus pulposus. Water is absorbed into the disc from the vertebral bodies via the hyaline cartilage. Under normal loading conditions a healthy nucleus pulposus is approximately 80% to 90% water (Kapandji 1974; Lindh 1989). The nucleus pulposus is enclosed within a virtually inextensible casing formed by the vertebral bodies and the annulus fibrosus. Consequently, as the volume of water absorbed into the nucleus pulposus increases, there is a simultaneous increase in the pressure within the nucleus, since its capacity for expansion is limited. The pressure in the nucleus pulposus is transmitted to the vertebral bodies and annulus fibrosus such that the thickness of the disc increases as more water is absorbed into the nucleus pulposus (figure 6.3). The flow of water into the nucleus pulposus ceases when the pressure exerted on the nucleus pulposus via the bony-fibrous casing is equal to the OSP.

When the body is in a recumbent position (lying down), the vertebral column does not have to support the weight of the upper body. In this situation the amount of water the nucleus pulposus of each disc can absorb is limited only by the extensibility of the corresponding annulus fibrosus. Consequently, during rest or sleep in a horizontal position, and any other condition where the compression load on the vertebral column is considerably reduced (such as during space flight), the nucleus pulposi absorb the maximum possible amount of water and the discs achieve maximum thickness. For this reason, most individuals are at their tallest just after getting out of bed in the morning.

During normal daily activities, with the trunk in a fairly upright position, the vertebral column has to support the weight of the upper body. Consequently, the discs are subjected to almost continuous compression loading. In this situation, the amount of water the nucleus pulposus of each disc can absorb is limited by the size and duration of the compression loads. In most individuals the compression loads

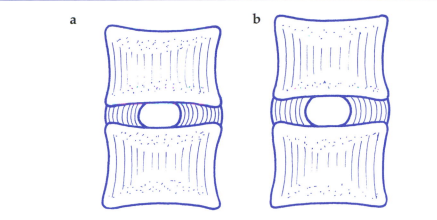

Figure 6.3   Effect of osmotic swelling pressure (OSP) on thickness of intervertebral disc; (*a*) disk thickness as a result of equilibrium between OSP and compression load on motion segment; (*b*) increased disc thickness following establishment of new state of equilibrium between OSP and reduced compression load on motion segment.

exerted on the nucleus pulposi due to maintaining an upright posture tend to exceed the OSP, at least during the first few hours after getting out of bed in the morning. Consequently, water is gradually forced out of the nucleus pulposi and back into the vertebral bodies such that the discs become thinner. In healthy joints the loss of water is gradual due to the low permeability of the hyaline cartilage layers. Nevertheless, the cumulative reduction in the thickness of the discs during a normal day may be as much as 2 cm (Kapandji 1974). This represents a reduction of approximately 15% in the combined maximum thickness of all the discs.

Due to the OSP of the nucleus pulposus, the annulus fibrosus of each disc and, therefore, the disc as a whole is constantly under load; that is, the annulus fibrosus is constantly under a certain degree of tension, which is analogous to the tension in the casing of an inflated ball. The greater the degree of inflation—comparable to the amount of water in the nucleus pulposus, the greater the tension in the casing. Since a healthy annulus fibrosus is under considerable load even when there is no external force acting on the disc, the disc is said to be permanently preloaded (Kapandji 1974). The preloaded condition, which can only be created when the nucleus pulposus and the annulus fibrosus function normally, is mechanically significant, as described below.

> *Each intervertebral disc consists of a peripheral annulus fibrosus surrounding a central nucleus pulposus. In a healthy disc the nucleus pulposus and annulus fibrosus continually exert an OSP that simultaneously increases the thickness of the disc and preloads the disc. The preloaded condition is necessary for normal disc function.*

Restoration of Normal Vertebral Column Alignment

When the vertebral column is flexed, extended, or laterally flexed with respect to its normal alignment, the vertebral bodies of each motion segment pivot about the corresponding nucleus pulposus, which, again, is comparable to an inflated beach ball. Consequently, as one side of the annulus fibrosus is compressed, the other side is stretched (subjected to greater than normal tension). In addition, the nucleus pulposus is displaced laterally in the direction of the stretched side of the joint; the greater the angular displacement in the motion segment, the greater the lateral displacement of the nucleus pulposus (figure 6.4). Consequently, the increased

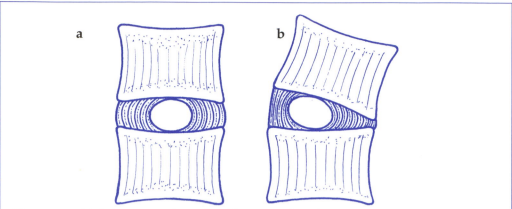

Figure 6.4   (*a*) Orientation of annulus fibrosus and nucleus pulposus in normal upright posture; (*b*) lateral displacement of nucleus pulposus in response to angular displacement of motion segment.

tension in the annulus fibrosus on the stretched side of the joint is due not only to the angular displacement of the vertebral bodies, but also to the pressure exerted by the laterally displaced nucleus pulposus. The net result of compression on one side of the disc and tension on the other produces considerable torque, which helps to restore the normal orientation of the annulus fibrosus and nucleus pulposus.

In the absence of a nucleus pulposus, the deformation of the annulus fibrosus tends to be asymmetric due to differences in the tensile and compressive strength of the annulus fibrosus. This, in turn, reduces the disc's restoring torque. It follows that the restoring torque of a disc diminishes whenever the disc's normal capacity to preload is impaired.

> *When the vertebral column is flexed, extended, or laterally flexed with respect to its normal alignment, the intervertebral discs are stretched on one side and compressed on the other. This action creates considerable torque, which helps to restore the normal alignment of the column.*

Protection of Vertebrae From Excessive Loading

In response to external loading imposed on the vertebral column by normal everyday activities, the discs function to cushion the effect of the external loading and protect the vertebrae. The type of deformation the discs experience depends on the rate at which the external load is applied. In response to impact loads—loads with a high rate of loading and short duration—the discs behave elastically. For example, in situations such as heel strike in walking and running, kicking a football, and hitting a tennis ball, there is insufficient time for water to be squeezed out of the discs. Consequently, the discs behave elastically; they deform instantaneously on loading and instantaneously recover their original shapes on unloading. The degree of strain experienced by a disc in response to impact loading depends on the magnitude of the load and the degree of preloading in the disc. Generally speaking, the higher the state of preloading, the lower the strain, and the better the cushioning effect of the disc. This cushioning effect of the discs in response to impact loading is often referred to as shock absorption.

In contrast to their elastic response to impact loading, discs behave viscoelastically in response to loads applied relatively slowly and for a prolonged period. For example, in response to prolonged compression loading, water is gradually squeezed out of the discs so that they become progressively thinner. This response is similar to water being squeezed out of a sponge or to air being squeezed out of an inflated plastic bag that has a number of pinholes. However, when the load on the discs is removed or reduced to a level below the OSP of the discs, they will begin absorbing water to restore their original size and shape.

The reduction in disc thickness due to supporting the weight of the upper body in upright postures, followed by the restoration of the thickness of the discs during sleep, exemplifies the viscoelastic behavior of the discs. The extent of the reduction in disc thickness in response to prolonged compression loading depends on the magnitude and duration of the load. Generally speaking, the greater the magnitude and duration, the thinner the discs become. However, as mentioned earlier in this chapter, in a healthy vertebral column the reduction in the thickness of the discs from normal everyday activity is unlikely to be greater than 15% of the combined maximum thickness of all the discs.

In addition to its mechanical importance in protecting the vertebrae from excessive loading, the viscoelastic behavior of the discs is important physiologically in terms of the nutrition of the discs. The alternate expulsion and absorption of water creates a circulation of water through the discs similar to the circulation of synovial fluid through articular hyaline cartilage. The circulation of water enables the discs' chondrocytes to receive nutrients and to get rid of waste products (Adams and Hutton 1985).

> In response to external loading by normal everyday activities, the intervertebral discs deform to cushion the load and thereby protect the vertebrae. In response to impact loads, the discs behave elastically; in response to nonimpact loads, the discs behave viscoelastically. The capacity of the discs to respond effectively to loading is reduced when the capacity of the discs to preload is impaired.

Facet Joints

The facet joints in each motion segment are synovial gliding joints formed by the inferior articular processes of the upper vertebra and the superior articular processes of the lower vertebra (figure 6.5). The joint capsule of each facet joint is supported by relatively thick capsular ligaments on the anterior and posterior aspects. The function of the facet joints is largely determined by the orientation of the articular surfaces, which gradually change throughout the length of the vertebral column (see figure

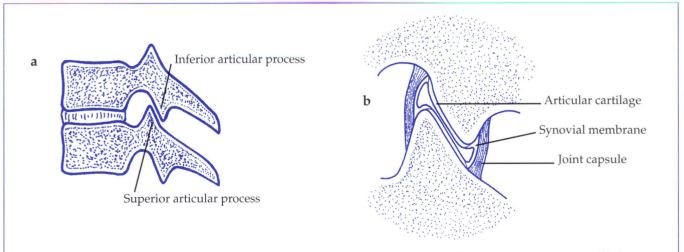

Figure 6.5 Facet joint; (*a*) orientation of articular surfaces in a facet joint of a typical thoracic motion segment; (*b*) close-up of a sagittal section through a facet joint.

3.10). In general, the facet joints assist the intervertebral joints to transmit loads between the vertebrae. Specifically, the facet joints contribute to load transmission in certain postures and restrict certain ranges of movement of the intervertebral joints in order to prevent injury to the intervertebral discs (Adams and Hutton 1983).

With respect to standing postures, the load on the facet joints depends on the thickness of the discs and the alignment of the vertebral column. In relaxed upright standing the compression load on the facet joints of a healthy vertebral column, with the discs at maximum thickness, is small (Adams and Hutton 1980). However, due to prolonged standing the thickness of the discs gradually decreases such that the inferior articular processes slide progressively downward relative to the superior articular processes, and the compression stress on the articular surfaces gradually increases (figure 6.6). For example, after standing upright for three hours the facet joints of the lumbar motion segments have been shown to support approximately 16% of the total vertical compression load on the motion segments (Adams and Hutton 1980). If disc thickness continues to decrease, eventually the tips of the articular processes impinge upon the vertebral arch; that is, the tips of the inferior articular processes impinge upon the laminae of the lower vertebra or the tips of the superior articular processes impinge upon the pedicles of the upper vertebra (figure 6.6c). This condition is referred to as **extra-articular impingement** (EAI).

EAI, which may involve impingement of the facet joint synovial membranes, is a common feature of motion segments in which the discs are badly degenerated. In such cases, the compression loads on the facet joints and areas of EAI may be very high. For example, in badly degenerated lumbar motion segments the facet joints and areas of EAI support approximately 70% of the total compression load on the motion segments in the upright standing position (Adams and Hutton 1980).

> *The facet joints contribute to load transmission between vertebrae in certain postures and restrict certain ranges of motion to prevent injury to intervertebral discs. In relaxed upright standing the compression load on the facet joints of a healthy vertebral column may be small. Prolonged standing decreases the thickness of the discs (due to expulsion of water), which, in turn, increases the compression load on the facet joints and possibly results in EAI.*

The loads on the facet joints between the fourth and fifth lumbar vertebrae, L4 and L5, and between L5 and the first sacral vertebra, S1, tend to be greater than the loads on facet joints in other regions of the column due to the orientation of the L4, L5, and S1 vertebrae. In relaxed upright standing the plane of the inferior end plate of L4 makes an angle (L4 angle) of approximately 15° with the horizontal, and the plane of

Extra-articular impingement (EAI): when the tips of the inferior articular processes impinge upon the laminae of the lower vertebra or the tips of the superior articular processes impinge upon the pedicles of the upper vertebra in a motion segment

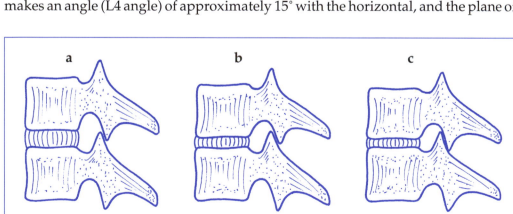

Figure 6.6   Effect of disc thickness on compression load on facet joints; (*a*) normal disc thickness; (*b*) reduced disc thickness resulting in increased compression load on facet joint; (*c*) extra-articular impingement.

the inferior end plate of L5 makes an angle (L5 angle) of approximately 25° with the horizontal (figure 6.7, *a* and *b*). Consequently, there is a tendency for the vertebral bodies of L4 and L5 to slide forward and downward on L5 and S1, respectively, and thereby subject the L4/L5 and L5/S1 discs to considerable shear loading.

However, in a healthy vertebral column, the facet joints and supporting ligaments prevent L4 and L5 from sliding (see next section; figure 6.7, *b* and *c*). In each vertebra the parts of the vertebral arch between the superior and inferior articular processes— at the junctions of the pedicles and laminae—are called the **pars interarticularis** (PI: interarticular parts) (figure 6.7*d*). In resisting the tendency of L4 and L5 to slip forward and downward, the pars interarticularis of the L4/L5 and L5/S1 motion segments are subjected to shear loading (figure 6.7*c*). As shown in figure 6.7*c*, F1 is the component of the load exerted by L4 on L5 perpendicular to the end plate; F2 is the component of the load exerted by L4 on L5 parallel to the end plate; and F3 and F4 are components of the load exerted by the sacrum on L5 perpendicular and parallel to the end plate, respectively. F2 and F4 constitute a shear load on the pars interarticularis of L5.

Pars interarticularis (PI): the parts of the vertebral arch between the superior and inferior articular processes

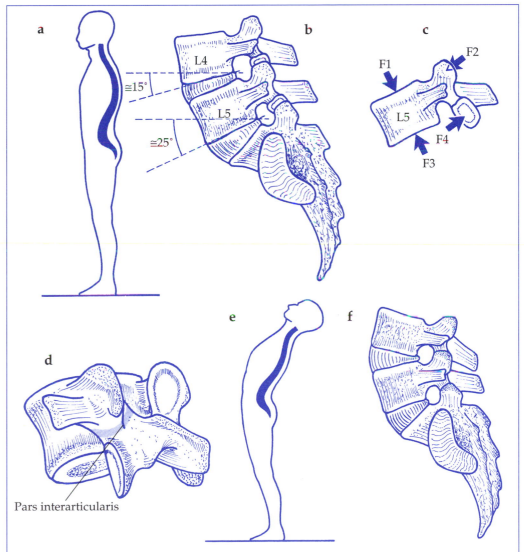

Figure 6.7   Orientation of fourth (L4) and fifth (L5) lumbar vertebrae and sacrum in upright postures; (*a*, *b*, and *c*) normal upright standing; (*d*) oblique view of lumbar vertebra showing left pars interarticularis; (*e* and *f*) trunk extension during upright standing.

With reference to the shape of the vertebral column in relaxed upright standing, extension of any part of the vertebral column tends to increase the compression loading on the articular surfaces of the facet joints in the extended region. The increase in compression loading is offset to a certain extent by a simultaneous increase in the area of contact between the articular surfaces so that the increase in compression stress on the articular surfaces is much less than the increase in compression load (figure 6.7, *e* and *f*). However, the increase in compression load on the articular facets may result in an increase in shear load on the pars interarticularis (Silver, Silver, and Godfrey 1986; Aspden 1987). Consequently, repeated forceful extension of any part of the vertebral column is likely to result in partial or complete fracture of the PI in the affected region (see discussion later in chapter).

Flexion of any part of the vertebral column with respect to the shape in relaxed upright standing decreases the compression load on the articular surfaces of the facet joints. For example, a very small amount of lumbar flexion has been shown to completely unload the lumbar facet joints (Adams and Hutton 1980). Since most relaxed sitting postures involve a certain amount of lumbar flexion, the lumbar facet joints likely are unloaded in these situations (figure 6.8). The only situation where the facet joints are likely loaded during flexion is when lifting a heavy weight with the trunk flexed forward and the legs kept fairly straight. In this situation, any tendency of trunk flexion to unload the facet joints is more than offset by the tendency of the vertebrae to slide forward and downward on each other, thereby increasing the compression loading on the facet joints.

Flexion of any part of the vertebral column tends to decrease the compression load on the affected facet joints, whereas extension tends to increase the compression load on the facet joints. The greater the compression load on the facet joints of a vertebra the greater the shear load on the pars interarticularis of the vertebra.

Ligaments of the Vertebral Column

The vertebral column is supported by three longitudinal ligaments, which run almost the whole of the length of the column, and by three groups of intersegmental ligaments (Williams et al. 1995). The longitudinal ligaments are the anterior longitudinal ligament, posterior longitudinal ligament, and supraspinous ligament. The anterior longitudinal ligament is a broad band running down the anterior aspect of the vertebral bodies from the anterior aspect of the occipital bone to the anterior aspect of the sacrum (figure 6.9). It is attached to the intervertebral discs and to the anterior aspects of the vertebral bodies except for the upper and lower edges. The posterior longitudinal ligament is a broad band running down the posterior aspect of the vertebral bodies from the axis to the posterior aspect of the sacrum. It is attached to the intervertebral discs and to the upper and lower edges of the vertebral bodies. In the lower thoracic and lumbar regions the ligament fans out at the level of each intervertebral joint to gain a broad attachment to the intervertebral disc (figure 6.10).

The supraspinous ligament is a thick cord that links the external occipital protuberance and the tips of the vertebral spines between the seventh cervical vertebra and the posterior aspect of the sacrum (see figure 6.9). In the cervical region the supraspinous ligament expands anteriorly in the median plane by means of a series of radiating fibers that attach onto the tips of the spines of the upper six cervical vertebrae. The series of radiating fibers constitute a membranous syndesmosis called the ligamentum nuchae (figure 6.11). The lateral aspects of the ligamentum nuchae provide areas of attachment for muscles of the neck. The ligamentum nuchae, and that part of the supraspinous ligament from which it arises, consists largely of regular

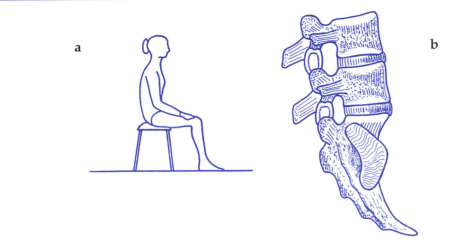

Figure 6.8   Effect of lumbar flexion on loading on facet joints during relaxed sitting on a stool; (*a*) relaxed sitting posture; (*b*) separation and unloading of facet joints as a result of lumbar flexion.

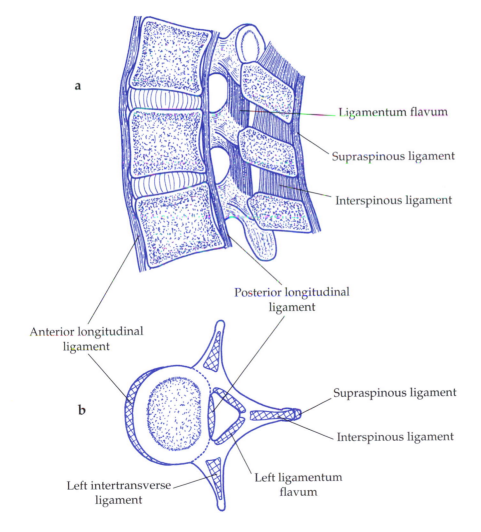

Figure 6.9   Ligaments of the vertebral column; (*a*) median section through lumbar region; (*b*) superior aspect of a typical vertebra showing location of ligaments.

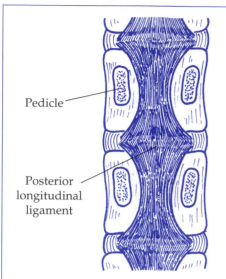

Figure 6.10 Vertical section through pedicles in lumbar region: posterior aspect of bodies of vertebrae showing location of posterior longitudinal ligament.

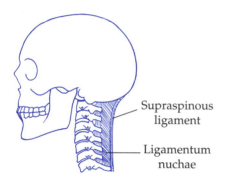

Figure 6.11   Ligamentum nuchae.

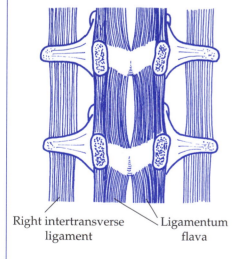

Figure 6.12   Vertical section through pedicles in lumbar region: anterior aspect of vertebral arch showing location of intertransverse ligaments and ligamentum flava.

elastic tissue. The elasticity of these ligaments helps to restore the head to its normal orientation following flexion of the neck and, in general, relieves the load on the extensor muscles of the neck. The ligamentum nuchae is extensive in animals such as the horse and giraffe that normally stand with the head suspended in front of the rest of the body.

The three groups of intersegmental ligaments are the ligamentum flava, interspinous ligaments, and intertransverse ligaments. The ligamentum flava link the laminae of adjacent vertebrae (see figures 6.9, a and b, and 6.12). In each motion segment there is a left and a right ligamentum flavum with a space between them in the median plane for the passage of blood vessels (figure 6.12). Each ligamentum flavum consists of a sheet of regular elastic tissue. The ligamentum flava, like the ligamentum nuchae, are stretched during flexion of the vertebral column and, therefore, help to restore the normal orientation of the column after flexion. In doing so the ligamentum flava relieve the load on the back extensor muscles.

The interspinous ligaments are membranous syndesmoses that link together the spines of adjacent vertebrae (see figure 6.9, a and b). They are continuous with the supraspinous ligament and provide areas of attachment for muscles on both lateral aspects. The intertransverse ligaments link together the transverse processes of adjacent vertebrae. They tend to become progressively broader from the top to the bottom of the column.

The vertebral column is supported by three longitudinal ligaments (anterior longitudinal ligament, posterior longitudinal ligament, supraspinous ligament) and three groups of intersegmental ligaments (ligamentum flava and interspinous and intertransverse ligaments).

Range of Movement in the Vertebral Column

Movement in the individual motion segments and in the whole vertebral column is limited by the combined effects of the longitudinal and intersegmental ligaments, the thickness of the intervertebral discs, the orientation of the facet joints, the shape of the vertebral spines, and, in the case of the thoracic region, the splinting effect of the ribs. The extensibility of the supraspinous ligament, interspinous ligaments, intertransverse ligaments, posterior longitudinal ligament, capsules of facet joints, and posterior aspects of the intervertebral discs limit flexion of the column. Extension is limited by the extensibility of the anterior longitudinal ligament and anterior aspects of the intervertebral discs and by impingement of the vertebral spines on each other. Lateral flexion, always associated with a certain amount of axial rotation, is limited by the extensibility of the supraspinous and interspinous ligaments and, on the convex side, by the intertransverse ligaments and lateral aspects of the intervertebral discs (Kapandji 1974). Axial rotation is limited by the extensibility of the supraspinous, interspinous, and intertransverse ligaments, torsion in the intervertebral discs, and, in the lumbar region, by the orientation of the facet joints.

Table 6.1   Normal Ranges of Movement (Degrees) in the Vertebral Column Relative to Relaxed Upright Standing

	Flexion		Extension		Lateral flexion		Axial rotation	
	A	B	A	B	A	B	A	B
Cervical	6.7	40	12.5	75	5.8	35	8.3	50
Thoracic	3.7	45	2.1	25	1.7	20	2.9	35
Lumbar	12.0	60	7.0	35	4.0	20	1.0	5
Total		145		135		75		90

A = approximate range of movement per intervertebral joint
B = approximate range of movement in each region

Based on Kapandji 1974

In addition to the limitations on movement imposed by the shapes of the vertebrae and the various fibrous supporting structures, the ranges of movement of the vertebral column about the three principal axes are also limited by the extensibility of the surrounding muscles. Whereas the range of movement in each motion segment is quite small, the cumulative ranges of movement are quite large (table 6.1). Similarly, the shock-absorbing capacity of the vertebral column is determined to a certain extent by the cumulative shock-absorbing capacity of the individual motion segments. However, the shock-absorbing capacity of the vertebral column also depends upon its curved shape in the median plane. Whereas a vertical metal rod will provide little or no shock absorption in response to a vertical impact load, the same piece of metal can be converted into an excellent shock absorber by making it into a helical spring (Adams and Hutton 1985). In the same way, the cervical, thoracic, and lumbar curves of the vertebral column considerably increase its shock-absorbing capacity in activities involving axial compression of the vertebral column such as landing from a jump.

> *Movement in the individual motion segments and in the vertebral column as a whole is limited by the combined effects of the discs, ligaments, orientation and size of the vertebral spines, and, in the thoracic region, by the splinting effect of the ribs. While the range of movement in each motion segment is quite small, the cumulative ranges of movement are quite large.*

Degeneration and Damage in the Vertebral Column

Back pain, particularly lower-back pain (pain in the lumbar region) is a common medical disorder in industrial countries and is the main cause of absence from work (Dwyer 1987; Lonstein and Wiesel 1987). The source of back pain is often difficult to diagnose (Nachemson 1992) and it is suggested that it may occur with or without structural damage (Plum and Ofeldt 1985; Luoto et al. 1995; Sparto et al. 1997). In the absence of structural damage the cause of back pain may be muscle fatigue arising from relatively low levels of muscle strength or endurance (Plum and Ofeldt 1985; Luoto et al. 1995; Sparto et al. 1997). However, persistent back pain is likely to be due to structural damage or degeneration in one or more motion segments (Nachemson 1992).

Muscle Fatigue

The maximum amount of force a muscle (or muscle group) can exert—the strength of the muscle—largely depends on the extent it is used in everyday physical activities; the greater the level of activity, the greater the strength of the muscle, and vice versa. When an individual's level of habitual physical activity decreases, his or her muscles atrophy—decrease in size—and, consequently, become weaker. As muscles weaken, the proportion of the strength needed to bring about or maintain a particular movement or posture progressively increases; that is, the intensity of the muscular effort progressively increases. The blood flow through a muscle (which supplies oxygen to and removes metabolic waste products from the muscle) depends on the intensity of muscular effort; the greater the intensity, the lower the blood flow and, consequently, the faster the muscle fatigues.

Muscle fatigue is characterized by the buildup of metabolic waste products including toxic substances such as lactic acid; this buildup results in pain. To avoid this type of pain the individual is likely to further restrict movement of the affected muscles and associated joints, which, in turn, may result in pain due to stiffness, cartilage degeneration, and increased muscle atrophy (Nelson et al. 1995) or local overload as a result of compensatory movements in response to muscle fatigue (Plum and Ofeldt 1985; Risch et al. 1993; Nelson et al. 1995).

Structural Damage and Degeneration

Structural damage to a motion segment or its supporting structures is likely to result in pain. Minor muscle tears are a common cause of temporary low-back pain (LBP) (Clarkson and Tremblay 1988). Minor muscle tears occur in a variety of everyday situations when the vertebral column is subjected to above-average loading such as when trying to lift a heavy load. With rest, minor muscle tears tend to heal quickly such that the LBP gradually goes away within a few days. Major muscle tears are usually associated with other types of damage, which prolong LBP.

Whereas minor muscle tears tend to result in temporary LBP, damage or degeneration in intervertebral discs may result in persistent LBP. The intervertebral discs have no nerve supply and, therefore, damaged or degenerated discs are not a direct source of pain. However, the mechanical integrity of the discs, especially the ability to preload to normal levels, profoundly influences the loads on other parts of the motion segments. Damage or degeneration in an intervertebral disc may cause overloading of other parts of the motion segment resulting in pain from a variety of sources.

> Low-back pain is one of the most common medical disorders in industrial countries and the main cause of absence from work. The source of back pain is often difficult to diagnose and it may occur with or without structural damage.

Degeneration and Damage in Intervertebral Discs

Degeneration and damage in intervertebral discs occur as a result of aging and mechanical overload.

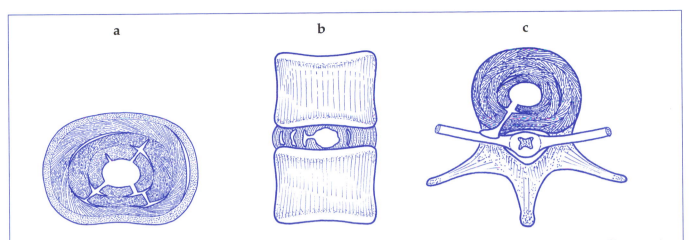

Figure 6.13   Degeneration and damage in an intervertebral disc; (*a*) superior aspect of disc showing fissures; (*b*) vertical section through disc showing displaced portions of the nucleus pulposus trapped in fissures; (*c*) superior aspect of intervertebral joint showing prolapsed disc and impingement of spinal nerve by part of nucleus pulposus.

Aging

After the second decade of life, degenerative changes that impair the capacity of a disc to preload are likely to occur in both the nucleus pulposus and annulus fibrosus (Brinckmann 1985). These degenerative changes bring about a gradual reduction in the OSP of the nucleus pulposus and a gradual softening and weakening of the annulus fibrosus, which includes the occurrence of fissures between and across the layers of fibrocartilage, especially those close to the nucleus pulposus (Adams and Hutton 1982) (figure 6.13). The fissures are similar to the circumferential and radial tears that occur in the menisci of the knee. In the medium to long term, these degenerative changes are likely to cause pain, either directly or indirectly, in the following three ways.

1. **Abnormal loading on end plates of vertebral bodies**. In each motion segment the central areas of the end plates of the vertebral bodies tend to be weaker than the surrounding circumferential areas. When the disc is healthy, most of the compression load on the motion segment is transmitted through the circumferential areas of the end plates. However, as the disc degenerates, the compression loads on the central areas tend to increase, leading to increased compression loading on the subchondral bone and, therefore, the likelihood of pain.

2. **Internal derangement of the nucleus pulposus**. During flexion, extension, and lateral flexion of the vertebral column, the vertebrae in each motion segment pivot about the nucleus pulposus, which is itself laterally displaced toward the stretched side of the joint (see figure 6.4). As the normal orientation of the vertebrae is restored the nucleus pulposus in a healthy disc moves back to its normal position. However, in a degenerated disc small portions of the nucleus pulposus may become trapped in the fissures of the degenerated annulus fibrosus— portions of the nucleus pulposus stick in an off-center position (figure 6.13*b*). This situation may result in unequal tension in the muscles on opposite sides of a motion segment, with the muscles under the greater tension going into spasm. The muscle spasm may stretch the adjacent spinal nerve resulting in pain (Nachemson 1992).

3. **Reduced thickness of the disc**. In degenerated discs the capacity to preload is impaired; the greater the degeneration the greater the impairment. Conse- quently, under normal loading conditions degenerated discs are thinner than healthy discs. Reduction in the thickness of a disc has a number of potentially

painful consequences (Adams and Hutton 1980; Letts et al. 1986; Rydevik, Brown, and Lundborg 1984):

- Impingement of the spinal cord, spinal nerves, and nerves of supporting ligaments and muscles due to reduced size of intervertebral foramen and protrusion of discs, especially at posterior and posterolateral aspects (figure 6.14)
- Increased loading on articular surfaces of facet joints resulting in pain from subchondral bone and progressive damage to articular cartilage
- Fractures of vertebral bodies and pars interarticularis due to reduced shock-absorbing capacity and increased shear load on the pars interarticularis
- Extra-articular impingement of the facet joints (see figure 6.6c)

Mechanical Overload

Repetitive or continuous high-level loading for long periods of time is referred to as **chronic loading**. The intervertebral discs are subjected to chronic loading, especially in the form of bending and torsion, during heavy manual work—activities involving fairly strenuous lifting, carrying, shoveling, pulling, and pushing. Chronic loading accelerates the degeneration of the discs as they age.

Fissures in the annulus fibrosus are the result of a combination of degeneration of the fibrocartilage and strain in response to loading. Given adequate rest, some healing of the fissures may occur. However, the supply of nutrients to the annulus fibrosus is relatively poor and the load on the discs is usually high, even when simply supporting the weight of the upper body; consequently, healing is usually slow. Under chronic loading conditions the repair process may be outpaced by the rate at which fissuring occurs so that the damage to the annulus fibrosus gradually accumulates. This form of cumulative damage is referred to as **fatigue damage** (Adams and Hutton 1982). **Fatigue failure** occurs when the annulus fibrosus becomes so badly damaged that the capacity of the disc to preload is severely and permanently impaired. The effects of fatigue failure are the same as those that result from aging, except that the various sources of pain that result from disc damage tend to occur earlier.

As the annulus fibrosus of a disc degenerates it becomes progressively weaker and, consequently, more vulnerable to damage from all types of loads. In particular, the annulus fibrosus is less able to withstand the tension loads imposed on it during

Chronic loading: repetitive or continuous high-level loading for long periods of time

Fatigue damage: cumulative structural damage due to chronic loading

Fatigue failure: failure of a structure due to fatigue damage

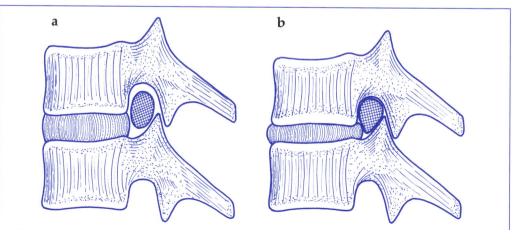

Figure 6.14 Impingement of spinal nerve; (*a*) normal orientation of vertebrae, disc, and spinal nerve; (*b*) impingement of spinal nerve by reduced intervertebral foramen and posterior protrusion of disc.

bending movements of the vertebral column. In these circumstances the nucleus pulposus may burst through the annulus fibrosus resulting in what is referred to as a **prolapsed**, herniated, or "slipped" disc, or an external derangement of the nucleus pulposus (Brinckmann 1985; Adams and Hutton 1982; see figure 6.13c). A prolapse usually occurs suddenly due to an **acute failure** of the annulus fibrosus—the annulus fibrosus suddenly bursts in a manner similar to a blowout of a tire.

A prolapse may occur in a previously healthy annulus fibrosus, but in the majority of cases it is likely that the disc was degenerated to a certain extent prior to the prolapse (Brinckmann 1985). In a slightly degenerated disc the nucleus pulposus still has a high OSP and can, therefore, still exert a high pressure on the annulus fibrosus. Consequently, a slightly degenerated disc may be more vulnerable to prolapse than a severely degenerated disc in which the nucleus pulposus has lost its osmotic swelling property.

Many people between 30 and 50 years of age have slightly to moderately degenerated discs. Since many people in this age group are still active and impose large loads on the vertebral column, it follows that the discs are particularly vulnerable to prolapse during this period. If disc degeneration does not occur until later in life, when the individual's level of activity is considerably lower than in the middle years, it is unlikely that a vulnerable disc will be subjected to the high loading necessary to cause prolapse (Brinckmann 1985).

Prolapses may occur in the cervical, thoracic, and lumbar regions of the vertebral column, but occur most frequently in the L4/L5 and L5/S1 discs (Adams and Hutton 1982; Brinckmann 1985). Prolapses in the lumbar region are usually the result of postures in which the lumbar region is heavily loaded while fully flexed, such as in stooping to lift an object from the floor while keeping the legs straight (figure 6.15b). In this situation the combined weight of the upper body and the object impose a considerable load on the lumbar region. The tension load on the stretched posterior aspect of a lumbar disc may cause the disc to prolapse, usually in a posterolateral direction, so that the nucleus pulposus shoots through the tear in the annulus fibrosus and rams into the adjacent spinal nerve causing a sharp severe pain (see figure 6.13c). After the prolapse, the extruded part of the nucleus pulposus may continue to impinge upon the spinal nerve, resulting in continuous pain. The posterior longitudinal ligament strengthens the posterior part of the disc (see figures 6.9 and 6.10) so that prolapses usually occur in the relatively weaker posterolateral aspects of the disc.

Prolapsed disc: acute failure of the annulus fibrosus of an intervertebral disc resulting in external derangement of the nucleus pulposus

Acute failure: sudden failure of a structure due to overload

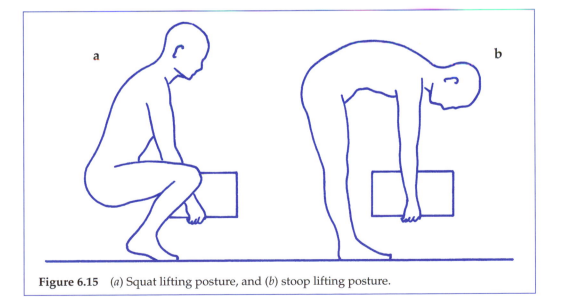

Figure 6.15   (a) Squat lifting posture, and (b) stoop lifting posture.

After the second decade of life degenerative changes that impair the capacity of the disc to preload are likely to occur in both the nucleus pulposus and annulus fibrosus. In particular, there is a gradual reduction in the OSP of the nucleus pulposus and a gradual softening and weakening in the annulus fibrosus. These degenerative changes may result in direct or indirect pain, in the following three ways:

- *Increased loading on subchondral bone of vertebral end plates*
- *Muscle spasm due to internal derangement of the nucleus pulposus*
- *Impingement of spinal cord and spinal nerves due to reduced thickness of the intervertebral disc and disc prolapse*

With regard to lifting postures, the load on the lumbar region increases as the inclination of the trunk increases. Consequently, lifting an object with a bent back and straight legs—a stoop lift—puts maximum load on the lumbar region, whereas lifting the same load with bent legs and a fairly straight back—a squat lift—minimizes the load on the lumbar region (figure 6.15a). In a squat lift, the back muscles hold the trunk steady while the extensor muscles of the legs do most of the work in lifting the object. However, in a stoop lift, the leg extensor muscles contribute little to the work of lifting the object; the extensor muscles of the lumbar region do most of the work of lifting the object, which, in turn, puts a considerable load on the lumbar structures.

The load on the lumbar region when lifting loads in the stooped posture depends not only on the overall inclination of the trunk, but also on the shape of the lumbar region. With respect to any particular trunk inclination, the greater the flexion of the lumbar region the shorter the moment arm of the back extensor muscles adjacent to the lumbar region and, consequently, the greater the force exerted by the muscles and the greater the reaction forces on the intervertebral joints. The greater the muscle forces and joint reaction forces, the greater the risk of injury. Maintaining a lumbar lordosis—posterior concavity in the lumbar region—as shown in figure 6.16, reduces the muscle forces and joint reaction forces and, consequently, the risk of injury (Aspden 1987).

The ability to maintain a lumbar lordosis in the stooped posture may depend to a certain extent on **intratruncal pressure**—the pressure developed in the thoracic and the abdominal cavities in response to all types of activity that impose large loads on the lumbar region (Troup 1970; Aspden 1987; McGill and Norman 1987). The thorax contains air and the abdomen contains liquid and semisolid material. By breathing in and then holding the breath, the pressure inside the thorax—intrathoracic pressure—can significantly increase above normal (figure 6.16). Similarly, in association with intrathoracic pressure, coordinated activity in the back extensor muscles and abdominal muscles considerably increases the pressure inside the abdomen—intra-abdominal pressure (IAP). The role of IAP in lifting mechanics is not yet clear (McGill and Norman 1987), but it is suggested that IAP, in conjunction with the force exerted by the back extensor muscles, stabilizes the lumbar lordosis and thereby reduces the likelihood of overloading the lumbar structures (figure 6.16) (Aspden 1987).

> *Intratruncal pressure may help to maintain a lumbar lordosis in lifting postures and thereby reduce the likelihood of overloading the lumbar structures.*

Intratruncal pressure: the pressure developed in the thoracic and abdominal cavities by coordinated activity of the muscles surrounding the thorax and abdomen

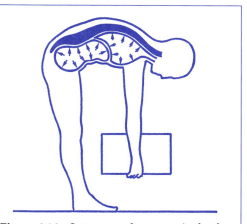

Figure 6.16 Intratruncal pressure in the thoracic and abdominal cavities.

Figure 6.17 summarizes the sources of back pain that may result from degeneration and damage in intervertebral discs.

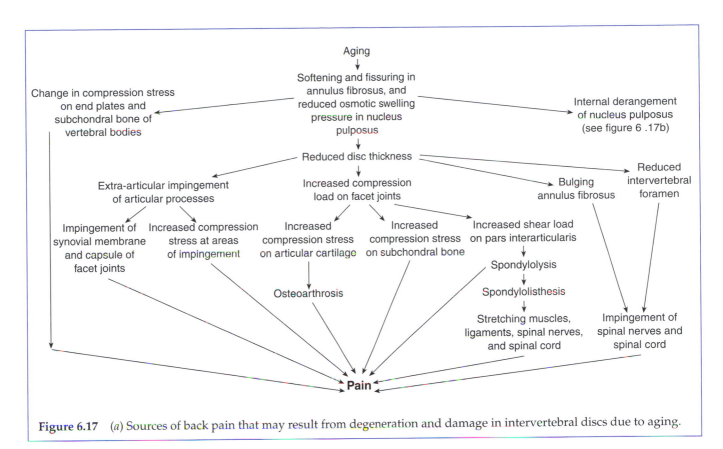

Figure 6.17   (*a*) Sources of back pain that may result from degeneration and damage in intervertebral discs due to aging.

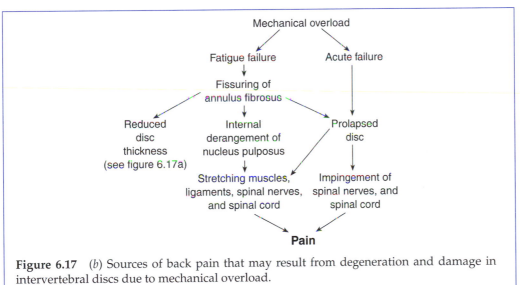

Figure 6.17   (*b*) Sources of back pain that may result from degeneration and damage in intervertebral discs due to mechanical overload.

Degeneration and Damage in Vertebrae

As with intervertebral discs, degeneration and damage in vertebrae occur due to aging and mechanical overload.

Aging

The main effect of aging on vertebrae is the gradual reduction in bone mass beginning at approximately 30 and 45 years of age in women and men, respectively (see chapter

4). The gradual loss in bone mass results in a gradual decrease in the strength of the vertebrae; for example, the compression strength of lumbar vertebral bodies in 60 to 70 year olds is approximately half of that found in the 20 to 30 year olds (Brinckmann 1985). Fractures of the vertebral bodies, vertebral arches, spines, and transverse processes are fairly common in aged individuals.

Mechanical Overload

In young and middle-aged individuals the most frequently reported form of damage to the vertebrae is fracture of the pars interarticularis (PI). Fractures of the PI occur most frequently in the cervical and lumbar regions and are usually associated with motor vehicle accidents or participation in sports that subject the vertebral column to high bending, torsion, and shear loads (Lowe et al. 1986; Letts et al. 1986; Stanitski 1982; Silver, Silver, and Godfrey 1986; Tall and Devault 1993). These sports include gymnastics, diving, trampolining, judo, wrestling, weightlifting, soccer, rugby football, and American football. Acute failure—sudden, complete fracture of the PI— may occur in previously undamaged PI. When this occurs it is usually the result of a sudden, violent hyperextension or hyperflexion. The cervical region is particularly vulnerable to damage from this type of movement and many neck injuries resulting from motor vehicle accidents (whiplash injuries) and trampolining (landing on the head) occur in this manner. Many of these injuries involve severe damage to the spinal cord and spinal nerves (Silver, Silver, and Godfrey 1986; Gozna and Harrington 1982).

Although acute failure of the PI is fairly common, it is likely that most complete fractures of the PI, like prolapsed intervertebral discs, are the result of progressive fatigue damage that eventually results in fatigue failure—complete fracture. Fatigue failure of the PI occurs in response to chronic overloading, when the rate of loading is such that the occurrence of microfractures outpaces the capacity of the healing processes to repair them. Fatigue failure of the PI is often the result of chronic bending loads—fairly high frequency flexion and extension of the lumbar region while it is under fairly high loading. This type of movement is a common feature of gymnastics and trampolining. The effect of this type of movement on the lumbar region as a whole and on the PI in particular is similar to that of repeatedly bending a piece of

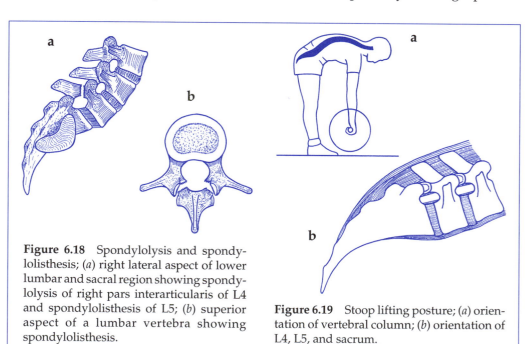

Figure 6.18 Spondylolysis and spondylolisthesis; (*a*) right lateral aspect of lower lumbar and sacral region showing spondylolysis of right pars interarticularis of L4 and spondylolisthesis of L5; (*b*) superior aspect of a lumbar vertebra showing spondylolisthesis.

Figure 6.19 Stoop lifting posture; (*a*) orientation of vertebral column; (*b*) orientation of L4, L5, and sacrum.

wire alternately in one direction then the other until it breaks (Hutton and Cyron 1978).

There are a number of stages in fatigue failure of the PI of a vertebra. In the first stage microfractures gradually accumulate until a stress fracture—partial fracture— is noticeable in one PI. The stress fracture progresses until the PI is completely fractured, but without separation of the fractured ends. This condition is referred to as **spondylolysis** (figure 6.18a). The same events occur in the other PI so that eventually both PI are completely fractured—bilateral spondylolysis. At this stage the fractured ends are held together by the supporting structures—ligaments, intervertebral discs, and muscles. However, if the chronic loading continues, the vertebral body may slide forward so the posterior part of the vertebral arch separates from the rest of the vertebra. This condition is called **spondylolisthesis** (figure 6.18).

> *In young and middle-aged individuals the most common form of damage to the vertebrae is fracture of the pars interarticularis. Acute failure of the pars interarticularis may occur without prior damage, but it is likely that most acute failures, like prolapsed discs, are the result of progressive fatigue failure. There are four main stages in fatigue failure of the pars interarticularis: stress fracture, unilateral spondylolysis, bilateral spondylolysis, and spondylolisthesis.*

Spondylolisthesis subjects the associated intervertebral disc to shear loading, stretches the supporting ligaments, and, depending on the degree of slippage, compresses or stretches the adjacent spinal cord and spinal nerves. In the general population, spondylolysis and spondylolisthesis are uncommon in children under approximately eight years of age. Between 8 and 15 years of age the incidence of spondylolysis rises to 5% and to 6% in adults (Letts et al. 1986). The incidence of spondylolisthesis in older children and adults is approximately 1.5%. Participation in high-level training and competition in a variety of sports has been found to result in an increased incidence of spondylolysis and spondylolisthesis. For example, in a group of female gymnasts the incidence of spondylolysis and spondylolisthesis was found to be 11% and 6%, respectively (Jackson, Wiltse, and Cirincione 1976). In a group of high school and university male sports participants from 20 different sports ranging from rifle shooting to American football, the incidence of spondylolysis was nearly 21% (Hoshina 1980).

In comparison to activities like gymnastics and trampolining, the frequency of flexion and extension of the lumbar region that occurs in weightlifting is much lower, but the load on the lumbar region tends to be much higher so that the lumbar region and the PI, in particular, are subjected to chronic bending loads. In addition, some of the static and quasi-static postures that occur in weightlifting subject the PI to considerable shear loading. To illustrate the types of loading on the lumbar structures during weightlifting, let's consider three postures (Troup 1970):

1. Just after the start of a clean and press lift (figure 6.19). This posture is basically the same as the stoop lifting position (see figure 6.15b). The lumbar region is in a flexed position and subject to considerable loading.

- All of the ligaments posterior to the vertebral bodies are likely to be stretched.

Spondylolysis: complete fracture of the pars interarticularis without separation of the fractured ends

Spondylolisthesis: separation of the vertebral arch from the vertebral body subsequent to bilateral spondylolysis

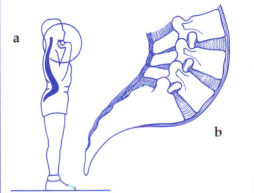

Figure 6.20 Upright posture when supporting a weighted barbell at shoulder height; (*a*) orientation of vertebral column; (*b*) orientation of L3, L4, L5, and sacrum.

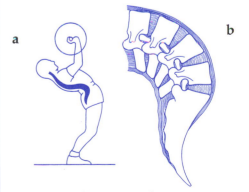

Figure 6.21 Posture involving an exaggerated lean-back for pressing a weighted barbell upwards from the shoulders; (*a*) orientation of vertebral column; (*b*) orientation of L2, L3, L4, L5, and sacrum.

- The posterior aspects of the discs are likely to be stretched, with the possibility of posterior prolapse.
- The shear load on the PI of the lower lumbar vertebrae is likely to be high, with the possibility of fatigue failure or acute failure of the PI, leading to spondylolisthesis.

2. Holding a barbell at the shoulder level or overhead with both feet in a frontal plane (figure 6.20). During the clean stage of a clean and press the heavily loaded lumbar region changes from a flexed to an extended position (see figures 6.19*a* and 6.20*a*). The lumbar region remains extended during the press stage and during the period when the weight is held overhead at the end of the lift. In the extended position the loads on the lumbar structures are considerable (figure 6.20*b*):

- The shear load on the PI of L5 may result in fatigue or acute failure.
- The spines of the lower lumbar vertebrae may impinge upon each other with the possibility of fracture of one or more spines.
- Extra-articular impingement (EAI) of the articular processes of the facet joints may occur.
- The anterior longitudinal ligament may be overstretched resulting in excessive tension in the anterior aspects of the intervertebral discs with the possibility of anterior prolapse of one or more discs.

3. A shoulder press involving an exaggerated lean-back (figure 6.21). This posture involves full extension of the lumbar region so that the lower thoracic to upper lumbar region is more or less horizontal:

- L2 tends to be displaced downward relative to L3. The L2/L3 disc, subjected to shear loading, resists this. The downward displacement of L2 is also resisted by most of the ligaments supporting the L2/L3 motion segment, including the ligaments and capsules of the facet joints, which are luxated in this posture.

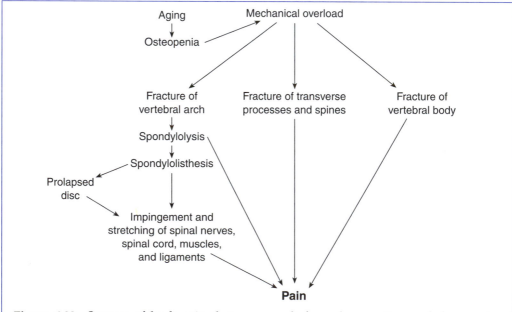

Figure 6.22 Sources of back pain that may result from degeneration and damage in vertebrae.

- The anterior longitudinal ligament may be overstretched, resulting in excessive tension in the anterior aspects of the intervertebral discs with the possibility of anterior prolapse.
- Whereas the L2/L3 facet joints are luxated, the L4/L5 and L5/S1 facet joints are heavily compressed, with the possibility of extra-articular impingement of the articular processes.
- The spines of the lumbar vertebrae may impinge upon each other with the possibility of fracture of one or more spines.

The three postures described are usually symmetrical about the median plane. If the postures are accompanied by axial rotation, a certain degree of twisting, the loads on the lumbar structures can be greater than in the symmetrical position. Figure 6.22 summarizes the sources of back pain that may result from degeneration and damage in vertebrae.

Normal Shape of the Vertebral Column

When the trunk is in an upright position, as in standing, the normal vertebral column has four moderate curves in the median plane, but is straight in the frontal plane (figure 6.23, *a* and *b*). The curvature in the median plane represents a compromise between mechanical and morphological requirements. Mechanically, the vertebral column provides stability and flexibility to support the weight of the upper body, act as a shock absorber in response to impact loads, and protect the spinal cord. Morphologically, the anterior concavities of the thoracic and sacral regions provide spaces for the thoracic and lower abdominal organs, respectively. In a rigid upright structure supported by a single pillar, the pillar is situated centrally to correspond to the line of action of the weight of the structure and thereby prevent any bending load on the supporting pillar (figure 6.23*c*). Since the vertebral column is curved, the weight of the upper body always exerts a bending load, which increases the curvature of the column. However, in normal upright posture the line of action of

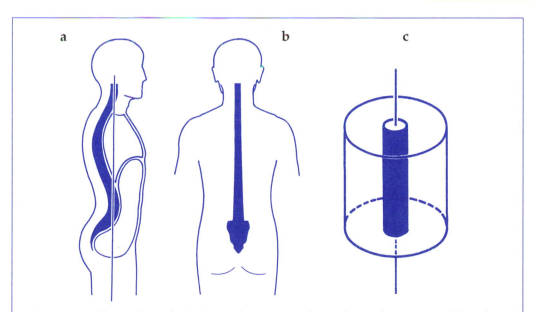

Figure 6.23   Shape of vertebral column during normal upright standing posture; (*a*) median plane; right lateral aspect showing line of action of upper body weight; (*b*) frontal plane; posterior aspect; (*c*) circular structure supported by a central pillar.

upper body weight passes through the cervical and lumbar vertebrae; this action minimizes the bending load on the column (Kapandji 1970).

> In the anatomical position the normal vertebral column has four moderate curves in the median plane (cervical, thoracic, lumbar, and sacral), but is straight in the frontal plane. The curvature in the median plane represents a compromise between mechanical and morphological requirements.

The vertebral column has three flexible regions—cervical, thoracic, and lumbar— and one rigid region, the sacrum. Since the four regions are linked together in a chain, a change in the orientation or curvature of one region results in compensatory changes in the orientation or curvature of the other regions. In normal upright posture the degree of curvature of the vertebral column is usually moderate—the alternating curves change direction gradually so that the anteroposterior depth of the column as a whole is relatively small. This distance increases as the curvature of the column increases.

Descriptions of correct upright posture may be functional or anatomical. Functional descriptions refer to the degree of muscular effort required to maintain a balanced position. For example, upright posture may be considered to be correct "if it is effortless, nonfatiguing and painless when the individual remains upright for reasonable periods" (Bullock-Saxton 1988, p94). Anatomical descriptions of correct upright posture refer to the orientation or curvature of the vertebral column. For example, in correct upright posture "the line of action of body weight lies in the median plane and passes through the mastoid processes, just in front of the shoulder joints, through or just behind the hip joints, through the knee joints and just in front of the ankle joints" (Basmajian 1965, p27) (figure 6.24a).

More detailed anatomical descriptions may refer to the degree of curvature of the different regions of the column (figure 6.24b). The curvature of a particular region is assessed by measuring, from an X ray, the angle subtended by the limits of the region. For example, the lumbar region is usually convex anteriorly and concave posteriorly. Consequently, lines drawn parallel with the superior end plate of L1 and the inferior end plate of L5 intersect posteriorly. The angle of intersection is referred to as the lumbar angle or the angle of lumbar lordosis (lordosis = bent back). The angles of curvature of the cervical region and thoracic region are measured in a similar manner. The method of measuring the angles is called the Cobb method and the angles are often referred to as Cobb angles (Greenspan et al. 1978). The Cobb method is also used to assess abnormal curvatures in the frontal plane (see next section).

Like the lumbar region, the cervical region is convex anteriorly and concave posteriorly. For this reason the cervical angle of curvature is sometimes referred to as the angle of cervical lordosis. In contrast to the cervical and lumbar angles, which are subtended posteriorly, the thoracic angle is subtended anteriorly since the thoracic region is usually concave anteriorly and convex posteriorly (see figure 6.24b). The thoracic angle is referred to as the angle of thoracic kyphosis (kyphosis = keel shaped). The sacral angle, lumbosacral angle, and angle of pelvic tilt may also be used in anatomical descriptions of upright posture. The sacral angle is the angle between the plane of the superior end plate of S1 and the horizontal. The lumbosacral angle is the angle of intersection between lines drawn through the geometric centers of the end plates of L5 and S1, respectively. The pelvic tilt angle is the angle between the horizontal and a line through the promontory of the sacrum and the anterior superior border of the pubic symphysis.

Whereas the angles of curvature (cervical, thoracic, and lumbar) and inclination (sacral, lumbosacral, and pelvic tilt) are relatively easy to measure on X rays, the relationship among the angles in terms of what constitutes correct upright posture has not yet been established (During et al. 1985; Walker et al. 1987). For example, an

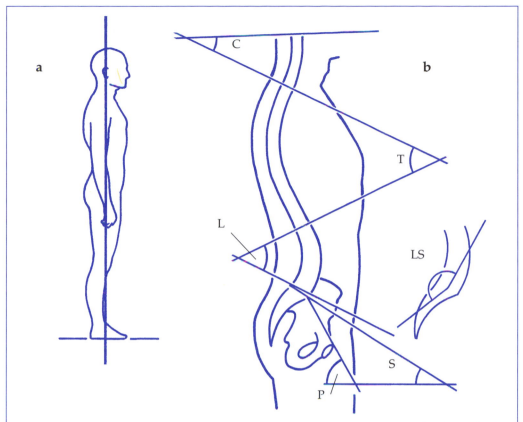

Figure 6.24   Anatomical descriptions of upright standing posture; (*a*) line of action of body weight in relation to anatomical landmarks; (*b*) angles of curvature of different parts of the vertebral column: C—cervical angle, T—thoracic angle, L—lumbar angle, S—sacral angle, LS—lumbosacral angle, P—pelvic tilt angle.

increase in the sacral angle will almost certainly result in an increase in lumbar lordosis. However, the actual increase in lumbar lordosis resulting from a specific increase in the sacral angle to maintain correct upright posture is not known. Similarly, the effect of an increase in the sacral angle on thoracic kyphosis and cervical lordosis is not known. Consequently, there is very little information on normal ranges for the various angles. When normal ranges are given, they are often broad; for example, normal ranges for thoracic kyphosis and the lumbosacral angle have been reported as 20° to 40° and 128° to 160°, respectively (Drummond 1987; Hollinshead 1962). It follows that in terms of the overall shape of the vertebral column, there is a broad range of shapes that may be considered correct upright posture with respect to the functional description given earlier.

A posture that may seem to an observer to be anatomically abnormal may, in fact, be functionally normal. This is often the case in childhood and early adulthood. During the growth period, the vertebral column, like the rest of the skeleton, models its size, shape, and structure in response to the loads imposed on it (see chapter 11). Consequently, the vertebral column may develop abnormal curvature (relative to the majority of individuals of the same age) in order to continue to function normally. In such cases, the shape of the vertebral column will, in the absence of any pathological conditions, eventually stabilize, that is, the vertebral column will have a consistent shape when in an upright posture. In middle to old age, the loading on the vertebral column may change due to changes in muscle strength and body weight. Since the vertebral column is no longer capable of modeling in response to the change in loading, it will become functionally abnormal, depending on the change in loading.

Functional abnormality may result in joint degeneration and pain (Bradford et al. 1974). After maturity, all joints are vulnerable to permanent changes in loading. However, the joints most vulnerable are anatomically abnormal weight-bearing joints (see chapter 11).

> *Descriptions of correct and incorrect upright posture tend to be functional or anatomical. Functional descriptions refer to the degree of muscular effort or load on the vertebral column in relation to the maintenance of a balanced position. Anatomical descriptions refer to the orientation or curvature of the regions of the vertebral column.*

Abnormal Curvature of the Vertebral Column

Abnormal curvatures of the vertebral column often occur prior to maturity in both the median and frontal planes. Whereas some forms of abnormal curvature are more common than others, the range of abnormal shapes is broad.

Abnormal Curvature in the Median Plane

The most common forms of abnormal curvature in the median plane are increased thoracic kyphosis and increased lumbar lordosis (Bullock-Saxton 1988). These two conditions, which may occur together, are often referred to as simply kyphosis and lordosis, but using the terms in this way may result in confusion since the terms are also used to describe normal curvature of the thoracic and lumbar regions and may be applied to other regions of the column to describe certain abnormalities. Increased thoracic kyphosis (i.e., an abnormally large thoracic angle) results in a pronounced

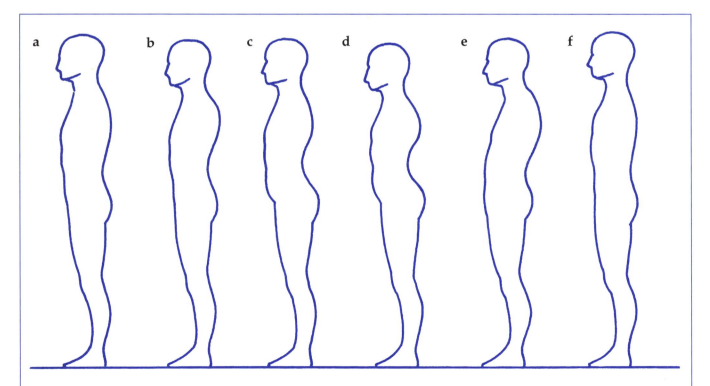

Figure 6.25   Abnormal curvature of the vertebral column in the median plane in upright standing posture; (*a*) normal posture; (*b*) roundback; (*c*) hollowback; (*d*) combined roundback and hollowback; (*e*) swayback; (*f*) flatback.

hump that projects posteriorly giving rise to the descriptive terms roundback, humpback, and, in extreme cases, hunchback. Figure 6.25 shows normal curvature of the vertebral column (6.25a) together with abnormal curvatures: **roundback** (figure 6.25b); **hollow-back** (figure 6.25c), combined roundback and hollow-back (figure 6.25d), swayback (figure 6.25e), and flatback (figure 6.25f).

There are two forms of roundback: postural roundback and Scheuermann's roundback (Drummond 1987; Bradford et al. 1974). Postural roundback is nonstructural and disappears when the individual makes a conscious effort to hold the trunk in a normal upright posture. It is caused by slouched sitting and standing postures that, over a period of time, result in abnormal lengthening of the muscles and ligaments on the posterior aspect of the trunk and abnormal shortening of the muscles and ligaments on the anterior aspect of the trunk (Bullock-Saxton 1988). This imbalance in the soft tissues maintains the postural roundback. Postural roundback is especially common in girls who start puberty early and become overly self-conscious of breast development; in this situation a girl is likely to adopt a round-shouldered posture to hide her breasts (Keim 1982). Normal thoracic curvature can usually be completely restored by a suitable exercise program.

In contrast to postural roundback, which is nonstructural, Scheuermann's roundback is a structural deformity involving abnormal growth of the vertebral bodies. Wedge-shaped vertebral bodies, in which the height of the bodies anteriorly is less than that posteriorly, and reduced disc space anteriorly characterize the condition. Scheuermann's roundback usually occurs between the ages of 11 and 17 years and is said to be present when at least three vertebrae show wedging of 5° or more (Drummond 1987) (figure 6.26, a and b). The degree of wedging is usually greatest at the apex of the hump. In addition to the wedging of the vertebrae, the anterior longitudinal ligament is tight resulting in a rigid hump that persists even when a conscious effort is made to flatten it.

Scheuermann's roundback is thought to be due to abnormal ossification of the vertebral bodies, which results in abnormally weak end plates. The line of action of

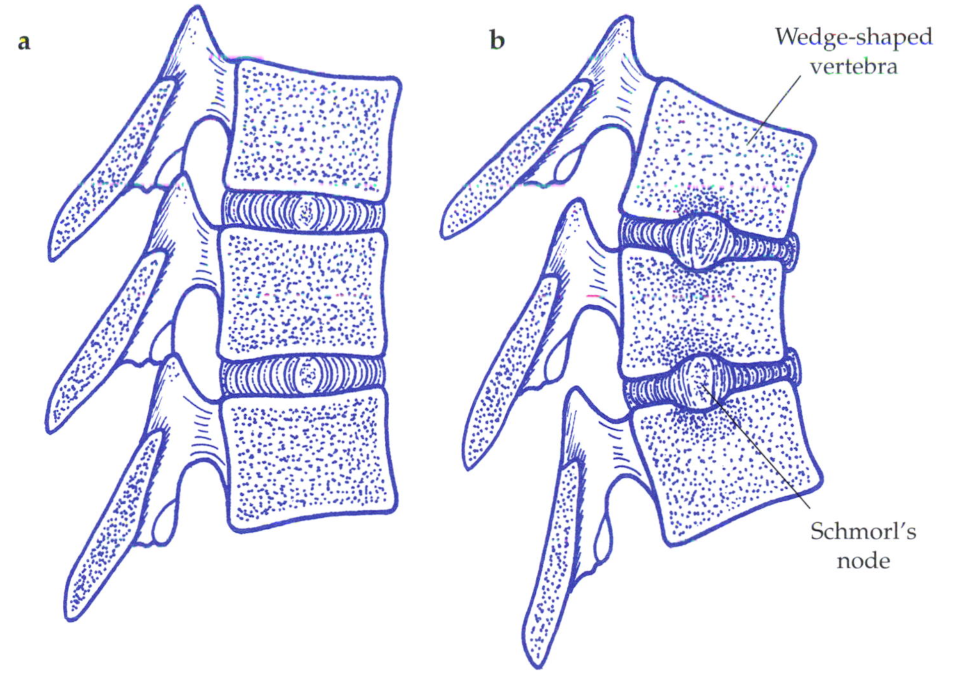

Roundback: an abnormally large thoracic angle resulting in a pronounced hump that projects posteriorly

Hollowback: an abnormally large lumbar angle resulting in a pronounced hollow in the lower back

Figure 6.26 Scheuermann's roundback; (a) normal shape and orientation of thoracic vertebrae; (b) Scheuermann's roundback showing wedging of vertebrae and Schmorl's nodes.

upper body weight passes in front of the thoracic region so that the thoracic region is normally subjected to an anterior bending load—compression loading on the anterior aspects of the vertebral bodies and tension on the posterior aspects. In response to this normal bending load, abnormally weak end plates grow abnormally and become wedged. The wedging increases the bending load on the affected region so that the condition becomes self-aggravating. In addition, the pressure exerted by the intervertebral discs may depress the central areas of the vertebral end plates; each small depression is called a Schmorl's node (see figure 6.26b). The incidence of Scheuermann's roundback has been reported as between 0.4% and 8.3% of the general population (Drummond 1987).

In children, Scheuermann's roundback may give rise to pain, probably as a result of increased compression stress on subchondral bone and the development of Schmorl's nodes. However, the incidence of pain from Scheuermann's roundback is greatest after maturity and the pain is often associated with an increase in the thoracic angle (Drummond 1987). Figure 6.27 shows a model of the development of Scheuermann's roundback and the likely sources of pain that result from it. Scheuermann's roundback usually results in compensatory changes in the curvature of the other regions of the vertebral column; for example, a marked increase in the thoracic angle is usually associated with a marked increase in the lumbar angle (see figure 6.25d).

Increased lumbar lordosis (i.e., an abnormally large lumbar angle) results in a pronounced hollow in the lower back, giving rise to the descriptive term hollowback (see figure 6.25c). Hollowback may occur in a similar manner to Scheuermann's roundback, that is, abnormal ossification of the vertebral end plates with subsequent abnormal growth of the vertebral bodies. Wedging of the vertebrae and Schmorl's nodes may occur, but not usually to the same extent as in roundback (Drummond

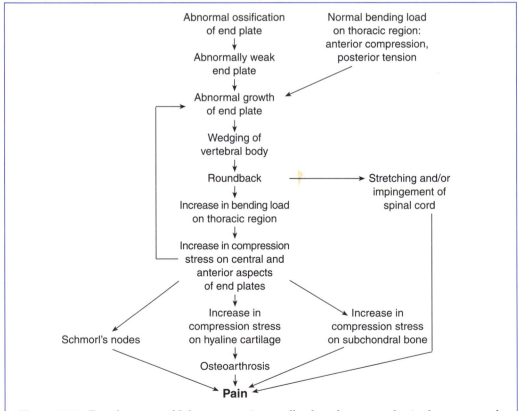

Figure 6.27 Development of Scheuermann's roundback and sources of pain that may result from the condition.

1987; Hollinshead 1962). The most frequent causes of hollowback appear to be

- a compensatory increase in the lumbar angle secondary to the development of roundback, and
- a strength imbalance between the muscles that control the position of the pelvis.

Weakness in the abdominal and buttock muscles may result in an increase in the pelvic tilt angle. This, in turn, causes an increase in the sacral angle and, therefore, an increase in the lumbar angle (figure 6.28). When an increase in the lumbar angle is due to muscle imbalance, normal lumbar curvature can usually be completely restored by a suitable exercise program (Keim 1982). Hollowback is often associated with pain that may arise from a variety of sources (figure 6.29).

In addition to the varieties of roundback and hollowback, there are two other distinct types of abnormal curvatures in the median plane: swayback and flatback (see figure 6.25, *e* and *f*). Swayback is characterized by a backward tilt of the trunk about the hips so that the abdomen protrudes anteriorly. It is usually associated with a forward head position and involves increases in cervical lordosis, thoracic kyphosis, lumbar lordosis, pelvic tilt angle, and sacral angle. Flatback is characterized by a flattening of the thoracic and lumbar curves and involves a decrease in thoracic kyphosis, lumbar lordosis, pelvic tilt angle, and sacral angle.

Abnormal Curvature in the Frontal Plane

In an upright posture the vertebral column is normally straight in the frontal plane (figure 6.30*a*). Abnormal curvature of the vertebral column in the frontal plane is referred to as **scoliosis**. A scoliosis may be unilateral (one sided) or bilateral (both sides). There are three main forms of unilateral scoliosis, namely, thoracic,

Scoliosis: abnormal curvature of the vertebral column in the frontal plane; it may be unilateral or bilateral

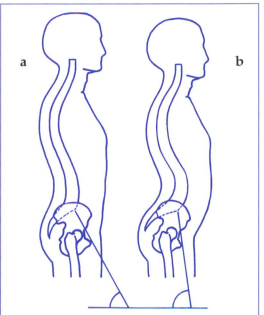

Figure 6.28   Relationship between degree of lumbar lordosis and pelvic tilt angle in upright standing posture; (*a*) normal orientation of vertebral column and pelvis; (*b*) increased degree of lumbar lordosis associated with an increase in pelvic tilt angle.

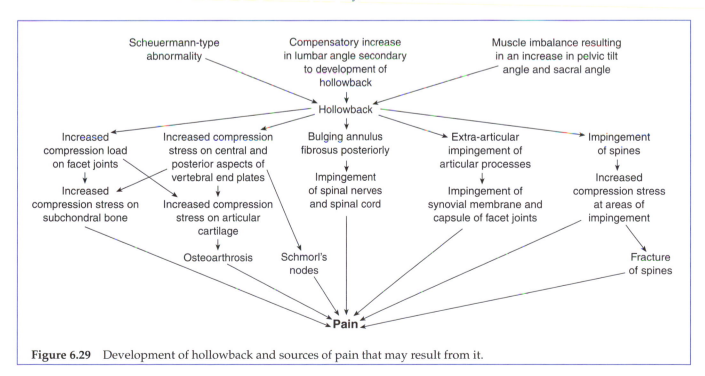

Figure 6.29   Development of hollowback and sources of pain that may result from it.

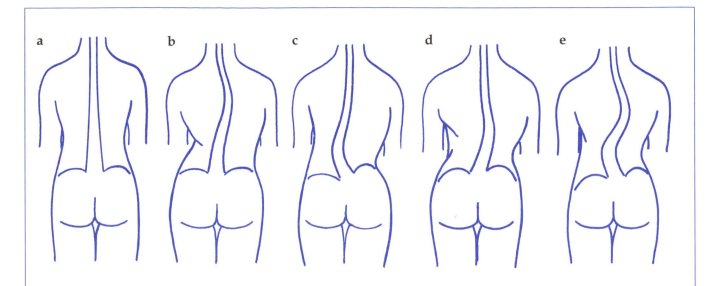

Figure 6.30   Forms of scoliosis; (*a*) normal orientation of vertebral column; (*b*) right thoracic scoliosis; (*c*) left lumbar scoliosis; (*d*) right thoracolumbar scoliosis; (*e*) bilateral scoliosis: right thoracic and left lumbar.

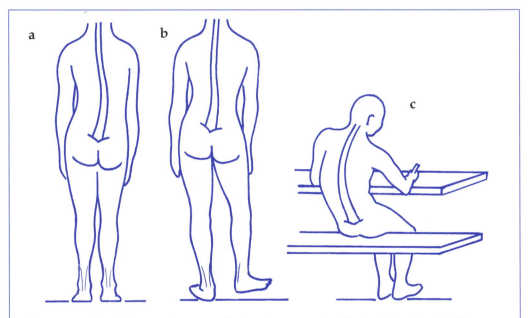

Figure 6.31   Possible contributors to the development of scoliosis; (*a*) leg length difference; (*b*) poor standing posture; (*c*) poor sitting posture.

lumbar, and thoracolumbar (figure 6.30, *b*, *c*, and *d*). In a bilateral scoliosis there are two compensatory curves, which bend in opposite directions to each other (figure 6.30*e*). Lumbar scoliosis and bilateral scoliosis are often associated with marked pelvis obliquity.

Scolioses are assessed by the Cobb method using the upper and lower vertebrae in the curve as the limits of the curve. In bilateral scoliosis both curves are assessed. A Cobb angle of less than 10° is regarded as within the bounds of normality. Approximately 2% of the adult population has some degree of scoliosis, but it exceeds 20° in less than 0.05% (Keim 1982). Scoliosis usually develops during the growth period and, whereas the incidence of scoliosis is similar in boys and girls, the incidence of severe scoliosis—a Cobb angle of 30° to 40°, is five to eight times greater in girls than in boys (Keim 1982).

In some cases of scoliosis the underlying cause may be a leg length inequality that tilts the pelvis in the frontal plane (Giles and Taylor 1981). This situation usually results in a unilateral lumbar or thoracolumbar scoliosis (figure 6.31a). In the absence of a leg length inequality, poor posture when standing and sitting may contribute to the development of scoliosis (Perdriolle and Vidal 1985) (figure 6.31, b and c). However, in 80% of cases of scoliosis the underlying cause is unknown—idiopathic (no known cause) scoliosis (Richter et al. 1985). In these cases the underlying cause is thought to be due to one of the following:

1. Abnormal ossification of the vertebral bodies, which results in abnormalities similar to those associated with Scheuermann's roundback, wedge-shaped vertebrae, and Schmorl's nodes (Keim 1982; Giles and Taylor 1982).

2. An imbalance in the levels of contraction in the muscles on opposite sides of the vertebral column due to malfunctioning in the nervous system (Haderspeck and Schultz 1981; Schultz, Haderspeck, and Takashima 1981; Ford et al. 1988).

3. A combination of skeletal abnormality and malfunctioning in the nervous system. It has been shown that in children with idiopathic scoliosis, the height/width ratio of the vertebral bodies is greater than in children without the condition; children with idiopathic scoliosis have more slender vertebral columns (Skoglund and Miller 1981). The increase in slenderness makes the vertebral column more vulnerable to slight changes in loading. For example, whereas a slight imbalance in the levels of contraction in the muscles on opposite sides of the vertebral column may have no effect on a normal vertebral column, the same imbalance may result in an excessive bending load on an abnormally slender vertebral column (Schultz, Haderspeck, and Takashima 1981).

A scoliosis tends to develop in three phases:

1. Occurrence of slight scoliosis

2. A period of rapid progression of the scoliosis

3. A period of more gradual progression of the scoliosis up to maturity when the scoliosis stabilizes (Perdriolle and Vidal 1985)

Scolioses are classified according to the age of the individual at the onset of phase 2—the period of rapid progression. There are three categories (Perdriolle and Vidal 1985):

1. Infantile scoliosis: phase two occurs in the period between birth to six years of age

2. Juvenile-pubertal scoliosis: phase two occurs in the period between six years of age and the onset of puberty

3. Pubertal (adolescent) scoliosis: phase two occurs in the period between the onset of puberty and maturity

In cases of juvenile-pubertal and pubertal scoliosis, there may be a considerable period of time between phases one and two in which the degree of scoliosis remains at a fairly low level. The degree of scoliosis at maturity depends on the age of the individual at the onset of phase two; the earlier the onset of phase two, the greater the amount of skeletal growth remaining, and the greater the final degree of scoliosis (Bunnell 1986). Consequently, infantile scoliosis often results in the greatest deformity and pubertal scoliosis tends to result in the least deformity. Fortunately, pubertal scoliosis is the most common form of scoliosis (Richter et al. 1985). Scoliosis often gives rise to pain, especially after maturity. Figure 6.32 summarizes the development of scoliosis and the sources of pain that arise from it.

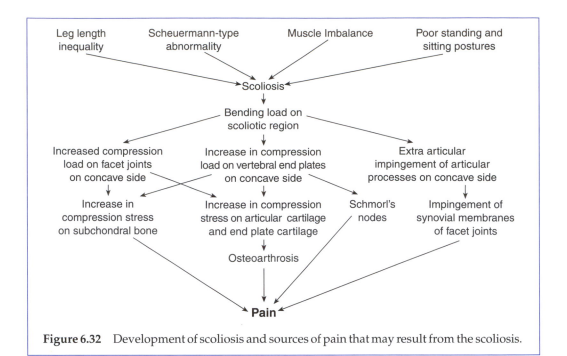

Figure 6.32   Development of scoliosis and sources of pain that may result from the scoliosis.

Joints of the Pelvis: Sacroiliac Articulations and Pubic Symphysis

The axial skeleton articulates with the lower limbs via the sacroiliac articulations—the joints between the sacrum and the innominate bones (see figure 3.31; figure 6.33). The sacroiliac articulations are partly synovial and partly syndesmosis. The joints between the auricular surfaces of the sacrum and innominate bones—usually referred to as sacroiliac joints—are synovial in structure. Located immediately posterior to each sacroiliac (SI) joint is a short, strong interosseous ligament between sacrum and ilium, which constitutes a syndesmosis (figure 6.34).

The sacrum is directly responsible for transmitting loads between the vertebral column and pelvis across the lumbosacral and SI articulations. Like adjacent motion segments in the vertebral column, the lumbosacral motion segment and SI articulations are functionally interdependent. Malfunction in one SI articulation is likely to overload not only the lumbosacral motion segment, but also the other SI articulation. Similar malfunction in the lumbosacral motion segment is likely to overload the SI articulations (Grieve 1976; Wilder, Pope, and Frymoyer 1980).

Sacroiliac Joints

The SI joints have many of the structural characteristics of a typical synovial joint—the auricular (articular) surfaces are covered with a form of articular cartilage, there is a small joint cavity containing synovial fluid, and the whole joint is lined on the inside with a synovial membrane (Bowen and Cassidy 1981). Each joint is also supported by a number of ligaments. However, there are a number of distinct differences between the SI joints and a typical synovial joint in terms of articular cartilage, articular surfaces, and range of movement.

The type of articular cartilage on the sacral surface of a SI joint is different from that on the iliac surface. The limited amount of information available suggests that the sacral articular cartilage resembles hyaline cartilage, whereas the iliac articular

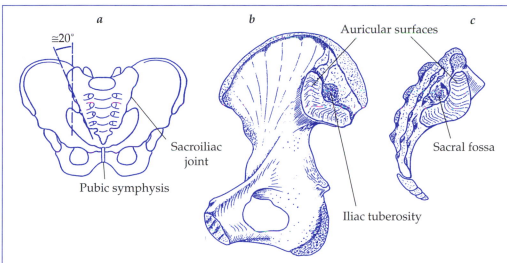

Figure 6.33   Sacroiliac joint; (*a*) anterior aspect of male pelvis showing angle of inclination of sacroiliac joint with respect to median plane; (*b*) medial aspect of right innominate bone; (*c*) right lateral aspect of sacrum.

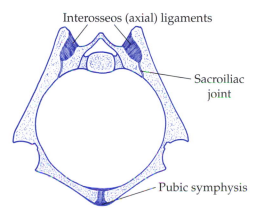

Figure 6.34   Oblique section through the pelvis—the pubic symphysis and angles of the sacroiliac joints—that shows the interosseous ligaments.

cartilage resembles fibrocartilage, although both may change with age (Bowen and Cassidy 1981). The sacral articular cartilage is about three times thicker than the iliac articular cartilage, 6 mm versus 2 mm in a normal adult.

At birth the articular surfaces are fairly smooth and flat, allowing sliding movements in virtually all directions. Consequently, at this stage the SI joints resemble synovial joints. During the first decade the surfaces remain fairly flat, but become less smooth as the difference in texture between the sacral and iliac cartilage becomes more pronounced. After puberty, each of the articular surfaces begins developing an irregular, uneven contour. Prior to the start of the third decade the sacral articular surface becomes depressed along its length, whereas the iliac articular surface develops a reciprocal elevation along its length (see figures 6.33, *b* and *c,* and 6.34). The sacral depression and iliac elevation are sometimes referred to as the sacral groove and iliac ridge (Bowen and Cassidy 1981). The actual articular surfaces gradually develop a complex irregular pattern of grooves, ridges, eminences, and depressions that reciprocate with each other only in part (Grieve 1976; Wilder, Pope and Frymoyer 1980). The irregular and uneven articular surfaces are atypical of ordinary synovial joints. Furthermore, whereas ordinary synovial joints are usually designed to allow a large range of movement, the partial interlocking of the SI articular surfaces severely restricts movement. In old age the SI joint cavity often becomes obliterated with a gradual increase in fibrocartilaginous adhesions, and synostosis may occur (Williams et al. 1995).

The overall shape of the auricular surfaces varies between a C shape and an L shape (see figure 6.33, *b* and *c*). For descriptive purposes, each auricular surface is often divided into cranial and caudal portions. With the sacrum in the anatomical position, the cranial portion points upward and, in some cases, slightly backward, while the caudal portion points backward and, in some cases, slightly downward (see figure 6.33, *b* and *c*). The region between the cranial and caudal portions is sometimes referred to as the angle of the auricular surface.

The sacrum is wedge shaped; it tapers from top to bottom such that on average the plane of each SI joint makes an angle of approximately 20° to the median plane when viewed from the front (figure 6.33*a*). The configuration of the SI joints is such that any downward force on the sacrum, such as the downward load of upper body weight when the trunk is upright, is resisted by the ligaments supporting the SI joints, which, in turn, stabilizes the SI joints.

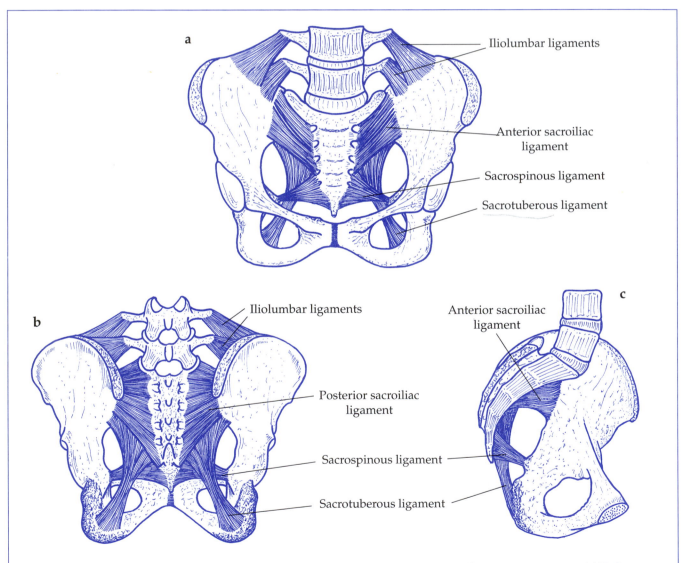

Figure 6.35   Supporting ligaments of the sacroiliac articulations; (*a*) anterior aspect; (*b*) posterior aspect; and (*c*) left aspect of medial section through the pelvis.

Syndesmoses and Supporting Ligaments of the SI Articulations

The syndesmosis of each SI articulation consists of a short but broad and, therefore, strong ligament called the interosseous ligament or axial ligament. The interosseous ligament spans the gap between the sacrum and ilium posterior to the SI joint (see figure 6.34). In this region the sacral surface is dominated by the large sacral fossa, and the iliac surface is dominated by the large iliac tuberosity (see figure 6.33). The area of articulation of the interosseous ligament is approximately 50% greater than that of the SI joint (22 cm^2 versus 14 cm^2 on average in adults) and reflects the importance of the interosseous ligament to the strength of the SI articulation (Miller, Schultz, and Anderson 1987). In addition to the interosseous ligament, there are three groups of ligaments that support the SI joints. The anterior and posterior SI ligaments provide direct support, whereas the iliolumbar ligaments, above the SI joint, and the sacrospinous and sacrotuberous ligaments, below the SI joint, provide indirect support. The anterior SI ligaments run downward and medially across the front of the SI joint. They arise from the anterior aspect of the ilium adjacent to the SI joint and attach onto the upper two-thirds of the anterior lateral aspect of the sacrum (figure 6.35a). The posterior SI ligaments run across the posterior aspect of the SI joint. The ligaments arise from the posterior medial aspect of the iliac crest and the adjoining notch between the posterior superior and posterior inferior iliac spines. The ligaments run medially and fan out to attach onto the tubercles of the intermediate and lateral sacral crests (figure 6.35b).

The iliolumbar ligaments run laterally downward and slightly forward from the transverse processes of L4 and L5 to the anterior medial superior aspect of the ilium above the SI joint (see figure 6.35). The sacrospinous and sacrotuberous ligaments join the sacrum to the ischium. The sacrospinous ligament is thin and triangular; it arises from the ischial spine and fans out medially superiorly and posteriorly to attach on to the lateral border of the coccyx and lower third of the sacrum (figure 6.35, a and c). The sacrotuberous ligament is very strong. Laterally, it is attached to the medial aspect of the tuberosity and ramus of the ischium. From this broad attachment the ligament converges to form a thick central band that passes behind the sacrospinous ligament. The central band then fans out to gain attachment to the superior medial aspect of the greater sciatic notch and lateral border of the sacrum below the SI joint. The fibers of the posterior sacroiliac, sacrospinous, and sacrotuberous ligaments blend with each other to a considerable extent.

> *The axial skeleton articulates with the lower limbs via the sacroiliac articulations—the sacroiliac joints and the interosseous syndesmoses. The sacroiliac articulations are supported by five groups of ligaments: anterior and posterior sacroiliac ligaments, iliolumbar ligaments, sacrospinous ligaments, and sacrotuberous ligaments.*

Pubic Symphysis

The pubic symphysis or interpubic joint is a symphysis joint that lies in the median plane (see figures 6.33a and 6.34; figure 6.36). The elliptical medial ends of the pubic bones (see figure 6.33b) are covered with hyaline cartilage, and a disc of fibrocartilage is sandwiched between the layers of hyaline cartilage (figure 6.36). In men the disc is approximately 4 mm thick and has a cross-sectional area in the median plane of approximately 5 cm^2. Consequently, the amount of movement allowed by deformation of the disc is very small. Strong ligaments on the superior, anterior, inferior, and posterior aspects of the joint further restrict movement in the joint. The superior ligament is attached to the pubic crest, that is, the region between the pubic tubercles.

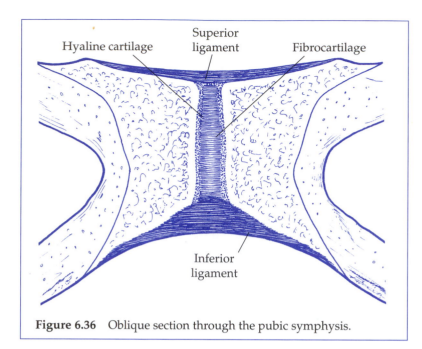

Figure 6.36   Oblique section through the pubic symphysis.

The posterior ligament runs across the posterior aspect of the joint and is fairly thin. The inferior or arcuate (arch) ligament, on the other hand, is thick. It attaches to the pubic arch and blends with the disc. The anterior ligament is also thick and blends with the tendons and aponeuroses of the abdominal muscles. The combination of relatively thin disc and strong ligaments produces a strong joint with a minimal range of movement.

The pubic symphysis stabilizes the pelvis and, in particular, the SI joints, and at the same time provides a certain amount of shock absorption in response to loads transmitted across the pelvis. Under normal circumstances the stability of the SI joints is maintained largely by the combined clamping effect of the pubic symphysis and the ligamentous support posterior to the SI joints (see figure 6.34). Thus, a decrease in the stability of the pubic symphysis reduces the clamping effect on the SI joints and, therefore, reduces the stability of the SI joints. Any reduction in the stability of the pubic symphysis results in abnormal transmission of loads across the SI joints. This, in turn, may result in pain as a consequence of impingement and stretching of spinal nerves or progressive degeneration of the pubic symphysis and SI joints.

In men and young women the pubic symphysis and, therefore, the SI joints are normally very stable. However, in adult women the stability of the pubic symphysis and SI joints is decreased during the latter part of the menstrual cycle, during the menopause, and during pregnancy (Grieve 1976; Don Tigny 1985; Colliton 1996). Approximately 7 to 10 days prior to the start of menstruation there is a hormone-induced softening and lengthening of the ligaments of the pubic symphysis and SI articulations. Consequently, the joints become less stable and more susceptible to injury. During menstruation, the ligaments regain their normal length and toughness to restore normal stability of the joints. During pregnancy the ligaments of the pubic symphysis and SI articulations gradually soften and lengthen. In addition, the cartilage of the joints gradually becomes thicker due to absorption of water. Consequently, the joints become much more flexible, an essential feature of normal vaginal childbirth, but the stability of the joints is considerably reduced. After delivery the stability of the joints is gradually restored over a period of 6 to 12 weeks (Grieve 1976).

The pubic symphysis stabilizes the pelvis, particularly the sacroiliac joints, and provides shock absorption in response to loads transmitted across the pelvis.

Movements of the Sacroiliac Joints

In men the SI joints are normally stable. As previously described, the stability of the SI joints in women is reduced at certain times by hormonal changes. However, at other times the stability of the SI joints in women is similar to that in men. The function of the SI articulations is to transmit loads between the sacrum and ilia. This is achieved most efficiently when the SI joints are stable and, as such, have a limited range of movement. A high level of stability not only facilitates shock absorption in response to impact loading, but also ensures that all of the main components of the SI articulations (articular cartilages, syndesmoses, supporting ligaments) participate fully in the transmission of load so that none of the components is overloaded. Inadequate stability of the SI joints can overload certain components of the SI articulations during load transmission.

In response to impact loads that result, for example, from heel strike during walking and running, the SI articulations act as shock absorbers in concert with the other joints of the vertebral column and the pubic symphysis to absorb the energy of the impact. Shock absorption in the SI articulations occurs in a manner similar to that in intervertebral joints—in response to impact loading a small amount of movement occurs in the SI joint, which is strongly resisted by the interosseous ligament and other supporting ligaments. Energy is absorbed by deformation of the articular cartilages and tension in the interosseous ligament and other supporting ligaments (Wilder, Pope, and Frymoyer 1980). Degeneration of the SI joints and consequent loss of shock-absorbing capacity may overload the lumbar intervertebral joints; this is reflected in the increased incidence of degenerative joint disease in the lumbar region following degeneration of the SI joints (Wilder, Pope, and Frymoyer 1980).

It is generally agreed that a small amount of movement takes place in the SI joints in response to nonimpact changes in loading (Wilder, Pope, and Frymoyer 1980; Colachis et al. 1963). For example, in a completely relaxed recumbent posture, as shown in figure 6.37a, upward pressure on the base of the sacrum rotates the sacrum backward relative to the ilia about a transverse axis; the coccyx moves upward and, to a lesser extent, the sacral promontory moves downward (relative to figure 6.37, a and b). In upright stance the weight of the upper body (head, arms, and trunk) is exerted downward on the sacrum via the lumbosacral joint rotating the sacrum forward about a transverse axis relative to the ilia; in moving from a recumbent position to upright standing, the sacral promontory moves forward and downward and the coccyx moves upward and backward (Kapandji 1974; Don Tigny 1985) (figure 6.37, c and d).

Authors disagree regarding the position of the axis of rotation of the sacrum; some regard the interosseous ligament (axial ligament) as the axis of rotation, hence the name of the ligament, whereas others maintain that the axis of rotation passes through the angle of the auricular surfaces of the SI joint (Bowen and Cassidy 1981; Kapandji 1974). Both assertions are probably correct to a certain extent since it is likely that the rotation of the sacrum is accompanied by some linear movement (Wilder, Pope, and Frymoyer 1980). In this case the axis of rotation is not fixed, but tends to move in the region between the axial ligament and the angle of the auricular surfaces.

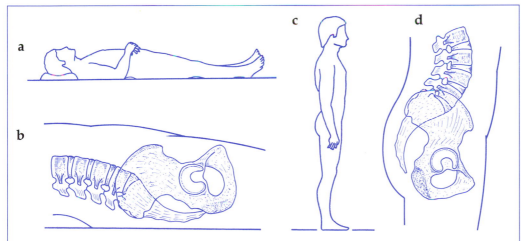

Figure 6.37 Movement of the sacrum relative to the ilia when moving between recumbent and standing postures; (a) recumbent posture; (b) position of sacrum relative to ilia in recumbent posture; (c) standing posture; (d) position of sacrum relative to ilia in standing posture.

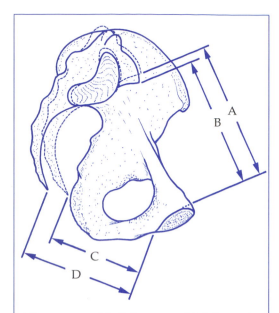

Figure 6.38 Medial aspect of left innominate bone and right lateral aspect of the sacrum showing the position of the sacrum relative to the innominate bone in nutation (solid outline) and counternutation (dotted outline) of the pelvis. A and B show the anteroposterior diameter of the pelvic brim in counternutation and nutation. C and D show anteroposterior diameter of the pelvic outlet in counternutation and nutation.

In view of the large variation in the size, shape, and contour of the SI joints among individuals, each individual probably has a unique pattern of SI movement consisting of certain amounts of linear and angular movement. Nevertheless, it is generally agreed that forward rotation of the sacrum is the dominant feature of SI joint movement in response to the downward force of upper body weight in the upright stance. The forward rotation of the sacrum results in movement of the innominate bones such that the posterior aspects of the ilia are drawn together and the ischial tuberosities move farther apart. The forward rotation of the sacrum and resulting movement of the innominate bones is referred to as nutation of the pelvis (Kapandji 1974) (nutare = to nod). Nutation results in a decrease in the anteroposterior diameter of the pelvic brim and an increase in the anteroposterior and transverse diameters of the pelvic outlet (figure 6.38). Nutation is usually severely restricted by the ligaments that support the SI joints. Counternutation is the reverse of nutation—backward rotation of the sacrum resulting in separation of the posterior aspects of the ilia and drawing together of the ischial tuberosities. Counternutation occurs in the recumbent supine position with the hips extended (figure 6.37, *a* and *b*). Counternutation results in an increase in the anteroposterior diameter of the pelvic brim and a decrease in the anteroposterior diameter of the pelvic outlet (figure 6.38). Like nutation, counternutation is usually severely restricted by the ligaments that support the SI joints.

The degree of nutation and counternutation increases considerably during pregnancy due to the decrease in stability of the SI and pubic symphysis joints. During childbirth the increased degree of nutation and counternutation facilitates the passage of the fetus along the birth canal. During the early stages of labor the woman is usually encouraged to adopt a recumbent supine position with the hips extended as in figure 6.37, *a* and *b*. The counternutation of the pelvis that results from this position enlarges the pelvic brim and favors the descent of the fetal head into the pelvis (Kapandji 1974). During the expulsive phase of labor the woman is often encouraged to flex the hips and knees. In this position, tension in the hamstring muscles in the back of the thighs rotates the innominate bones backward relative to the sacrum thereby resulting in nutation of the pelvis. The effect of nutation is to enlarge the pelvic outlet (Kapandji 1974).

The sacroiliac joints are usually very stable. However, small amounts of movement (combination of linear and angular motion) occur in the sacroiliac joints in response to changes in loading. These movements are referred to as nutation and counternutation of the pelvis.

Abnormalities of the Sacroiliac Joints and Pubic Symphysis

The anterior branches of the fourth and fifth lumbar nerves, which form part of the sacral plexus, pass in front of the SI joints (figure 6.39). Branches from these nerves innervate the joint capsules of the SI joints. Excessive movement of an SI joint stretches the joint capsule, which, in turn, stretches the capsular nerves, resulting in pain. Excessive movement in an SI joint may also cause pain due to stretching and impingement of the sacral spinal nerves. Pain from these sources is referred to as sacroiliitis and is often misdiagnosed as some form of lower back pain (Don Tigny 1985). Sacroiliitis can occur whenever the stability of the SI joints is decreased. For reasons described earlier, adult women are much more vulnerable to sacroliitis than men. This is reflected in a study of 222 sport-related hip and pelvic injuries in 204 patients; the incidence of sacroiliitis was found to be 15.6% and 6.4% in women and men, respectively (Lloyd-Smith et al. 1985). The pelvic pain that accompanies pregnancy and especially labor and childbirth is largely due to sacroiliitis resulting from the increased range of SI joint movement (Hainline 1994; Endresen 1995).

The contours of the auricular surfaces of the SI joints are irregular and in the normal stable condition partially interlock with each other. In this situation, the movement of the SI joints that occurs in response to normal levels of impact loading and nonimpact changes of loading is usually insufficient to disrupt the normal pattern of interlocking or to cause sacroiliitis. However, in an unstable joint, excessive movement may cause the contours of the auricular surfaces to interlock in an abnormal position, which will be reinforced by tension in the supporting ligaments created by the excessive movement. This situation, a form of dislocation of the SI joint, may result in continuous pain (chronic sacroiliitis). The most common form of this condition is called anterior dysfunction of the SI joint in which the SI joint becomes locked in an abnormal position with the innominate bone rotated anteriorly on the sacrum, that is, in a position of counternutation (Don Tigny 1985).

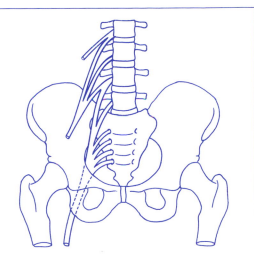

Figure 6.39 Anterior aspect of the pelvis showing the locations of the lumbar and sacral nerves in relation to the sacroiliac joints.

Anterior dysfunction occurs as a result of standing postures in which the trunk is inclined forward; for example, lifting with straight legs, making a bed, ironing, washing dishes, and shaving. In these situations the innominate bones rotate forward on the sacrum (counternutation), but the amount of rotation is usually small due to the stability of the SI joints and the abdominal and gluteal muscles, which stabilize the innominate bones. However, when the SI joints are unstable, the amount of rotation may be excessive, especially if the abdominal and gluteal muscles are weak. In these circumstances one or both SI joints may become locked in an abnormal position resulting in unilateral (one SI joint) or bilateral (both joints) anterior dysfunction. Anterior dysfunction can cause chronic sacroiliitis. In addition, since the amount of movement in an SI joint affected by anterior dysfunction is virtually nil, the ability of the joint to assist in shock absorption is severely impaired. Consequently, the adjacent lumbar motion segments are overloaded, which increases the risk of low-back pain. In some cases a change in posture may dislodge the anterior dysfunction in the affected joint(s). However, in other cases the anterior dysfunction may be aggravated by certain events, for example:

1. Women are particularly susceptible to anterior dysfunction just prior to and during menstruation. If the anterior dysfunction persists after menstruation the likelihood of self-correction is considerably reduced, since the ligaments of the SI articulations and pubic symphysis shorten, thereby tending to stabilize the anterior dysfunction.

2. After anterior dysfunction has occurred, there is an increased likelihood of the sacrum slipping downward on the ilium in the affected joint. When this happens the innominate bone is prevented from rotating backward on the sacrum, which prevents self-correction of the anterior dysfunction. This condition occurs when there is a sudden increase in the vertical load exerted on the SI joint as, for example, when lifting an object, stepping down heavily, or sitting down heavily.

Coughing and sneezing may increase the pain of sacroiliitis when anterior dysfunction persists. Manipulation therapy is usually required to correct persistent anterior dysfunction.

Instability in the pubic symphysis reduces the stability of the SI joints and, therefore, increases the likelihood of anterior dysfunction. Except for when hormonal changes in women reduce SI joint stability, the pubic symphysis is usually a very stable joint in women and men and dislocation of the pubic symphysis rarely occurs (Kapandji 1974). Consequently, when dislocation or separation of the pubic symphysis does occur, it is, like a prolapsed intervertebral disc, a serious injury that is difficult to treat and that may not heal completely.

The pubic symphysis and ipsilateral SI joint are subjected to considerable shear loading in any activity involving a one-legged stance such as walking, running, jumping, and landing. It is not surprising that repetitive high-level shear loading on the pubic symphysis can cause microtears in the disc and supporting ligaments, giving rise to a painful condition called osteitis pubis (Hanson, Angevine, and Juhl 1978). Just as sacroiliitis is often misdiagnosed as low-back pain, osteitis pubis is often misdiagnosed as muscle strain. In the study of sport-related hip and pelvic injuries by Lloyd-Smith et al. (1985), the incidence of osteitis pubis in men and women was found to be 7.9% and 4.2%, respectively. During the second decade of life, a median cleft often develops in the rear part of the disc of the pubic symphysis, which seems to be the long-term effect of repetitive high-level shear loading.

> *Excessive movement of a sacroiliac joint stretches the joint capsule and impinges sacral spinal nerves, resulting in sacroiliitis. The pelvic pain accompanying pregnancy, and especially labor and childbirth, is largely due to sacroiliitis. Excessive movement in the pubic symphysis will likely cause microtears in the disc and supporting ligaments resulting in osteitis pubis.*

Summary

This chapter described the structure and function of the joints of the axial skeleton and pelvis. These joints are well designed—both in terms of the structure of individual joints and the shape of the vertebral column—to transmit large forces and to provide shock absorption at the expense of flexibility. The next chapter describes the structure and function of the joints of the appendicular skeleton.

Review Questions

1. Differentiate between a motion segment and an intervertebral joint.
2. Describe the structure and function of an intervertebral joint and a facet joint.
3. Describe the variation in size and shape of intervertebral discs.
4. Explain how disc thickness is affected by external loading and osmotic swelling pressure.
5. Describe the response of the annulus fibrosus and nucleus pulposus of an intervertebral disc to movement (about a transverse axis) in a motion segment.
6. Describe the effects of flexion and extension of the vertebral column on loading in facet joints.
7. Describe the ligaments of the vertebral column.
8. Describe the various factors that limit flexibility in different regions of the vertebral column.
9. Explain how back pain may arise in the absence of structural damage.
10. Describe the effects of aging and chronic loading on intervertebral discs and the sources of pain that may result.
11. Differentiate between spondylolysis and spondylolisthesis.
12. Differentiate between functional and anatomical descriptions of upright posture.
13. Describe the structure and functions of the sacroiliac articulations and the pubic symphysis.
14. Differentiate between counternutation of the pelvis and anterior dysfunction of a sacroiliac joint.

Chapter 7

Joints of the Appendicular Skeleton

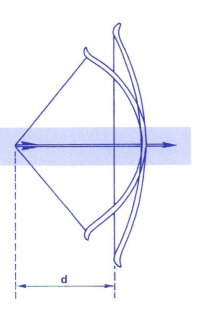

The larger joints of the axial skeleton (intervertebral and pelvic joints) are mainly cartilaginous and designed primarily to provide shock absorption at the expense of flexibility. In contrast, virtually all of the joints of the appendicular skeleton are synovial, or mainly synovial, and designed primarily to facilitate flexibility at the expense of shock absorption. This chapter describes the structure and function of the joints of the appendicular skeleton.

Objectives

After reading this chapter you should be able to do the following:

1. Define or describe the key terms.
2. Differentiate between joints and joint complexes.
3. Describe the structure and function of the shoulder complex.
4. Describe the structure and function of the elbow complex.
5. Describe the structure and function of the wrist complex.
6. Describe the structure and function of the hip joint.
7. Describe the structure and function of the knee complex.
8. Describe the structure and function of the ankle complex.

Joints and Joint Complexes

Most whole body movements involve simultaneous movement in many different joints. Whereas the range of movement patterns adopted by different individuals to carry out a particular task may be broad, each movement pattern reflects a certain degree of functional interdependence between the joints used; movement in a particular joint is associated with simultaneous movements in other joints, especially those in the same skeletal chain such as the hip, knee, and ankle joints, and the shoulder, elbow, and wrist joints. For some groups of joints, such as the intervertebral joints discussed in chapter 6, the functional interdependence is clear, that is, the movement of the vertebral column always involves simultaneous movement in a number of motion segments. For others, such as the group consisting of the shoulder, acromioclavicular, and sternoclavicular joints, the degree of functional interdependence is less clear. For example, whereas small movements of the shoulder joint, such as in writing at a desk, may occur without simultaneous movement in the acromioclavicular and sternoclavicular joints, most movements of the arm about the shoulder joint involve simultaneous movement in all three joints.

Joint complex: a group of joints with a relatively high degree of functional interdependence

A group of joints with a relatively high degree of functional interdependence is called a **joint complex** (Peat 1986). With respect to the following descriptions of joints and joint complexes in the appendicular skeleton, the values given for ranges of movement refer to normal healthy untrained individuals. As described in chapter 5, the flexibility of a joint largely depends on the length range in which the muscles that control the joint normally function. Consequently, trained individuals may have much greater ranges of movement in certain joints than untrained individuals.

Shoulder Complex

Shoulder complex: the shoulder, acromioclavicular, and sternoclavicular joints

The **shoulder complex** consists of the shoulder (glenohumeral), acromioclavicular, and sternoclavicular joints, which link together the humerus, scapula, clavicle, and sternum in a chain (figure 7.1a). Loads transmitted from the upper limb toward the axial skeleton are transmitted directly via the shoulder, acromioclavicular, and sternoclavicular joints, and indirectly by ligaments and muscles supporting the shoulder complex and by muscles linking the shoulder complex to the axial skeleton.

The shoulder complex, in association with the muscles that link it to the axial skeleton, provides the upper limb with a range of movement exceeding that of any other joint or joint complex (Peat 1986). The large range of movement is due to four sources of relative motion in the shoulder complex: the shoulder, acromioclavicular and sternoclavicular joints, and the **scapulothoracic gliding mechanism**. The latter refers to the ability of the scapula to glide and rotate relative to the posterior aspect of the rib cage. The scapulothoracic gliding mechanism, combined with movements in the acromioclavicular and sternoclavicular joints, enables the glenoid fossa to follow the head of the humerus and, thereby, maintain maximum congruence of the shoulder joint in movements of the humerus relative to the axial skeleton; this includes all movements of the arm that involve medium- to full-range extension, flexion, abduction, and adduction of the shoulder complex. Consequently, the range of movement of the humerus relative to the axial skeleton is normally much greater than the range of movement of the shoulder joint (the range of movement of the humerus relative to the scapula). Normal movement of the humerus and, therefore, of the upper limb, depends on normal functional interrelationships between the four sources of relative motion. Restriction of normal movement in any of the four sources of relative motion can impair normal function of the upper limb.

The shoulder complex and the scapulothoracic gliding mechanism provide the upper limb with a range of movement exceeding that of any other joint or joint complex.

Scapulothoracic gliding mechanism: the system of muscles that enables the scapula to glide and rotate relative to the posterior aspect of the rib cage

Shoulder Joint

The shoulder joint is a synovial ball and socket joint. The head of the humerus is almost hemispherical and articulates with the glenoid fossa, which is shallow and relatively small; the articular surface of the glenoid fossa is approximately one-third the size of that of the head of the humerus (Williams et al. 1995).

Due to the relatively small and shallow contact area between the articulating surfaces of the shoulder joint and the relatively large loads transmitted across the shoulder joint in forceful actions of the upper limb, the shoulder joint is vulnerable to dislocation. Dislocations of the joint, especially anterior dislocation (dislocation of the head of the humerus forward of the glenoid fossa), are common in contact sports (Hawkins and Mohtadi 1994). The size of the area of articulation is increased by a ring of fibrocartilage called the glenoid labrum (labrum = lip) around the edge of the glenoid fossa (figure 7.1b). The glenoid labrum considerably deepens the glenoid fossa and, thus, increases the congruence and stability of the joint.

The joint is enclosed by a joint capsule strengthened on its anterior aspect by three capsular ligaments—the superior, middle, and inferior glenohumeral ligaments (figure 7.1a). The superior aspect of the joint is supported by the coracohumeral ligament, which consists of two bands linking the lateral aspect of the base of the coracoid process to the superior aspects of the lesser tuberosity (anterior band) and greater tuberosity (posterior band) (figure 7.1c). Although the coracohumeral and glenohumeral ligaments, together with the rest of the capsule, help stabilize the joint in certain joint positions, the stability of the joint is largely maintained by four muscles collectively referred to as the rotator cuff (figure 7.2). The muscles originate from the subscapular fossa (subscapularis), supraspinous fossa (supraspinatus), and infraspinous fossa (infraspinatus and teres minor) and converge on the proximal end of the humerus to attach onto the lesser tuberosity (subscapularis) and greater tuberosity (supraspinatus, infraspinatus, and teres minor). In addition to stabilizing the shoulder joint, the rotator cuff assists other muscles that move the shoulder joint and shoulder girdle.

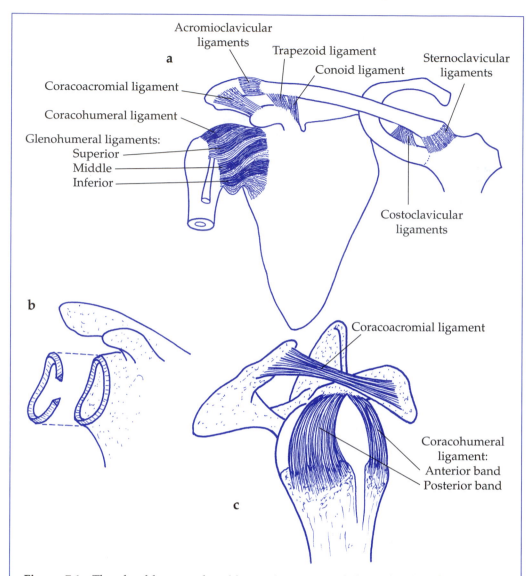

Figure 7.1   The shoulder complex; (*a*) anterior aspect of the right shoulder complex showing the main ligaments; (*b*) glenoid labrum; (*c*) lateral aspect of right shoulder showing coracoacromial ligament and coracohumeral ligaments.

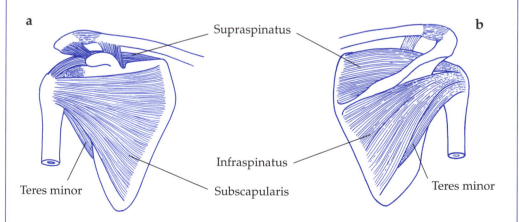

Figure 7.2   (*a*) Anterior aspect and (*b*) posterior aspect of the right rotator cuff.

The shoulder joint is a synovial ball and socket joint. The glenoid fossa is small and shallow relative to the head of the humerus, but the area of articulation is increased by the glenoid labrum. The joint capsule is strengthened on its anterior aspect by the glenohumeral ligaments and on its superior aspect by the coracohumeral ligament, but joint stability is largely maintained by the rotator cuff.

The coracoacromial arch restricts shoulder joint movements. The arch is formed by the acromion process, coracoid process, and the coracoacromial ligament linking the two bony processes (see figures 7.1, *a* and *c*). Impingement of the surgical neck of the humerus on the coracoid process and anterior aspect of the coracoacromial ligament restricts flexion. Extension is restricted by impingement of the surgical neck of the humerus on the acromion process. Abduction is restricted by impingement of the greater tuberosity of the humerus on the coracoacromial ligament.

Acromioclavicular Joint

The acromioclavicular joint is a synovial gliding joint and usually has an incomplete articular disc or one or two menisci (figure 7.3). The capsule of the acromioclavicular joint is strengthened superiorly by a capsular ligament called the superior acromioclavicular ligament. The acromioclavicular joint is indirectly stabilized by the coracoclavicular ligaments—the trapezoid and conoid ligaments, which join the clavicle to the coracoid process close to the acromioclavicular joint (see figure 7.1*a*). The fibers of the trapezoid ligament run downward and medially from the inferior aspect of the clavicle to the superior aspect of the coracoid process. The fibers of the conoid ligament run vertically from the inferior aspect of the clavicle to the superior aspect of the base of the coracoid process.

Due to the oblique orientation of the articular surfaces of the acromioclavicular joint (figure 7.3*a*), forces transmitted across the shoulder joint toward the axial skeleton often dislocate the acromioclavicular joint by driving the acromion process under the lateral end of the clavicle. Such forces, which occur, for example, as a result of a punching action or falling on an outstretched hand, are normally strongly resisted by the coracoclavicular ligaments, especially the trapezoid. However, dislocation of the acromioclavicular joint is not uncommon, especially in contact sports, usually due to a fall or a hit (Taft, Wilson, and Oglesby 1987).

The acromioclavicular joint is a synovial gliding joint. The joint is indirectly stabilized by the coracoclavicular ligaments, which resist forces transmitted across the shoulder joint that tend to dislocate the acromioclavicular joint. Dislocation of the acromioclavicular joint is not uncommon, especially in contact sports.

Sternoclavicular Joint

The sternoclavicular joint is a synovial joint that has structural and functional characteristics of both saddle and gliding joints. Basically, the sternoclavicular joint has two degrees of freedom: rotation in the coronal plane, as in raising the shoulders, and rotation in a transverse plane, as in moving the shoulders forward. The capsule of the sternoclavicular joint is strengthened on its posterior, superior, and anterior aspects by capsular ligaments. The joint is indirectly stabilized by the costoclavicular ligaments, which join the clavicle to the costal cartilage of the first rib close to the sternoclavicular joint (see figure 7.1*a*).

The fibers of the anterior band of the costoclavicular ligament run downward and medially from the inferior aspect of the clavicle to the costal cartilage and adjoining bony portion of the first rib. The fibers of the posterior band of the costoclavicular

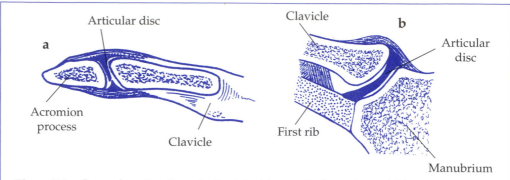

Figure 7.3   Coronal section through the right (*a*) acromioclavicular and (*b*) sternoclavicular joints.

ligament run downward and laterally from the inferior aspect of the clavicle to the costal cartilage and adjoining bony portion of the first rib. The sternoclavicular joint has a complete articular disc, which is attached superiorly to the superior aspect of the medial end of the clavicle, inferiorly to the costal cartilage of the first rib, and to the joint capsule for the remainder of its circumference (Williams et al. 1995; see figure 7.3*b*). This attachment enables the disc, assisted by the joint capsule and the posterior band of the costoclavicular ligament, to strongly resist forces transmitted along the clavicle toward the axial skeleton, which can dislocate the sternoclavicular joint by driving the medial end of the clavicle over the manubrium. The strength of this mechanism is such that dislocation of the sternoclavicular joint is rare; fracture of the clavicle is more common (Williams et al. 1995).

> *The sternoclavicular joint is a synovial joint with characteristics of both saddle and gliding joints. The joint has a complete articular disc, which, together with the joint capsule and costoclavicular ligaments, strongly resists forces that could dislocate the joint. Dislocation of the sternoclavicular joint is rare.*

Movements During Abduction of the Upper Limb

Due to the various restrictions on the movement of the joints of the shoulder complex, most upper limb movements that involve more than 30° of abduction of the arm with respect to the trunk are brought about by a combination of movement in two or more of the four sources of movement in the shoulder complex (Peat 1986). Consequently, lack of flexibility in any of these components can impair the ability to abduct the arm. Abduction is an essential feature of many overhead arm actions such as reaching upward to a high shelf, changing a bulb in a ceiling light, putting on your hat, washing and combing your hair, pitching a baseball, throwing a football, serving in tennis, throwing a javelin, spiking and blocking in volleyball, bowling in cricket, and all of the major swimming strokes (Miniaci and Fowler 1993; Copeland 1993).

The first 30° of abduction of the arm is brought about almost entirely by abduction of the shoulder joint (figure 7.4, *a* and *b*). The next phase of abduction of the arm—from 30° to approximately 100°—utilizes a combination of shoulder abduction and rotation of the scapula and clavicle as a single unit about an oblique axis passing through the sternoclavicular joint and the region of the medial end of the spine of the scapula. During this phase, shoulder abduction contributes about 30° and rotation of the shoulder girdle (sternoclavicular joint and scapulothoracic mechanism) contributes about 40° (figure 7.4*c*).

At this point the costoclavicular ligaments become taut and prevent further upward rotation of the clavicle about the sternoclavicular joint. The final phase of abduction of the arm—from 100° to approximately 180°—uses a combination of shoulder abduction, rotation of the scapula about an anteroposterior axis through the acromioclavicular joint, and axial rotation of the clavicle (counterclockwise with respect to the right clavicle when viewed from a lateral aspect). Rotation of the scapula about the acromioclavicular joint is limited to about 20° by the coracoclavicular ligaments, which then become taut.

Consequently, the other 60° in this final phase is brought about by a combination of shoulder abduction, axial rotation of the clavicle, and scapulothoracic movement (rotation and sliding). During this final phase, shoulder abduction is accompanied by external rotation of the shoulder so the articular surfaces of the joint remain in close contact with each other and the greater tuberosity of the humerus is prevented from impinging on the coracoacromial arch (figure 7.4d).

Impingement of any supporting structures of the shoulder joint (capsule, ligaments, and rotator cuff) on the coracoacromial arch results in a painful condition generally referred to as **shoulder impingement** (Miniaci and Fowler 1993). Shoulder impingement can occur in any form of overhead arm movement when flexibility in one or more components of the shoulder complex is limited (Brunet, Haddad, and Porche 1982; Silliman and Hawkins 1991). Whereas tight shoulder ligaments may restrict flexibility in these joints, shoulder impingement is more likely to occur due to lack of extensibility in some of the muscles of the shoulder complex, especially those concerned with the scapulothoracic mechanism. Not surprisingly, shoulder impingement is fairly common in sports such as swimming and tennis, but it can also affect everyday movements such as washing and combing one's hair (Miniaci and Fowler 1993).

> *Lack of flexibility in the shoulder complex and lack of extensibility in the muscles that control the scapulothoracic gliding mechanism can impair abduction of the upper limb and result in shoulder impingement.*

Shoulder impingement: a painful condition that results from impingement of one or more of the supporting structures of the shoulder joint (capsule, ligaments, and rotator cuff) on the coracoacromial arch

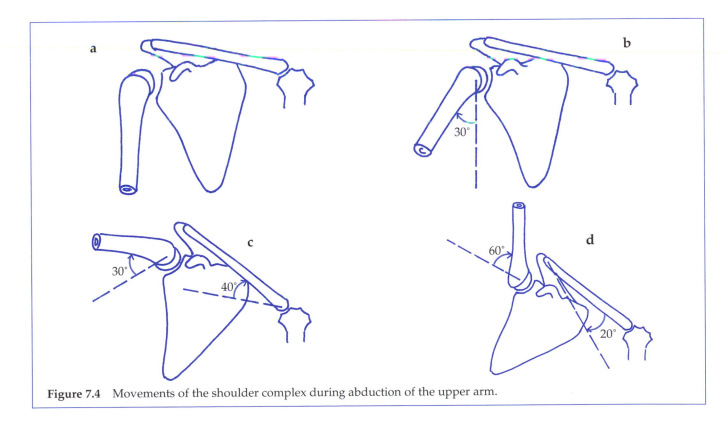

Figure 7.4   Movements of the shoulder complex during abduction of the upper arm.

Elbow Complex

Elbow complex: the humero-ulnar joint and the humeroradial joint

The **elbow complex** is synovial and consists of the humero-ulnar joint (between the trochlea and trochlea notch) and the humeroradial joint (between the capitulum and the head of the radius) (see figures 3.25 and 3.26). The humero-ulnar joint is a typical uniaxial hinge joint designed for flexion and extension. The humeroradial joint is biaxial—it can rotate about two axes:

- **A vertical axis**, the long axis of the radius, as in pronation and supination of the forearm. In this type of movement the head of the radius spins on the capitulum.
- **A transverse axis**, as in flexion and extension of the elbow. In this type of movement the head of the radius slides on the capitulum.

The elbow complex is enclosed within a joint capsule strengthened on all aspects by capsular ligaments (figure 7.5). The fibers of the capsule blend with the annular ligament, which holds the head of the radius against the radial notch of the ulna (figure 7.5a). The anterior aspect of the capsule helps prevent hyperextension of the elbow and is strengthened by the lateral and medial oblique ligaments (figure 7.5a). The posterior aspect helps prevent hyperflexion and is strengthened by transverse, lateral oblique, and medial oblique ligaments (figure 7.5b). Fan-shaped medial and lateral ligaments strengthen the medial and lateral aspects of the capsule, respectively. The medial ligament arises from the medial epicondyle of the humerus and is characterized by three distinct bands—the anterior, intermediate (middle), and posterior bands (figure 7.5c). A transverse ligament also strengthens the medial aspect of the capsule. The lateral ligament arises from the lateral epicondyle of the humerus and, like the medial ligament, is characterized by anterior, intermediate (middle), and posterior bands (figure 7.5d). The medial and lateral ligaments to maximize congruence of the elbow complex in all positions of the flexion and extension range and, thereby, prevent abduction and adduction of the elbow. The normal range of flexion and extension in the elbow is approximately 140°.

The elbow complex is synovial and consists of the humero-ulnar joint (uni-axial), which facilitates flexion and extension of the elbow, and the humeroradial joint (biaxial), which facilitates flexion and extension of the elbow and pronation and supination of the forearm. Capsular ligaments strengthen all aspects of the elbow complex capsule.

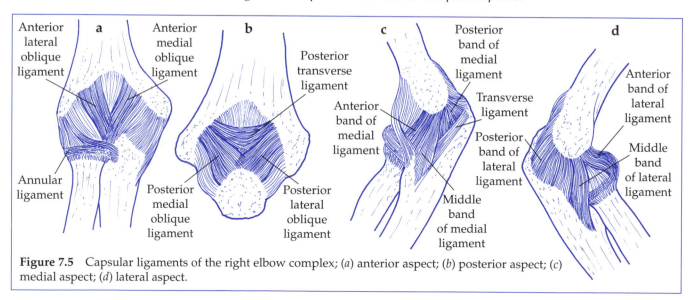

Figure 7.5   Capsular ligaments of the right elbow complex; (*a*) anterior aspect; (*b*) posterior aspect; (*c*) medial aspect; (*d*) lateral aspect.

Wrist Complex

The **wrist complex** consists of the wrist joint and the midcarpal joint. The wrist joint, a synovial ellipsoid joint, is between the distal ends of the radius and ulna and the proximal surfaces of the scaphoid, lunate, and triquetrum (figure 7.6*a*). The radius articulates directly with the scaphoid and lunate. However, an articular disc separates the ulna from the lunate and triquetrum. The articular disc is attached to the anterior lateral aspect of the styloid process of the ulna and the inferior medial aspect of the radius just below the ulna notch. The articular surface of the radius is continuous with the lower surface of the articular disc; together they form a roughly elliptical biconcave surface that articulates with the reciprocal biconvex surface formed by the scaphoid, lunate, and triquetrum. Another articular disc, sometimes referred to as a meniscus, links the styloid process of the ulna, the triquetrum, and the pisiform. The midcarpal joint consists of the series of synovial gliding joints between the proximal and distal rows of carpals (figure 7.6*a*).

The capsule of the wrist joint follows the line of the joint and is separate from that of the midcarpal joint. The capsule of the midcarpal joint is more extensive than that of the wrist joint; it has a very irregular shape and incorporates not only the intercarpal joints of the midcarpal joint, but also the intercarpal joints directly linked to the midcarpal joint.

Stability of the Wrist

The wrist complex is stabilized by

- interosseous ligaments (ligaments that link adjacent carpals) between some of the carpals,
- more extensive ligaments that span various sections of the carpus, and
- muscles that move the carpus.

On the anterior aspect of the carpus the main ligaments are the anterior (palmer) radiocarpal and ulnocarpal ligaments (figure 7.6*b*). The anterior radiocarpal ligament has two oblique bands. The inferior band passes downward and medially from the anterior inferior border of the styloid process of the radius to attach onto the capitate. The superior band passes downward and medially from the anterior lateral border of the distal end of the radius (medial to the attachment of the inferior band) to attach onto the triquetrum. The anterior (palmer) ulnocarpal ligament also has two bands, which originate from the anterior lateral aspect of the styloid process of the ulna. The superior and inferior bands fan out to attach onto the lunate and capitate, respectively.

On the posterior aspect of the carpus the main ligaments are the posterior (dorsal) radiocarpal, posterior (dorsal) transverse, medial, and lateral ligaments (figure 7.6*c*). The posterior radiocarpal ligament consists of superior and inferior bands arising from a common origin on the triquetrum. The bands fan upward and laterally to attach onto the posterior border of the distal end of the radius. The transverse ligament arises from the triquetrum; it divides into three bands as it passes laterally to attach onto the scaphoid, trapezoid, and trapezium. The medial ligament originates from the inferior aspect of the styloid process of the ulna; it fans out as it passes almost directly downward to attach onto the pisiform and triquetrum. The lateral ligament originates from the posterior inferior aspect of the styloid process of the radius; it fans out as it passes almost directly downward to attach onto the trapezoid and trapezium.

With respect to the anatomical position, the lateral border of the wrist joint is inferior to the medial border such that lateral displacement of the carpus at the wrist

Wrist complex: the wrist and midcarpal joints

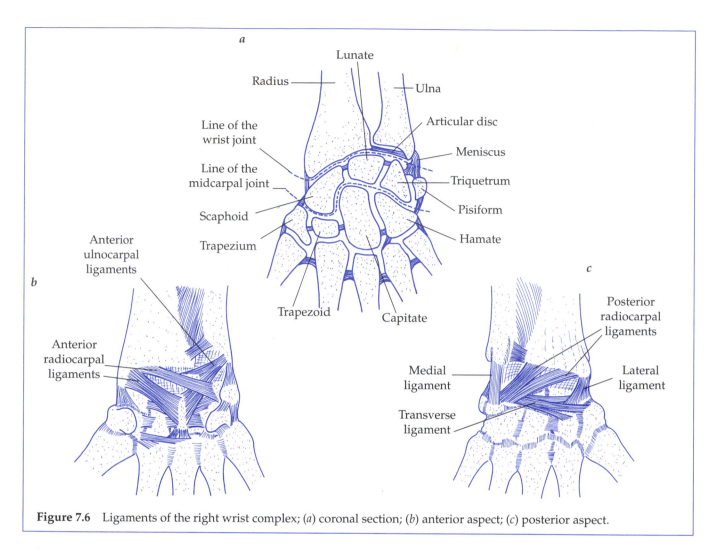

Figure 7.6 Ligaments of the right wrist complex; (*a*) coronal section; (*b*) anterior aspect; (*c*) posterior aspect.

joint is considerably restricted by the styloid process of the radius (figure 7.6*a*). In comparison, medial displacement of the carpus at the wrist joint is relatively unrestricted. Stability of the carpus in this direction appears to be one of the main functions of the obliquely oriented anterior and posterior radiocarpal ligaments, which restrict movement of the carpus medially (and downward) with respect to the wrist joint.

Movements of the Wrist

The main movements in the wrist complex are flexion, extension, abduction, adduction, and circumduction. All of these movements occur as a result of a complex combination of movements in the wrist and midcarpal joints. With respect to the anatomical position, the range of flexion and extension are both limited to approximately 85°. The posterior ligaments restrict flexion, and the anterior ligaments restrict extension. Abduction is restricted to approximately 15° by the oblique orientation of the wrist joint and by the medial ligament. In comparison, adduction is relatively unrestricted and has a 45° range of motion. Whereas movements of the wrist complex beyond any of the normal ranges can result in injury, injury occurs most often as a result of excessive abduction, usually in combination with extension. Such movements may result in fracture of the scaphoid or styloid process of the radius (Recht, Burk, and Dalinka 1987).

The wrist complex is synovial and consists of the wrist joint (basically ellipsoid) and midcarpal joint (series of gliding joints). The wrist complex is stabilized by interosseous ligaments between some of the carpals, more extensive ligaments that span various sections of the carpus, and muscles that move the carpus. It facilitates circumduction of the hand.

Hip Joint

The hip joint is a synovial ball and socket joint. The head of the femur is slightly more than a hemisphere and articulates with the relatively deep acetabulum. The acetabular notch is spanned by a transverse ligament that is flush with the acetabular rim. The acetabular rim and transverse ligament form a ring-shaped articular surface with the inner border of the ring recessed with respect to the outer border like a countersunk metal washer. The recessed acetabular fossa forms the center of the ring. The acetabulum is deepened considerably by a ring of fibrocartilage around its edge and inferior border of the transverse ligament. The ring of fibrocartilage is called the acetabular labrum and is structurally and functionally similar to the glenoid labrum in the shoulder. The hip joint is enclosed within a cylindrical joint capsule that is strengthened by capsular ligaments. The anterior aspect is strengthened by three ligaments, namely, the superior iliofemoral, inferior iliofemoral, and pubofemoral ligaments (figure 7.7a). The posterior aspect of the capsule is strengthened by the ischiofemoral ligament (figure 7.7b). Unlike the shoulder joint, the hip joint has an intracapsular ligament called the ligamentum teres (figure 7.7c), which links

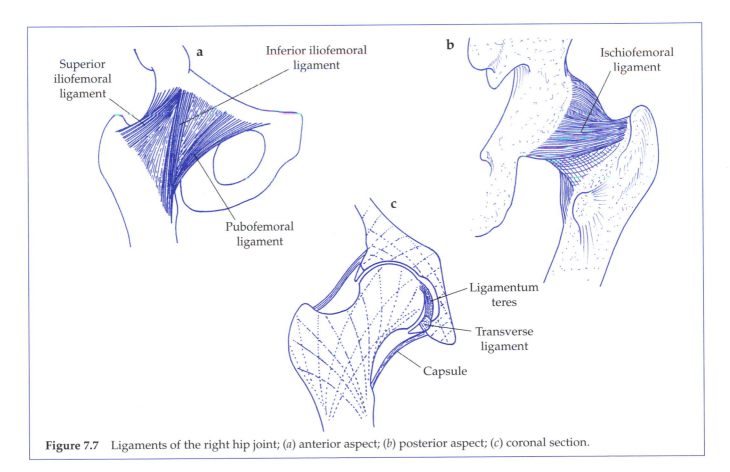

Figure 7.7 Ligaments of the right hip joint; (*a*) anterior aspect; (*b*) posterior aspect; (*c*) coronal section.

together the head of the femur (at a point called the fovea) and the transverse ligament. As the head of the femur slides on the acetabular rim, the ligamentum teres moves within the acetabular fossa.

The shape of the articular surfaces together with the capsule, capsular ligaments, and intracapsular ligament makes the hip joint a very stable joint. In comparison to the shoulder, the hip joint is far more stable, but much less flexible. With respect to the anatomical position, all of the capsular ligaments of the right hip joint run in a clockwise direction from the innominate bone to the femur when viewed from the lateral aspect. Similarly, all of the capsular ligaments of the left hip joint run in a counterclockwise direction from the innominate bone to the femur when viewed from the lateral aspect. This is thought to be the result of evolution from a quadruped posture to an upright posture (Williams et al. 1995). The orientation of the ligaments is such that they quickly become taut during hip extension and slacken during hip flexion. Consequently, the ligaments restrict hip extension but not hip flexion. However, the range of flexion and extension of the hip also depends on the angle of the knee joint. The hamstring and quadriceps muscles cross over both the hip and knee joints (see figure 2.5). With the knee extended, hip extension is limited to approximately 30° by tightness in the quadriceps as well as in the capsular ligaments, and hip flexion is limited to approximately 70° by tightness in the hamstrings. With the knee flexed, hip extension is reduced to virtually nil by tightness in the quadriceps, and hip flexion is increased to approximately 120° due to relaxation of both muscle groups as well as relaxation of the ligaments. The range of hip flexion is usually much greater than that of hip extension regardless of the angle at the knee. In this respect the hip joint is similar to the shoulder joint in that the flexion range is much greater than the extension range (with respect to the anatomical position).

During abduction of the hip the pubofemoral and ischiofemoral ligaments become taut and the superior iliofemoral ligament slackens off; the normal range of hip abduction is approximately 40°. During adduction of the hip the pubofemoral and ischiofemoral ligaments slacken and the superior iliofemoral ligament becomes taut; the normal range of hip adduction is approximately 30°. During lateral rotation of the hip all of the anterior ligaments become taut and the ischiofemoral ligament

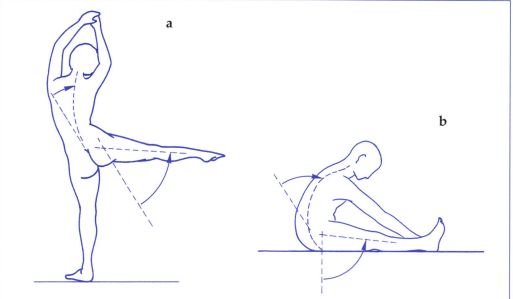

Figure 7.8 Actions involving movement in the hip joint and vertebral column; (*a*) hip abduction and lateral flexion of the vertebral column; (*b*) hip flexion and flexion of the vertebral column.

slackens; the normal range of lateral rotation is approximately 45°. During medial rotation of the hip all of the anterior ligaments slacken and the ischiofemoral ligament becomes taut; the normal range of medial rotation is approximately 30°.

Just as movements involving the shoulder joint are associated with movements of the shoulder girdle, movements of the hip joint are associated with movements of the vertebral column. For example, abduction of the hip is usually associated with lateral flexion of the trunk, as in standing on one leg with the free leg abducted (figure 7.8a). Similarly, hip flexion is usually associated with trunk flexion as in a sit-up exercise (figure 7.8b). Consequently, reduced flexibility in the hip is partially compensated by movement in the vertebral column. To this extent, the hip joints and joints of the vertebral column may be regarded as a functional group similar to the shoulder complex.

Injuries to the hip joint and its ligamentous supporting structures occur far less frequently than do injuries to the shoulder, knee, and ankle (Lloyd-Smith et al. 1985; Weiker and Munnings 1994). This probably reflects, at least in part, the high level of stability of the hip joint relative to the shoulder, knee, and ankle. Musculoskeletal pain from the hip region is usually the result of injury to structures associated with hip joint movement rather than the hip joint itself; these injuries include sacroiliitis, osteitis pubis, stress fractures of the femoral neck, bursitis, and muscle or tendon strains (Brody 1980).

> *The hip joint is a synovial ball and socket joint and the area of articulation is increased considerably by the acetabular labrum. The hip joint is supported internally by the ligamentum teres and externally by a cylindrical joint capsule strengthened anteriorly by the iliofemoral and pubofemoral ligaments, and posteriorly by the ischiofemoral ligament. The hip joint is more stable, but less flexible, than the shoulder joint.*

Knee Complex

The **knee complex** consists of two joints: the tibiofemoral joint and the patellofemoral joint (figure 7.9). Both joints are synovial and share the same joint capsule. The movement of the tibiofemoral joint largely determines patellofemoral joint movement. Abnormal movements in the tibiofemoral joint tend to cause abnormal movements in the patellofemoral joint. In situations involving dynamic weight bearing, the load on the two joints is usually high, and even slightly abnormal movements can overload parts of the joints (Huberti and Hayes 1984). Consequently, the incidence of injury to the knee complex is relatively high in certain groups of individuals such as those involved in games and sports and in manual occupations that require lifting heavy weights (LaBrier and O'Neill 1993).

Knee complex: the tibio-femoral joint and patello-femoral joint

> *The knee complex is synovial and consists of the tibiofemoral joint (modified hinge) and the patellofemoral joint (gliding joint), which share the same joint capsule.*

Tibiofemoral Joint

Whereas the tibial condyles are virtually flat, the femoral condyles have a convex pulley shape. Consequently, from a skeletal point of view the tibiofemoral joint is very unstable. However, this lack of skeletal stability is compensated to a considerable extent by two menisci, strong extracapsular ligaments, and powerful muscles. The shape and orientation of the menisci and ligaments enable the tibiofemoral joint to function as a modified hinge joint with flexion and extension as the principal plane of movement.

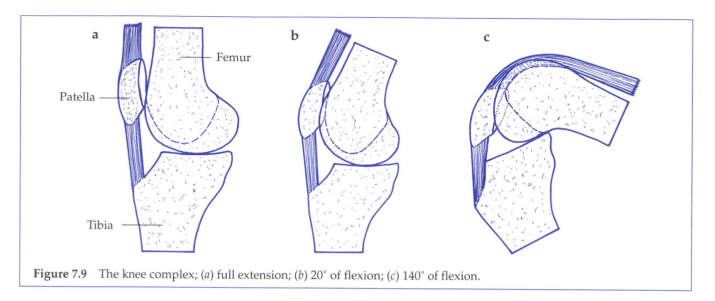

Figure 7.9   The knee complex; (*a*) full extension; (*b*) 20° of flexion; (*c*) 140° of flexion.

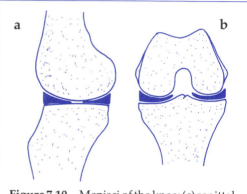

Figure 7.10   Menisci of the knee; (*a*) sagittal section; (*b*) coronal section.

Menisci of the Knee

At any particular joint position the area of contact between the articular surfaces of the femoral and tibial condyles is relatively small (see figure 7.9). However, the effective area of contact is normally increased by three to four times by the presence of two C-shaped menisci, one medial and one lateral, which form wedges between the peripheral incongruent articular surfaces of the femoral and tibial condyles (figure 7.10). The horns (or ends) of the menisci are attached to the intercondylar areas of the tibial table. The anterior horn of the lateral meniscus is attached to the posterior lateral aspect of the anterior intercondylar area (figure 7.11*a*). The posterior horn of the lateral meniscus is attached to the anterior lateral aspect of the posterior intercondylar area. The anterior horn of the medial meniscus is attached to the anterior medial aspect of the anterior intercondylar area. The posterior horn of the medial meniscus is attached to the anterior aspect of the posterior intercondylar area. The two menisci are joined anteriorly by a transverse ligament (figure 7.11*b*). The periphery of each meniscus is attached to the inner wall of the joint capsule. However, the articular surfaces of the menisci are free; they are in contact with the articular surfaces of the femoral and tibial condyles, but free to slide on them. The ability of the menisci to slide on both articular surfaces is essential for normal joint function. The menisci have four main functions:

1. To maintain congruence between the articular surfaces in all positions of the joint. During joint movement the menisci deform in response to the changing curvature of the femoral condyles and thus maintain congruence between the articular surfaces in all positions of the joint. The increase in congruence provided by the menisci results in an increased joint stability and also minimizes the compressive stress on the articular cartilages.

2. To provide shock absorption in the joint. The menisci deform under loading to help absorb the shock of impact loads on the joint.

3. To maintain a circulation of synovial fluid through the articular cartilages. The structure of articular cartilage is similar to that of a sponge, with a dense network of tiny interconnecting pores. During joint movement the joint reaction force compresses the articular cartilages, squeezing synovial fluid out of the

cartilage. During periods of relaxation, when the joint reaction force is reduced, the viscoelasticity of the cartilage restores it to its unloaded size and shape and synovial fluid flows into the cartilage.

Even in joints that do not normally have menisci or articular discs, the articular surfaces are not perfectly congruent. Consequently, at any particular joint position some parts of the articular surfaces are under greater compression load than others. Furthermore, the parts of the articular surfaces under the greatest compression load change as the joint position changes. In this way joint movement creates a circulation of synovial fluid through the articular cartilages; this enables the cartilage cells to receive nutrients and to get rid of waste products. The menisci of the knee help to maintain the circulation of synovial fluid through the articular cartilages of the tibial and femoral condyles.

4. To help bring about normal movement between the articular surfaces. Knee flexion from full extension involves two main phases: rolling only followed by simultaneous rolling and sliding (Kapandji 1970; Norkin and Levangie 1992). The two phases are reversed when the joint is extended. For the medial condyle, the rolling-only phase corresponds approximately to the first 15° of flexion, whereas

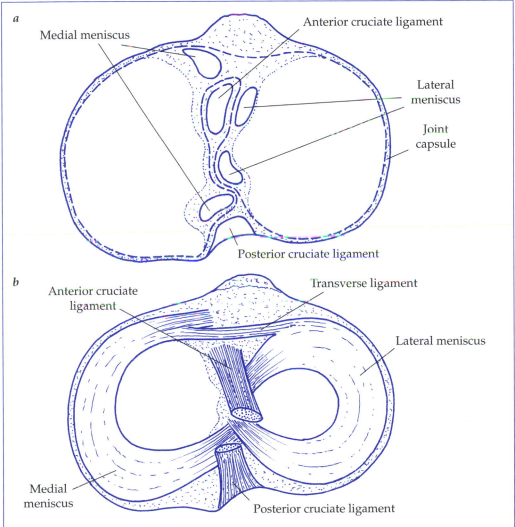

Figure 7.11   Menisci and cruciate ligaments; superior aspect of right tibial table showing the location of (*a*) the attachments of the menisci and cruciate ligaments and the outline of the joint capsule; (*b*) the menisci and cruciate ligaments.

for the lateral condyle, the rolling-only phase corresponds approximately to the first 20° of flexion. Consequently, the lateral condyle rolls farther than the medial condyle, resulting in the lower leg being internally rotated through about 20° during the rolling-only phase. The rolling-only phase at the start of knee flexion, which corresponds to the normal range of flexion and extension in ordinary walking, is sometimes referred to as unlocking the knee. The reverse movement, external rotation of the lower leg during the rolling-only phase of knee extension, is sometimes referred to as locking the knee or the screw-home mechanism.

The rolling-only phase at the start of knee flexion is followed by a phase in which rolling and sliding occur simultaneously, with sliding becoming more important as flexion progresses. The total range of knee flexion and extension is normally in the region of 160°. The ability of the menisci to deform and slide in response to the loads exerted on them during joint movement is essential for the maintenance of the normal pattern of rolling and sliding between the articular surfaces. In turn, the normal pattern of rolling and sliding is essential for maintaining normal congruence, shock absorption, and circulation of synovial fluid through articular cartilage.

The femoral condyles are convex and the tibial condyles are fairly flat. However, their effective area of contact is normally increased three to four times by medial and lateral menisci, which have four main functions:

- *To maintain congruence between the articular surfaces in all positions of the joint*
- *To provide shock absorption in the joint*
- *To maintain a circulation of synovial fluid through the articular cartilages*
- *To help bring about normal movement between the articular surfaces*

Injuries to menisci are common in the general population and especially in activities such as dancing (modern and ballet) and sports such as football (all types), basketball, baseball, wrestling, and skiing (Baker et al. 1985; Silver and Campbell 1985). The incidence of injury to the medial meniscus is approximately four times higher than for the lateral meniscus. In comparison to that of the lateral meniscus, the movement of the medial meniscus is rather restricted due to its firm attachment to the medial ligament via the joint capsule and the relatively large distance between its attachments to the tibial table. This restricted range of movement is thought to be responsible for the increased incidence of injury to the medial meniscus. Damage to the menisci may result from all types of abnormal movements of the knee, but especially from twisting the knee, that is, excessive axial rotation of the lower leg.

There are two main types of meniscal injury:

1. Tearing the attachments of the menisci to the tibial table and joint capsule
2. Crushing the menisci between the femoral and tibial condyles, which produces circular (bucket handle) and radial tears

These injuries are illustrated in figure 7.12. A meniscus or part of a meniscus may also become trapped in the intercondylar notch, which frequently results in the knee locking—the individual is unable to fully extend the knee. The only blood supply to the menisci is via the synovial membrane at the periphery. Furthermore, the menisci are usually under considerable load, especially when bearing weight, which may keep the torn parts of a damaged meniscus apart from each other. Consequently,

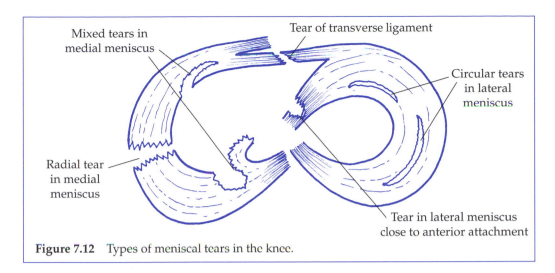

Mixed tears in
medial meniscus

Tear of transverse ligament

Circular tears
in lateral
meniscus

Radial tear
in medial
meniscus

Tear in lateral meniscus
close to anterior attachment

Figure 7.12 Types of meniscal tears in the knee.

repair of a damaged meniscus tends to be slow and may not take place at all. For this reason it used to be normal medical practice to remove a badly torn meniscus. However, in recent years, improved surgical techniques and a greater awareness of the importance of the menisci to knee function have resulted in a move toward partial rather than total meniscectomy whenever possible (Yates and Jackson 1984; Silver and Campbell 1985).

Capsule of the Knee

The tibiofemoral and patellofemoral joints share a common capsule with a complicated shape. The capsule is attached above to the femur, below to the tibia, and anteriorly to the patella (figure 7.13). To accommodate the extremes of full flexion and full extension, the anterior part of the capsule is folded upward during extension and the posterior part of the capsule is folded downward during flexion. During knee flexion the anterior part of the joint capsule straightens and pushes synovial fluid backward where it collects in a pouch formed by the slackened posterior part of the capsule. This pouch, only present during knee flexion, is called the gastrocnemius bursa since it intervenes between the posterior aspect of the knee and the gastrocnemius muscle. A bursa is a flattened sac of synovial membrane containing synovial fluid. Bursas minimize friction between structures that slide across each other during normal movement (see chapter 8).

During knee extension the posterior part of the joint capsule straightens—the gastrocnemius bursa gradually becomes smaller and disappears at full extension. As this occurs, the synovial fluid in the gastrocnemius bursa is pushed forward where it collects in a pouch formed by the slackened anterior part of the capsule. This pouch, only present during knee extension, is called the suprapatellar bursa. The suprapatellar bursa intervenes between the quadriceps tendon and the anterior aspect of the femur just proximal to the patellar surface.

The suprapatellar bursa is separated from the femur by a pad of fat called the suprapatellar fat pad. Lying in the space bounded by the upper two-thirds of the posterior aspect of the patellar ligament, the anterior intercondylar area of the tibia, and the anterior inferior aspect of the articular surface of the femoral condyles is a pad of fat called the infrapatellar fat pad. It is roughly triangular in sagittal cross section and is suspended superiorly from a fibro-adipose band called the infrapatellar fold. The infrapatellar fold is attached anteriorly to the inferior pole of the patella and posteriorly to the anterior border of the intercondylar notch (figure 7.13). The infrapatellar fat pad is attached anteriorly to the posterior aspect of the patellar ligament and extends at both sides of the patellar ligament. The infrapatellar fat pad

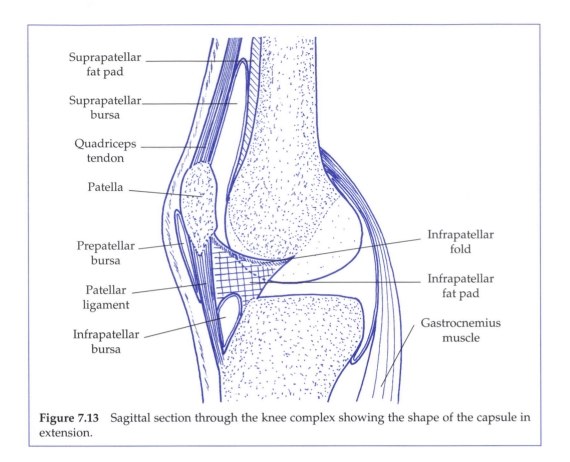

Figure 7.13   Sagittal section through the knee complex showing the shape of the capsule in extension.

also extends narrow branches halfway up each side of the patella; the branches are referred to as alar folds (alar = winglike). The infrapatellar fat pad cushions the patellar ligament and lower part of the patella during movements of the knee.

Knee Ligaments

Four extracapsular ligaments support the tibiofemoral joint. The lateral ligament (also referred to as the lateral collateral ligament and the fibular collateral ligament) is attached superiorly to the lateral epicondyle of the femur and inferiorly to the head of the fibula (figure 7.14*a*). The medial ligament (also referred to as the medial collateral ligament and the tibial collateral ligament) is attached superiorly to the medial epicondyle of the femur and inferiorly to the medial aspect of the tibia below the tibial condyle (figure 7.14*b*). The posterior fibers of the medial ligament blend with the joint capsule at the level of the medial meniscus and, as such, the medial ligament is not completely extracapsular. The arrangement of the medial and lateral ligaments and the curvature of the femoral condyles are such that both ligaments are relatively slack when the knee is flexed, but become progressively more taut as the joint extends. In a normal joint the ligaments are fully taut when the joint is fully extended. In this position of the joint both ligaments help to prevent hyperextension and lateral rotation of the tibia relative to the femur. In addition, the medial ligament and lateral ligament help to prevent abduction and adduction, respectively (see figures 5.23 and 5.24).

With the knee fully extended, the range of abduction and adduction is normally zero. Since the medial and lateral ligaments are relatively slack when the knee is flexed, a certain amount of axial rotation and abduction and adduction of the lower leg can occur when the knee is in a flexed, nonweight-bearing position. For example, when sitting on a table with the lower leg hanging freely, the ranges of internal and

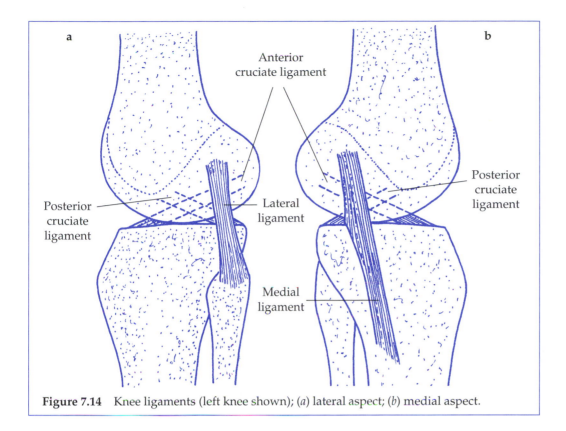

Figure 7.14   Knee ligaments (left knee shown); (*a*) lateral aspect; (*b*) medial aspect.

external rotation are normally around 30° and 40°, respectively (Norkin and Levangie 1992). However, the ranges of abduction and adduction are normally very small, approximately 2° to 5°.

Injuries to the lateral and, in particular, the medial ligaments are common in sports (Reider 1996). The classic cause of medial ligament injury is a blow to the lateral aspect of the knee while the foot is in contact with the ground resulting in abduction of the knee. This situation is common in sports such as American football, soccer, and rugby. The excessive loading on the medial ligament in such situations can result in serious injury (Indelicato 1995). Medial ligament injuries of a less severe but, nonetheless, persistent nature are common in swimmers, especially breaststroke swimmers. The whipkick leg action in breaststroke tends to simultaneously abduct and laterally rotate (tibia with respect to femur) the knee resulting in strain of the medial ligament and associated medial supporting structures (Vizsolyi et al. 1987).

The other two extracapsular ligaments that support the tibiofemoral joint are the anterior and posterior cruciate ligaments, which cross over each other in the intercondylar notch (cruciate = cross) (see figure 7.14). The distal end of the anterior cruciate ligament is attached to the posterior aspect of the anterior intercondylar area of the tibial table, medial to the attachment of the anterior horn of the lateral meniscus (see figure 7.11). The proximal end is attached to the posterior medial aspect of the lateral femoral condyle. The distal end of the posterior cruciate ligament is attached to the posterior aspect of the posterior intercondylar area of the tibial table, and the proximal end is attached to the anterior inferior lateral aspect of the medial femoral condyle. The anterior cruciate ligament runs laterally to the posterior cruciate ligament. Although the cruciate ligaments are located in the middle of the joint, they are extracapsular; the capsule is invaginated into the intercondylar notch where it encloses the menisci, but excludes the cruciate ligaments (see figure 7.11*a*).

The cruciate ligaments function to bring about normal movement between the articular surfaces of the femoral and tibial condyles. For example, during the rolling-sliding phase of knee flexion, the anterior cruciate ligament pulls on the femur so that the femoral condyles slide forward as they roll backward. Similarly, during the rolling-sliding phase of knee extension, the posterior cruciate ligament pulls on the femur so that the femoral condyles slide backward as they roll forward. The posterior cruciate ligament restricts forward movement of the femoral condyles and thus helps to prevent forward dislocation of the femur on the tibia (see figure 7.14). Similarly, the anterior cruciate ligament restricts backward movement of the femoral condyles and therefore helps to prevent posterior dislocation of the femur on the tibia. At full knee extension both cruciate ligaments become fully taut and therefore help the medial and lateral ligaments to prevent hyperextension. In addition, the orientation of the cruciate ligaments, the anterior ligament running laterally to the posterior ligament, is such that they help to prevent medial rotation of the tibia relative to the femur in full knee extension (see figure 7.11b). Consequently, the cruciate ligaments work with the lateral and medial ligaments to provide rotational stability of the knee. Damage to these ligaments causes excessive laxity in the tibiofemoral joint and considerably increases the likelihood of damage to the menisci (Gollehon, Torzilli, and Warren 1987).

> *The tibiofemoral joint is supported by four extracapsular ligaments: lateral, medial, anterior cruciate, and posterior cruciate. In a normal knee, all of the ligaments help to provide rotational stability and prevent hyperextension; in addition, the medial ligament and lateral ligament help to prevent abduction and adduction, respectively. Damage to the ligaments, especially the cruciate ligaments, causes excessive laxity and increases the likelihood of damage to the menisci.*

Patellofemoral Joint

The patellofemoral joint is a synovial gliding joint in which the patella slides up and down on the patellar surface and condyles of the femur during extension and flexion of the knee. In full knee extension, the superior half of the articular surface of the

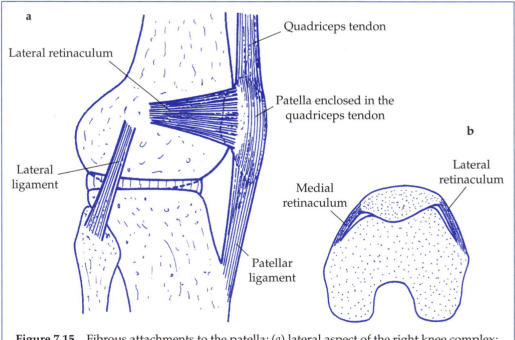

Figure 7.15   Fibrous attachments to the patella; (*a*) lateral aspect of the right knee complex; (*b*) transverse section through the patellofemoral joint.

patella rests against the suprapatellar bursa (see figure 7.13). During knee flexion, the patella slides down the intercondylar groove formed by the patellar surface and condyles of the femur. The area of contact between the patella and femur gradually increases up to about 90° of flexion, and then starts to decrease as the patella moves into the widest part of the intercondylar notch (see figure 7.9). From full knee extension to full knee flexion the patella slides on the femur for a distance of approximately 8 cm in the adult. The movement of the patella on the femur is determined by the shape of the articular surfaces of the patella and intercondylar groove and the various fibrous attachments to the patella. The articular surface of the patella is V-shaped and articulates reasonably well with the reciprocally shaped intercondylar groove of the femur for most of the range of flexion and extension of the knee. Consequently, in terms of the articular surfaces, normal movement of the patella involves sliding in a more or less sagittal plane.

There are four main fibrous attachments to the patella: quadriceps tendon, patellar ligament, and the medial and lateral retinacula (figure 7.15). The quadriceps tendon

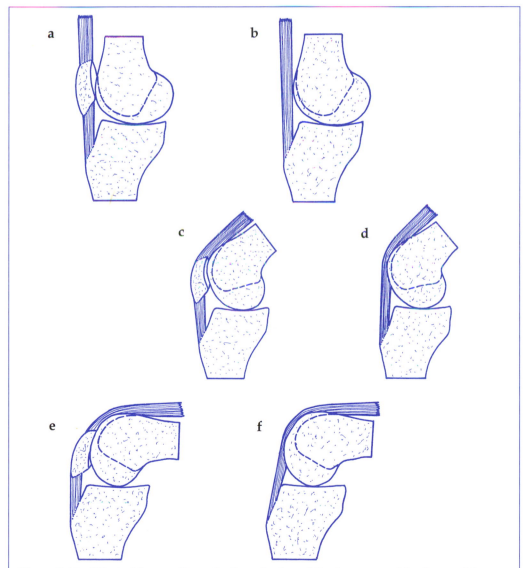

Figure 7.16   Effect of the patella on the line of action of the force exerted by the quadriceps; (*a, c*, and *e*) with patella; (*b, d*, and *f*) without patella; (*a* and *b*) full knee flexion; (*c* and *d*) knee flexed at 45°; (*e* and *f*) knee flexed at 90°.

and patellar ligament are continuous with each other and enclose the nonarticular parts of the patella. The distal end of the patellar ligament is attached to the tibial tuberosity of the tibia. The four quadriceps muscles, which form the bulk of the soft tissue in the anterior aspect of the thigh, attach onto the quadriceps tendon. The quadriceps muscles mainly function to extend the knee, and the patella functions to increase the mechanical efficiency of the quadriceps muscles. The quadriceps can still extend the knee if, for some reason, the patella has to be removed. However, in this case knee extension is relatively weak due to a reduction in the moment arm of the quadriceps. In addition, the absence of the patella results in the quadriceps tendon being compressed against the intercondylar groove (figure 7.16), which not only subjects the quadriceps tendon to friction during sliding, but also increases the compression stress on the intercondylar groove because of the reduced area of contact with the quadriceps tendon relative to contact with the patella. Consequently, the patella has three main functions:

1. To increase the mechanical efficiency of the quadriceps muscle group
2. To prevent contact and, therefore, friction between the quadriceps tendon and intercondylar groove
3. To minimize compression stress on the intercondylar groove

Normal and Abnormal Tracking of the Patella

The medial retinaculum attaches the medial border of the patella to the medial epicondyle of the femur. The lateral retinaculum attaches the lateral border of the patella to the lateral epicondyle of the femur. The retinacula blend with the capsule and, together with the quadriceps tendon and patellar ligament, normally maintain maximum congruence and normal movement of the patellofemoral joint throughout the range of flexion and extension of the knee. The movement of the patella along the intercondylar groove is usually referred to as **tracking**, and normal tracking corresponds to the maintenance of maximum congruence (figure 7.17a). Any change in the normal pattern of forces exerted on the patella by the various fibrous attachments may displace the patella laterally or medially in the intercondylar groove, which is likely to reduce congruence in the patellofemoral joint and produce abnormal tracking of the patella in the intercondylar groove during knee flexion and extension. Lateral tracking and medial tracking refer to abnormal tracking of the patella on the lateral and medial sides of the patellofemoral joint, respectively (figure 7.17, b and c).

Lateral and medial tracking of the patella stretches the joint capsule and retinacula, and also increases the compression stress on the articular surfaces due to the decrease in the area of contact between them. If prolonged, abnormal tracking can result in a dull pain around and beneath the patella. The pain can be exacerbated by exercise involving flexion and extension of the knee while weight bearing, such as running, jumping, and ascending and descending stairs. Pain may also occur when

Patellofemoral tracking: the movement of the patellofemoral joint

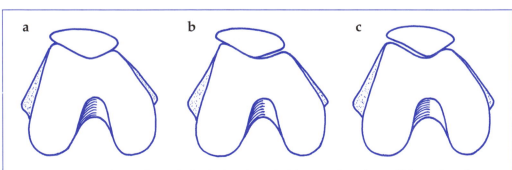

Figure 7.17 Tracking of patella (right leg shown); (*a*) normal tracking; (*b*) lateral tracking; (*c*) medial tracking.

sitting for long periods with the knee flexed such as in a car, airplane, or theater (Callaghan and Oldham 1996). In older adults the condition is often associated with damage to the articular cartilage of the patella and, as such, may be referred to as chondromalacia patellae (chondro = cartilage, malacia = softening) (Casscells 1982). However, in children and young adults the condition usually occurs without any obvious damage to the articular cartilage; these cases use the general term **patellofemoral pain syndrome** (or a similar term such as patellofemoral compression syndrome, patellofemoral malalignment syndrome, or patellagia) (Galea and Albers 1994; Percy and Strother 1985). The pain is thought to be due to stretching the fibrous supporting structures (joint capsule and retinacula) and associated synovial membrane, and compressive stress on the subchondral bone. Unlike the articular cartilage, the fibrous supporting structures, synovial membrane, and subchondral bone all have a rich supply of pain receptors. A high incidence of patellofemoral pain syndrome and, to a lesser extent, chondromalacia patellae is common among theatrical dancers (Reid 1987; Winslow and Yoder 1995) and participants in sports such as long distance running, volleyball (Clement et al. 1981; Ferretti, Papandrea, and Conteduca 1990), and training via jogging (Kujala et al. 1986). Patellofemoral pain syndrome also affects a high proportion of the nonathletic population (McConnell 1986).

> *The loads exerted on the patella by the quadriceps tendon, patellar ligament, and retinacula normally maintain maximum congruence and normal tracking of the patellofemoral joint. Any change in the normal pattern of these loads may result in abnormal tracking. If prolonged, abnormal tracking can result in patellofemoral pain syndrome.*

Causes of Abnormal Tracking

There are four main causes of abnormal tracking that often interact with each other: skeletal abnormalities, strength imbalance in the quadriceps, strength imbalance in fibrous supporting structures, and compensatory movements of the knee in response to abnormal movement in the foot.

Skeletal Abnormalities   A variety of skeletal abnormalities may contribute to abnormal tracking; these include the size of the Q angle, genu varum and genu valgum, patella alta and patella infera, and the size of the lateral border of the patellar surface.

- **The Q angle**. The position of the tibial tuberosity affects the line of pull of the quadriceps and, therefore, the tracking of the patella. The normal position of the tibial tuberosity is such that in the anatomical position the patellar ligament runs downward and slightly laterally (figure 7.18*a*). However, if the tuberosity is located more lateral than normal, there is a tendency to lateral tracking. Similarly, if it is located more medial than normal, there is a tendency to medial tracking. The orientation of the patellar ligament to the quadriceps tendon can be assessed by measuring the Q angle—the angle between the line joining the anterior superior iliac spine and the center of the patella, and the line joining the center of the patella and the midpoint of the tibial tuberosity (figure 7.18*b*). In men the mean Q angle is approximately 14°. In women it is approximately 17°, due to the relative increase in the width of the hips compared to men (Aglietti, Insall, and Cerulli 1983). Increases and decreases of the Q angle, with respect to these normal values, are associated with an increased incidence of patellofemoral pain syndrome (Huberti and Hayes 1984).
- **Genu varum and genu valgum**. Genu varum (bowlegs) and genu valgum (knock-knees) may lead to abnormal tracking of the patella. When these

Patellofemoral pain syndrome: persistent dull pain around the patella, frequently associated with abnormal tracking, which is thought to be due to stretching the fibrous supporting structures (joint capsule and retinacula) and associated synovial membrane and compressive stress on subchondral bone

conditions develop during childhood, it is usually the result of normal modeling in response to abnormal patterns of loading imposed on the knees by the combined effect of muscle forces and body weight (see chapter 11). The skeletal adaptations of genu varum and genu valgum ensure normal transmission of loads across the knee; the knees are functionally normal even though they appear to be abnormally aligned. The joints are only likely to become functionally abnormal when the balance between body weight and muscle forces changes. This applies to all individuals, but those with genu varum and genu valgum are more likely to develop functionally abnormal knees as a result of changes in the balance between muscle forces and body weight than individuals with normally aligned knees.

With increase in age, muscle weakness is likely to increase the degree of varus in an individual with genu varum and to increase the degree of valgus in someone with genu valgum. When this happens the joints become incongruent, which results in an increase in compression stress on the parts of the articular cartilage in contact—the lateral condyles of the femur and tibia in someone with genu valgum or the medial condyles of the femur and tibia in someone with genu varum. If the situation persists the overloaded cartilage may erode. As the cartilage becomes thinner, the degree of misalignment between femur and tibia increases. Removal of either the lateral or medial meniscus may lead to the same chain of events: excessive stress on one side of the joint, gradual thinning of the cartilage, and a gradual increase in the degree of varus or valgus (Yates and Jackson 1984). Clearly, any increase in the degree of varus or valgus of the knees during adulthood can cause abnormal tracking of the patella; an increase in varus can cause medial tracking and an increase in valgus can cause lateral tracking.

- **Patella alta and patella infera**. The length of the patellar ligament largely determines the height of the patella relative to the tibiofemoral joint. Normally

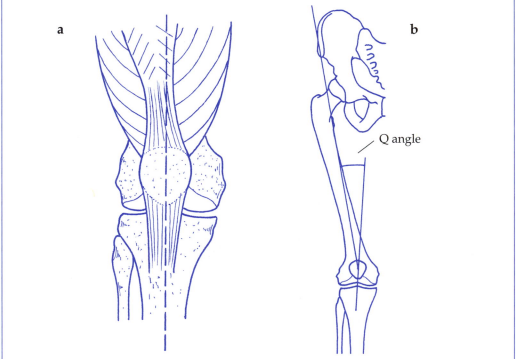

Figure 7.18   The Q angle; (*a*) anterior aspect of the right knee showing the alignment of the quadriceps tendon and the patellar ligament; (*b*) the Q angle.

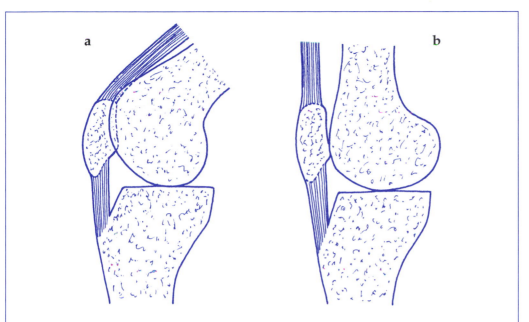

Figure 7.19   Effect of the knee joint angle on apposition of the patella; (*a*) knee flexed at 45°; (*b*) full extension of knee.

the ratio of the length of the patella to that of the patellar ligament is approximately 1.0; a variation of more than 20% in this ratio is regarded as abnormal (Insall and Salvati 1971). A ratio of 0.8 or less is referred to as patella alta—a high patella, and a ratio of 1.2 or greater is referred to as patella infera—a low patella. Patella alta and patella infera are associated with an increased incidence of patellofemoral pain syndrome, which may be due in part to abnormal tracking of the patella (Insall and Salvati, 1971; Lancourt and Cristini 1975; Kujala et al. 1987).

- **Lateral border of the patella surface**. Passive and active mechanisms restrain the movement of the patella. The passive mechanisms are the retinacula, their capsular attachments, and the shape of the patellofemoral joint. The patellofemoral joint is V-shaped in transverse section, which prevents side-to-side movement of the patella in the intercondylar groove (see figure 7.17*a*). The active mechanism is the appositional force exerted on the patella—the force pressing the patella into the intercondylar groove as a result of contraction of the quadriceps muscles. The greater the force exerted by the quadriceps and the greater the degree of knee flexion, the greater the appositional force (figure 7.19*a*). However, as the knee extends, the angle between the quadriceps tendon and patellar ligament decreases such that the appositional force on the patella decreases. At full knee extension the appositional force may be close to zero and, as such, the patella may displace laterally (figure 7.19*b*). In this situation one of the main restraints that prevents lateral displacement of the patella is the lateral border of the patellar surface, which is more prominent anteriorly than the medial border (see figure 7.17*b*). If the lateral border of the patellar surface is less prominent than normal, the patella may dislocate laterally at full extension of the knee (Kapandji 1970). Frequent lateral dislocation of the patella is referred to as recurrent dislocation of the patella.

Strength Imbalance in the Quadriceps   The four quadriceps muscles (rectus femoris, vastus lateralis, vastus intermedius, vastus medialis) pull on the patella in different directions, but in normal individuals the net effect of all four muscles is a more or less vertical pull that brings about normal tracking of the patella (figure 7.20).

The tendency of the vastus lateralis and, to a lesser extent, the rectus femoris and vastus intermedius to pull the patella laterally is normally balanced by the vastus medialis. The vastus medialis is most active during the last 20° of knee extension. If the full range of knee extension is not used, a strength imbalance may develop between the vastus medialis and vastus lateralis; the vastus medialis may become relatively weaker resulting in lateral displacement and, therefore, lateral tracking of the patella. A long distance runner who does not fully extend his knees during the propulsion phase of each stride is prone to develop this condition. Weakness of the vastus medialis is associated with an increased incidence of patellofemoral pain syndrome and recurrent lateral dislocation of the patella (Kujala et al. 1986; Bose, Kanagasuntheram, and Hosman 1980; Cash and Hughston 1988).

Strength Imbalance in Fibrous Supporting Structures The patellar ligament, quadriceps tendon, and the medial and lateral retinacula are all joined by the joint capsule and fascia of the muscles controlling the movement of the knee. In some individuals the fibrous structures on one side of the joint, usually the lateral side, become thicker or shorter than those on the other side of the joint and the patella is pulled to one side, resulting in abnormal tracking (Silver and Campbell 1985; Micheli and Stanitski 1981).

Compensatory Movement in the Knee Due to Abnormal Movement in the Foot In activities involving weight bearing, foot pronation (see next section) may result in internal rotation of the tibia with respect to the femur, and foot supination may result in external rotation of the tibia with respect to the femur. Internal rotation of the tibia can result in simultaneous medial rotation and medial displacement of the patella. If, as in many sports, foot pronation is associated with forceful activity in the quadriceps, abnormal tracking of the patella may occur due to a combination of medial rotation and medial displacement of the patella. If such movements occur with high frequency as, for example, in distance running, or in jumping and landing in volleyball, patellofemoral pain syndrome may occur.

It is not surprising that patellofemoral pain syndrome is the most common injury in joggers and long distance runners (Clement et al. 1981). Most runners contact the ground with the posterior lateral aspect of the heel. Immediately after heel strike

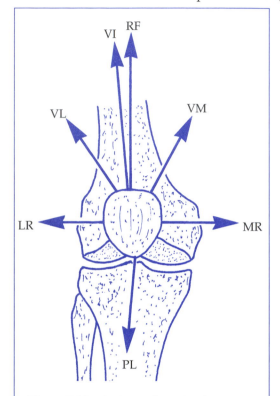

Figure 7.20 Active and passive forces acting on the patella. Active forces include the quadriceps: RF—rectus femoris; VL—vastus lateralis; VM—vastus medialis; VI—vastus intermedius. Passive forces include the PL—patellar ligament; LR—lateral retinaculum; MR—medial retinaculum.

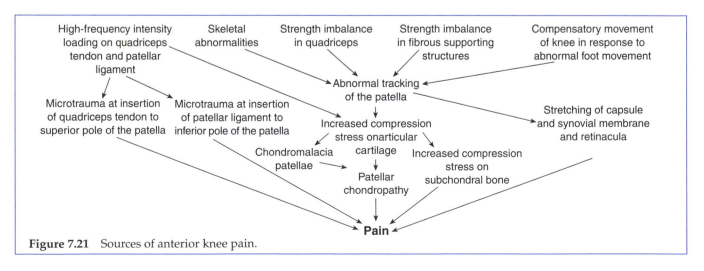

Figure 7.21 Sources of anterior knee pain.

the foot pronates and the arches of the foot depress to absorb the shock of the impact. In normal function the foot quickly recoils into a less-depressed and less-pronated position such that abnormal tracking of the patella is unlikely to occur. However, in many runners the ground contact period is associated with excessive or prolonged pronation, which considerably increases the likelihood of abnormal tracking and the occurrence of patellofemoral pain syndrome (Cavanagh 1980; Hontas, Haddad, and Schlesinger 1986). The excessive or prolonged pronation may be the result of extrinsic factors such as inadequate footwear, but it may also be the result of intrinsic factors such as fatigue in the muscles that control pronation.

Patellofemoral pain syndrome is associated with abnormal tracking of the patella, especially in activities involving high-frequency intensity loading of the quadriceps tendon-patellar ligament unit. In such activities, anterior knee pain can also occur in the form of jumper's knee even when patella tracking is normal. Jumper's knee is characterized by pain at the insertion of the quadriceps tendon into the upper pole of the patella and at the insertion of the patellar ligament into the inferior pole of the patella (Ferretti et al. 1984). The various sources of pain arising from jumper's knee and patellofemoral pain syndrome are summarized in figure 7.21.

> *There are four main causes of abnormal tracking of the patella that often interact with each other: skeletal abnormalities, strength imbalance in the quadriceps, strength imbalance in fibrous supporting structures, and compensatory movements of the knee in response to abnormal movement in the foot.*

Ankle Complex

Many of the 26 bones in each foot articulate with two or more other bones such that there are approximately 40 joints in each foot. Consequently, most movements of the foot involve a large number of joints and, therefore, are quite complex and difficult

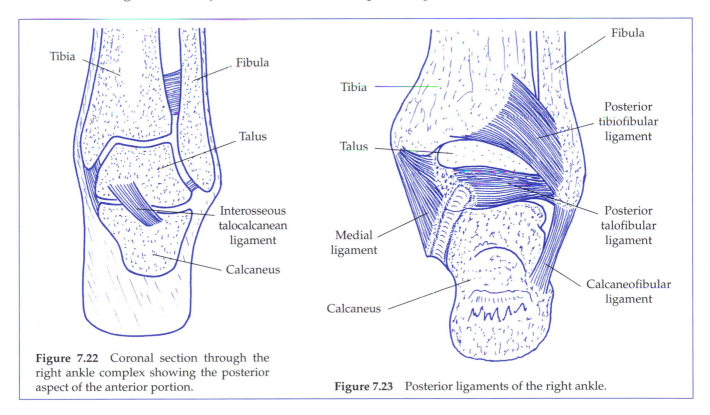

Figure 7.22 Coronal section through the right ankle complex showing the posterior aspect of the anterior portion.

Figure 7.23 Posterior ligaments of the right ankle.

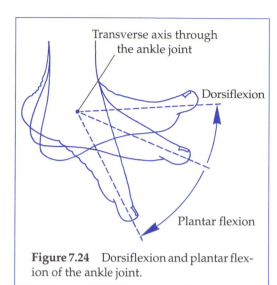

Figure 7.24 Dorsiflexion and plantar flexion of the ankle joint.

Plantar flexion: rotation of the foot about a transverse axis through the ankle joint (or subtalar joint as part of supination) in which the plantar surface of the foot is pushed away from the shin (pointing the toes)

Dorsiflexion: rotation of the foot about a transverse axis through the ankle joint (or subtalar joint as part of pronation) in which the dorsal surface of the foot is drawn closer to the shin

Supination of the foot: rotation of the foot about the subtalar joint involving simultaneous adduction, plantar flexion, and inversion

Pronation of the foot: rotation of the foot about the subtalar joint involving simultaneous abduction, dorsiflexion, and eversion

Ankle complex: the ankle joint and the subtalar joint

to describe. There are four basic movements of the foot; **plantar flexion** and **dorsiflexion** of the ankle joint (or talocrural joint), and **supination** and **pronation** of the subtalar joint (or talocalcanean joint). The **ankle complex** consists of the ankle joint and the subtalar joint.

Ankle Joint

The ankle joint is the joint between the tibia, fibula, and talus (figure 7.22). The distal ends of the tibia and fibula are joined at the distal tibiofibular joint; this is a syndesmosis that is supported by strong anterior and posterior tibiofibular ligaments. The fibers of the anterior tibiofibular ligament run downward and laterally from the anterior lateral aspect of the tibia just above the ankle joint to the anterior aspect of the lateral malleolus. The fibers of the posterior tibiofibular ligament run downward and laterally from the posterior lateral aspect of the tibia just above the ankle joint to the posterior aspect of the lateral malleolus (figure 7.23). The inferior aspect of the anterior tibiofibular ligament overlaps the anterior lateral aspect of the ankle joint, and the inferior aspect of the posterior tibiofibular ligament overlaps the posterior lateral aspect of the ankle joint.

The ankle joint is a synovial hinge joint allowing plantar flexion and dorsiflexion of the foot (figure 7.24). Dorsiflexion and plantar flexion of the ankle joint occur about a more or less transverse axis. In dorsiflexion—sometimes referred to as true flexion of the ankle—the dorsal (superior) surface of the foot is drawn closer to the shin. In plantar flexion—sometimes referred to as extension of the ankle—the plantar (inferior) surface of the foot is pushed farther away from the shin (pointing the toes).

The articular surfaces of the distal tibia and fibula form a mortise on the talus, which restricts side-to-side movement of the talus in the ankle joint (see figure 7.22). However, the degree of restriction depends on the angle of plantar flexion and dorsiflexion. When viewed from above, the trochlea surface of the talus is wedge-shaped with the broad end in front (figure 7.25). Consequently, the mortise is tightened during dorsiflexion, which increases the congruence and stability of the joint. However, during plantar flexion the mortise is loosened, which reduces the congruence and stability of the joint.

Subtalar Joint

The subtalar joint is the joint between the talus and calcaneus (see figure 7.22). It is part synovial and part syndesmosis. The anterior synovial part of the joint is separated from the posterior synovial part of the joint by a space called the sinus tarsi. The sinus tarsi is occupied by a strong interosseous talocalcanean ligament, which holds the talus and calcaneus together.

The subtalar joint is a uniaxial joint that facilitates supination and pronation (Norkin and Levangie 1992). Supination and pronation occur about an oblique axis through the subtalar joint as shown in figure 7.26. The orientation of the axis varies considerably among individuals (Inman 1976). Inman found that the mean angle of the axis, when projected onto a sagittal plane, was in the region of 42° with respect to the horizontal, with a range of 20° to 70° (figure 7.26a). Inman also found that the mean angle of the axis, when projected onto a horizontal plane, was in the region of 23° with respect to the sagittal plane, with a range of 5° to 47° (figure 7.26b). With respect to reference axes through the subtalar joint (figure 7.27), pronation involves simultaneous abduction (vertical axis), dorsiflexion (transverse axis) and eversion (anteroposterior axis) (figure 7.28, a and b). Similarly, supination involves simultaneous adduction, plantar flexion, and inversion (figure 7.28, b and c). It should be

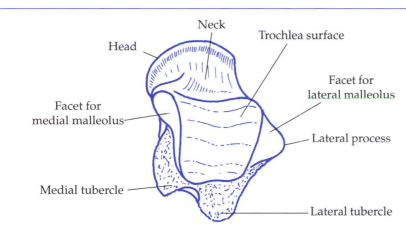

Figure 7.25 Superior aspect of the right talus.

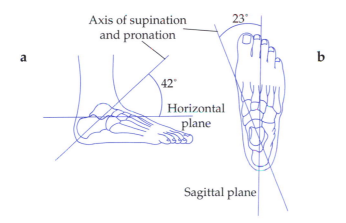

Figure 7.26 Axis of supination and pronation of the subtalar joint; (*a*) sagittal plane; (*b*) horizontal plane.

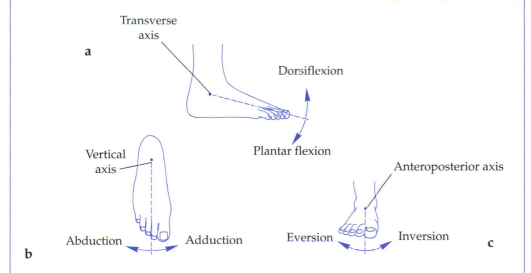

Figure 7.27 Movements of the subtalar joint involved in supination and pronation; (*a*) dorsiflexion and plantar flexion with respect to a transverse axis; (*b*) abduction and adduction with respect to a vertical axis; (*c*) eversion and inversion with respect to an anteroposterior axis.

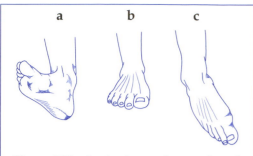

Figure 7.28   Supination and pronation of the subtalar joint (viewed from the front); (*a*) pronation; (*b*) neutral position; (*c*) supination.

clear that plantar flexion and dorsiflexion of the foot occur at the ankle joint and as components of supination and pronation (Edington, Frederick, and Cavanagh 1990).

> *The ankle joint is a synovial hinge joint allowing plantar flexion and dorsiflexion. The subtalar joint is part synovial and part syndesmosis. It has a complex shape and structure that reflects its function in facilitating supination and pronation of the foot.*

The movements of supination and pronation as previously described refer to movements of the subtalar joint when the foot is nonweight bearing. When the foot is weight bearing, these movements are constrained depending on the magnitude and distribution of the ground reaction force acting on the plantar part of the foot. Under weight-bearing conditions the most noticeable movements of the foot occur about an anteroposterior axis through the foot (similar to inversion and eversion). For this reason, in describing the movement of the foot under weight-bearing conditions the terms supination and inversion are sometimes used synonymously, as are the terms pronation and eversion. However, the actual movements of the foot under weight-bearing conditions are modifications of basic supination and pronation and, as such, involve simultaneous movement about vertical, transverse, and anteroposterior axes through the subtalar joint, as well as dorsiflexion and plantar flexion at the ankle joint (Lundberg and Svensson 1988; Stacoff, Kalin, and Stussi 1991).

Ligaments of the Ankle Complex

The capsule of the ankle joint is separate from that of the subtalar joint. Both capsules follow the lines of the joints and blend with the associated intertarsal ligaments. The ankle complex is supported by a number of noncapsular ligaments. The medial aspect of the complex is supported by the deltoid ligament (also referred to as the medial collateral ligament; figure 7.29). The deltoid ligament is a strong ligament that fans out from the anterior, medial, and posterior aspects of the medial malleolus to

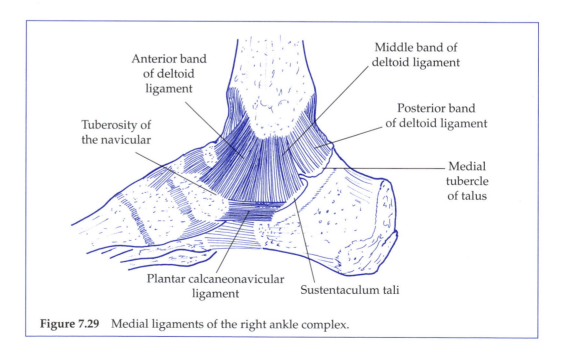

Figure 7.29   Medial ligaments of the right ankle complex.

attach onto a more or less continuous arc formed by the tuberosity of the navicular, the plantar calcaneonavicular ligament (which spans the gap between the tuberosity of the navicular and the sustentaculum tali of the calcaneus), the sustentaculum tali, and the medial tubercle of the talus. The deltoid ligament as a whole strongly resists abduction of the ankle and subtalar joints. The fibers of the anterior band become taut at the limit of plantar flexion and the fibers of the posterior band become taut at the limit of dorsiflexion.

The lateral aspect of the ankle joint complex is supported by the anterior talofibular ligament, the lateral talocalcanean ligament, the calcaneofibular ligament, and the posterior talofibular ligament (figure 7.30). The anterior talofibular ligament runs medially and slightly forward and downward from the anterior inferior aspect of the lateral malleolus to the anterior lateral aspect of the talus immediately in front of the lateral articular surface of the talus. The lateral talocalcanean ligament runs downward and slightly backward from the lateral process of the talus (see figure 7.25) to the adjacent lateral aspect of the calcaneus. The calcaneofibular ligament runs medially downward and backward from the inferior aspect of the lateral malleolus to the lateral aspect of the calcaneus. The posterior talofibular ligament runs medially and slightly backward from the posterior inferior aspect of the lateral malleolus to the lateral aspect of the talus immediately behind the lateral articular surface and the adjacent lateral tubercle of the talus (see figure 7.25). All of the lateral ligaments resist adduction of the ankle and subtalar joints. The posterior talofibular ligament becomes taut at the limit of dorsiflexion, and the anterior talofibular ligament becomes taut at the limit of plantar flexion.

Ligament injuries of the ankle are common, especially in sports (Boruta et al. 1990). The most common ankle injury is an inversion sprain, which results in damage to one or more of the lateral ligaments and, in some cases, fracture of the medial malleolus or the lateral malleolus (Lassiter, Malone, and Garrick 1989) (figure 7.31). Chronic, that is, persistent lateral ankle instability occurs in 10% to 20% of individuals following a severe inversion sprain (Perlman et al. 1987; Peters, Trevino, and Renström 1991). Most inversion sprains involve plantar flexion as well as inversion. As described previously, the ankle joint is least stable in plantar flexion due to loosening of the tibiofibular mortise.

Furthermore, the ligament likely to be under the most strain during plantar flexion and inversion is the anterior talofibular ligament, which is reported to be the weakest of the lateral ligaments (Siegler, Block, and Schneck 1988). Consequently, the ankle is vulnerable to injury during movements of the foot that involve rapid and forceful plantar flexion or inversion; these movements may occur, for example, when walking, running, or landing on uneven surfaces.

In addition to the possibility of ligament injuries, movements involving extreme plantar flexion also can result in injury due to impingement of the posterior lateral border of the trochlea surface of the tibia on the lateral tubercle of the talus or impingement of the posterior border of the medial malleolus on the medial tubercle of the talus (figure 7.32, *a* and *b*). This is a common cause of injury in theatrical dancers (Hardaker, Margello, and Goldner 1985). Extreme dorsiflexion also can result in injury due to impingement of the anterior border of the trochlea surface of the tibia on the neck of the talus (figure 7.32, *a* and *c*).

> *The medial aspect of the ankle complex is supported by the deltoid ligament and the lateral aspect is supported by the anterior and posterior talofibular ligaments, the lateral talocalcanean ligament, and the calcaneofibular ligament. Ligament injuries of the ankle are common, especially in sports. The most common ankle injury is an inversion sprain. Injuries may also occur due to impingement of the tibia on the talus in movements involving extreme plantar or dorsiflexion.*

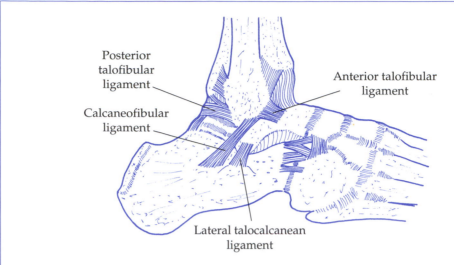

Figure 7.30   Lateral ligaments of the right ankle complex.

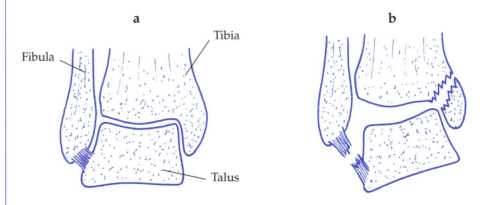

Figure 7.31   Inversion sprain injury; (*a*) normal orientation of the tibia, fibula, and talus; (*b*) torn lateral ligaments and fractured medial malleolus due to excessive and forceful inversion.

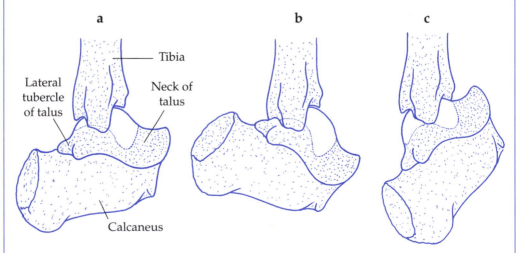

Figure 7.32   Medial aspect of the left foot: impingement of the tibia and talus; (*a*) neutral orientation of the tibia and talus; (*b*) impingement of the posterior border of the trochlea surface of the tibia on the lateral tubercle of the talus due to extreme plantar flexion; (*c*) impingement of the anterior border of the trochlea surface of the tibia on the neck of the talus due to extreme dorsiflexion.

Function of the Foot

The foot's main function is to transmit loads between the lower leg and the ground. In static situations such as standing upright on level ground, the size of the load transmitted, that is, body weight, is relatively small and transmission of this size of load could be satisfactorily achieved by a much less sophisticated structure than the foot. However, in dynamic situations such as walking, running, jumping, and landing, the foot is subjected to large loads, which, unless effectively transmitted, would be likely to excessively overload not only the foot, but also other parts of the musculoskeletal system.

In dynamic situations the foot is required to act as both a shock absorber (see chapter 10) and a propulsive mechanism. For example, during the first part of ground contact in running, the arches of a normal foot are depressed momentarily, rather like a compressed spring, to cushion the impact (figure 7.33). Having done so, the function of the foot rapidly changes from that of shock absorber to that of propulsive mechanism to assist the propulsive drive of the leg as a whole in projecting the body forward into the next stride. The foot often performs the functions of shock absorption and propulsion on a variety of support surfaces. Whereas floor surfaces tend to be firm and level, there are many other situations, such as in cross-country running, where the surface of the ground is neither firm nor level, but continually changes in terms of slope, evenness, and hardness. By being able to perform the functions of shock absorption and propulsion on a wide variety of support surfaces, the foot represents a very complex, sophisticated piece of mechanical engineering. The ability of the foot to perform these functions is largely due to the relatively large range of motion in the foot and, in particular, the ability of the foot to supinate and pronate.

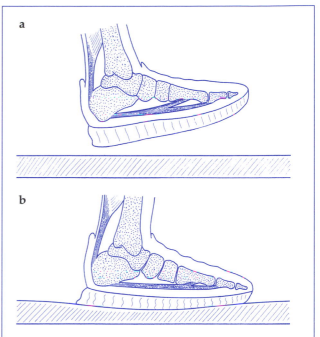

Figure 7.33 Compression of the arches of the foot during the ground contact phase in running; (*a*) just prior to ground contact; (*b*) during the first part of ground contact.

In dynamic situations the foot acts as a shock absorber and a propulsive mechanism. The ability of the foot to perform these functions is largely due to its relatively large range of motion and its ability to supinate and pronate.

Summary

This chapter described the structure and function of the joints of the appendicular skeleton. In contrast to the joints of the axial skeleton—designed primarily to provide shock absorption—the joints of the appendicular skeleton are designed primarily to facilitate relatively large ranges of movement; this design is reflected in the wide variety of different types of synovial joints present in the appendicular skeleton. All of the joints in the appendicular skeleton have a certain degree of functional interdependence with other joints, especially those in the same joint complex or skeletal chain, and therefore abnormal or restricted movement in a joint can result in abnormal movement in other joints in the joint complex or skeletal chain. Chapters 3 to 7 have been largely concerned with the structure and function of the skeletal system. Chapter 8 covers the neuromuscular system, which actively controls the movement of the skeletal system.

Review Questions

1. Describe the coracoacromial arch and its effect on movements of the shoulder joint.
2. Describe the role of the coracoclavicular and costoclavicular ligaments in the stability of the acromioclavicular and sternoclavicular joints.
3. Describe the movements of the shoulder complex during abduction of the upper limb to the overhead position.
4. Describe the ligaments of the elbow complex and their functions.
5. Explain why the normal range of wrist adduction is approximately three times that of wrist abduction.
6. Compare and contrast the structure and functions of the shoulder and hip joints.
7. Describe the structure and functions of the menisci of the tibiofemoral joint.
8. Describe the changes in the shape of the capsule of the knee complex during flexion and extension of the knee.
9. Describe the structure and functions of the collateral ligaments and cruciate ligaments of the knee.
10. Explain why knee flexion increases the range of hip flexion and decreases the range of hip extension.
11. Describe the functions of the patella.
12. Describe the main causes of abnormal tracking of the patella.
13. Describe the possible sources of pain characterized by patellofemoral pain syndrome.
14. Describe the effect of dorsiflexion and plantar flexion on the stability of the ankle joint.
15. Describe the types of ankle injury likely to result from extreme dorsiflexion and plantar flexion.

Chapter 8

The Neuromuscular System

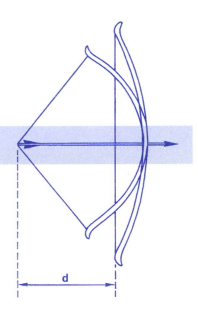

The muscular system is the interface between the nervous and skeletal systems. The muscles provide the force to move the levers of the skeletal system, but the nervous system determines the level and timing of force production. The nervous system constantly monitors and interprets information from the various senses concerning body position and body movement, including information from muscles and joint supporting structures, and on the basis of this information sends instructions to the muscles to coordinate body movement. Those parts of the nervous and muscular systems responsible for bringing about coordinated movement are referred to as the neuromuscular system. This chapter describes the structure and function of the neuromuscular system. Before reading this chapter you may want to review the sections in chapter 1 on work, energy, and power.

Objectives

After reading this chapter you should be able to do the following:

1. Define or describe the key terms.
2. Differentiate between the cerebrospinal nervous system and the autonomic nervous system.
3. Differentiate between the central nervous system and the peripheral nervous system.
4. Differentiate among unipolar, bipolar, and multipolar neurons.
5. Differentiate between myelinated and nonmyelinated nerve fibers.
6. Describe the events resulting in transmission of an impulse along a nerve fiber and from one nerve fiber to another.
7. Describe the general organization of nerve tissue in the brain, spinal cord, and spinal nerves.
8. Describe the sequence of events within the nervous system in a typical voluntary movement and a reflex action.
9. Describe Seddon's classification of injuries to nerve fibers.
10. Differentiate between pennate and nonpennate muscles.
11. Describe the structure of a muscle fiber.
12. Describe the sliding filament theory of muscular contraction.
13. Describe the composition of a motor unit.
14. Differentiate between fast twitch and slow twitch muscle fibers.
15. Differentiate between kinesthetic sense and proprioception.
16. Describe the structure and function of muscle spindles.
17. Describe the length-tension relationship in a sarcomere, muscle fiber, and muscle tendon unit.
18. Describe the force-velocity relationship in skeletal muscle.
19. Describe the stretch-shorten cycle in skeletal muscle.

Neuron: a nerve cell

Glial cell: a specialized connective tissue cell found only in the nervous system

Cerebrospinal nervous system: one of two functional divisions of the nervous system; it is concerned with consciousness and mental activities and control of skeletal muscle

Autonomic nervous system: one of two functional divisions of the nervous system; it is concerned with the control of the visceral muscles, the heart, and the exocrine and endocrine glands

Body movement is brought about by the musculoskeletal system under the control of the nervous system. The nervous system continuously monitors and interprets information from the various senses and uses this information to program muscular activity to bring about voluntary and involuntary (reflex) movements.

With regard to voluntary movements, improvements in effectiveness—the extent to which the objective of a particular movement is achieved—and efficiency—the extent to which chemical energy in the muscles is converted to useful mechanical energy in the control of joint movements—depend on changes in the operation of the neuromuscular system. The ability of a teacher, coach, or therapist to bring about positive changes in effectiveness and efficiency depends to a considerable extent on her understanding of the structure and function of the neuromuscular system.

The Nervous System

The nervous system consists of approximately 13,000 million nerve cells called **neurons**, and an equally large number of specialized connective tissue cells called **glial cells**. Neurons are specialized to conduct electrical impulses rapidly throughout the body to coordinate all the essential biological functions. The cells of the nervous system are organized into two functional divisions and two structural divisions (Williams et al. 1995). The functional divisions are the **cerebrospinal nervous system** and the **autonomic nervous system**.

The cerebrospinal nervous system, also known as the somatic, craniospinal, or voluntary nervous system, is under voluntary control except for reflex movements. A reflex movement provides protection by rapidly removing part of the body from a source of danger without conscious effort. The cerebrospinal nervous system includes those parts of the nervous system concerned with consciousness and mental activities and control of skeletal muscle. The autonomic nervous system, also known as the visceral or involuntary nervous system, is not under voluntary control; it includes those parts of the nervous system that control the visceral muscles, the heart, and the exocrine and endocrine glands.

The two structural divisions of the nervous system are the **central nervous system** and the **peripheral nervous system**. The central nervous system consists of the brain and spinal cord; the peripheral nervous system consists of 43 pairs of nerves (bundles of nerve fibers), which arise from the base of the brain and the spinal cord (figure 8.1). Twelve of the pairs of nerves arise from the base of the brain and are called **cranial nerves**. The other 31 pairs arise from the spinal cord and are called **spinal nerves**. The cranial and spinal nerves convey information between the central nervous system and the rest of the body.

Central nervous system: one of two structural divisions of the nervous system; it consists of the brain and spinal cord

Peripheral nervous system: one of two structural divisions of the nervous system; it consists of 43 pairs of nerves (bundles of nerve fibers), which arise from the base of the brain and the spinal cord

Cranial nerves: 12 pairs of peripheral nerves that arise from the base of the brain

Spinal nerves: 31 pairs of peripheral nerves that arise from the spinal cord

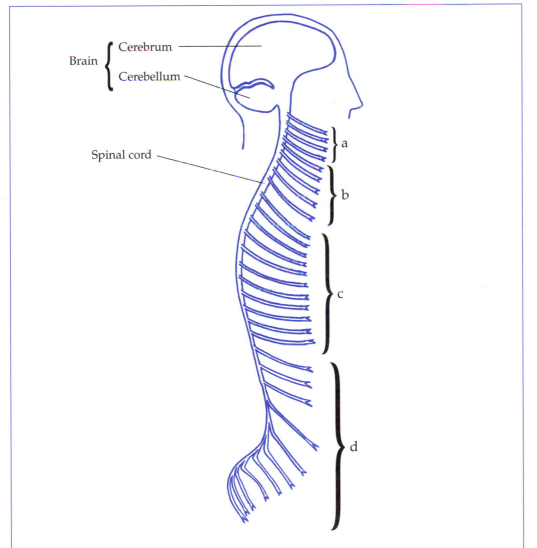

Figure 8.1   Peripheral nerves supplying: (*a*) the neck; (*b*) the upper limbs; (*c*) the thorax and abdomen; and (*d*) the lower limbs.

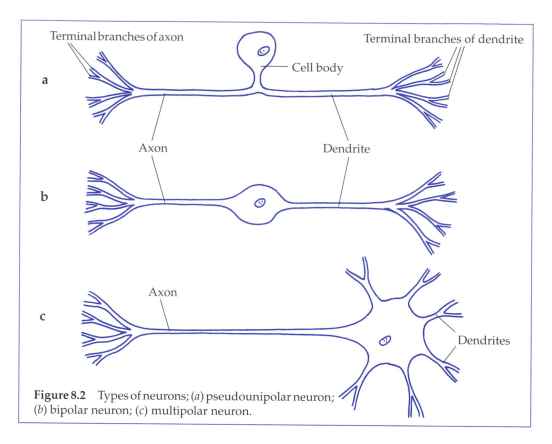

Figure 8.2   Types of neurons; (*a*) pseudounipolar neuron; (*b*) bipolar neuron; (*c*) multipolar neuron.

Neurons

The nervous system consists of neurons that conduct electrical impulses rapidly throughout the body to coordinate essential biological functions. Neurons differ in size and shape, but they all have three common features: a cell body, processes of varying length that extend from the cell body, and specialized sites for communicating with other neurons, with specialized receptors such as pain receptors, or with specialized effectors such as motor end plates in muscle. Neurons can be classified by the direction in which they conduct impulses in relation to the brain. **Sensory** or **afferent neurons** conduct impulses toward the brain, and **motor** or **efferent neurons** conduct impulses away from the brain.

Nerve Fibers, Dendrites, and Axons

Sensory neuron: a neuron that conducts impulses towards the brain; also called an afferent neuron

Motor neuron: a neuron that conducts impulses away from the brain; also called an efferent neuron

Nerve fibers: the processes that extend from the cell body of a neuron

Dendrite: a nerve fiber that conducts impulses toward the cell body

Axon: a nerve fiber that conducts impulses away from the cell body

The processes that extend from the cell bodies of neurons are called **nerve fibers**. Nerve fibers vary in length from a few millimeters to more than one meter. There are two types of nerve fibers, **dendrites** and **axons**. Dendrites, or afferent fibers, conduct impulses toward the cell body. Axons, or efferent fibers, conduct impulses away from the cell body. In addition to the sensory and motor classification, neurons are classified on the basis of the number of processes arising from the cell body into pseudounipolar, bipolar, and multipolar.

In a pseudounipolar neuron there appears to be one process arising from the cell body that quickly divides into an afferent fiber and an efferent fiber (figure 8.2*a*). Sensory neurons in the peripheral nervous system are pseudounipolar neurons. A bipolar neuron has two distinct processes, one afferent and one efferent (figure 8.2*b*). Bipolar neurons are found in the sensory areas of the eye, ear, and nose. Multipolar neurons have numerous relatively short dendrites with a single axon that may branch at various points (figure 8.2*c*). Most of the neurons in the brain and spinal cord are multipolar neurons.

Myelinated and Nonmyelinated Nerve Fibers

Glial cells provide mechanical and metabolic support to neurons. In the central nervous system there are a variety of glial cells including astrocytes, which provide support for blood vessels, and oligodendrocytes, which provide support for nerve fibers. In the peripheral nervous system there is only one type of glial cell, the **Schwann cell** (Gamble 1988). All nerve fibers of the peripheral nervous system are enveloped by Schwann cells, which provide the same type of mechanical support as the oligodendrocytes provide for nerve fibers in the central nervous system. The Schwann cells around some fibers produce a fatty substance called myelin, which is deposited around the fibers as a multilayered myelin sheath. The sheath is in the form of a spiral with up to 100 regularly spaced layers of myelin separated by folds of Schwann cell membrane. The outer fold of the Schwann cell membrane is referred to as the neurilemma (figure 8.3a).

The greater the number of layers of myelin in the sheath, the faster the speed of impulse transmission along the nerve fiber. Nerve fibers that have a myelin sheath are referred to as **myelinated** or **medullated** nerve fibers and those nerve fibers that do not are referred to as **nonmyelinated** or **nonmedullated** nerve fibers. Each myelinated nerve fiber is enclosed within a chain of Schwann cells and each Schwann cell envelops approximately 1 mm of nerve fiber. Between adjacent Schwann cells the myelin sheath is interrupted such that a small section of the nerve fiber is exposed; these regions are referred to as nodes of Ranvier (figure 8.3a). The nodes of Ranvier facilitate intercellular exchange between the nerve fibers and the surrounding

Schwann cell: the only type of glial cell in the peripheral nervous system

Myelinated nerve fiber: a nerve fiber that has a myelin sheath

Nonmyelinated nerve fiber: a nerve fiber that does not have a myelin sheath

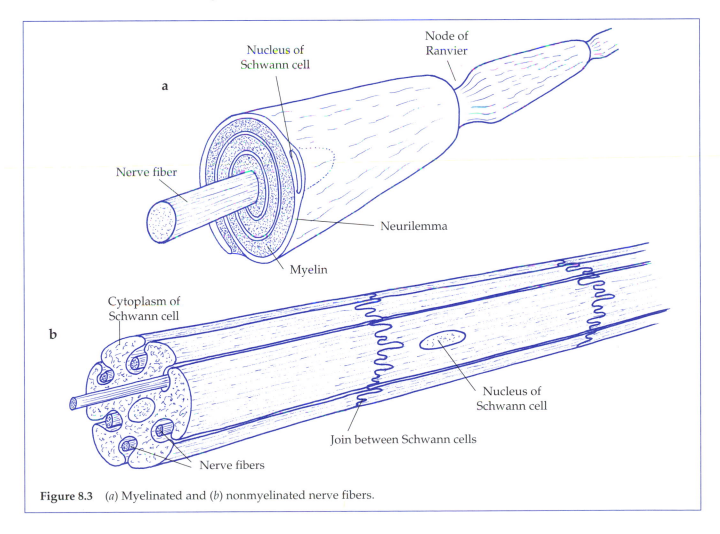

Figure 8.3 (*a*) Myelinated and (*b*) nonmyelinated nerve fibers.

extracellular fluid, which is important for the nutrition of the nerve fiber and for the transmission of impulses along the nerve fiber. Myelinated nerve fibers in the central nervous system are different from those in the peripheral nervous system in that they are not surrounded by Schwann cells (there are no Schwann cells in the central nervous system). It is thought that oligodendrocytes are responsible for the formation of the myelin sheath around these fibers at the embryonic stage (Williams et al. 1995).

Nonmyelinated nerve fibers of the peripheral nervous system are also enclosed within chains of Schwann cells. However, in comparison to myelinated fibers, there are no nodes of Ranvier in nonmyelinated fibers and there may be as many as nine nerve fibers enveloped within the folds of the Schwann cells in the chain (figure 8.3*b*).

Nerve Fiber Endings

The dendrites and axons of all types of neurons have a large number of terminal branches or nerve endings devoid of Schwann cells and myelin sheath. There are three types of nerve endings (**sensory, motor,** and **synapse**):

1. **Sensory:** where the nerve ending is in contact with a specialized receptor organ such as a pain receptor
2. **Motor:** where the nerve ending is in contact with a specialized effector organ such as a motor end plate in muscle
3. **Synapse:** where the nerve ending is in contact with another neuron

Sensory and motor nerve endings are referred to as **end organs.** Neurons that only have synapses at their nerve endings are called **association neurons** or **interneurons.**

Electrical Impulse Transmission

The cytoplasm of a neuron and the extracellular fluid surrounding the neuron contain many different ions (electrically charged atoms). These include positively charged inorganic ions such as sodium (Na^+) and potassium (K^+), negatively charged inorganic ions such as chloride (Cl^-), and various organic anions (negatively charged amino acids and proteins [A^-]).

Like most membranes in the body, a nerve fiber membrane is semipermeable; it has a large number of tiny holes through which ions and small molecules (aggregations of ions) pass from one side of the membrane to the other. The movement of ions through the nerve fiber membrane depends on the permeability of the membrane (the number and size of the holes in the membrane) and the force tending to drive the ions through the membrane. The driving force has electrical, chemical, and, in the case of Na^+ and K^+, mechanical components. The electrical component depends on the polarity of the ions; like charges repel each other and unlike charges attract each other. The chemical component depends on the concentration of ions in different regions; ions move from an area of high concentration to areas of lower concentration. The mechanical component results from specialized regions of the membrane collectively referred to as the Na^+-K^+ pump. The Na^+-K^+ pump transports Na^+ out of the cytoplasm and into the extracellular fluid and transports K^+ in the opposite direction.

Under resting conditions, when the nerve fiber is not transmitting an impulse, the net effect of the membrane permeability and the driving forces on the various ions is that the electrical charge on the outside of the fiber membrane is approximately 70 mV (millivolts) higher than on the inside; the potential difference across the membrane is approximately 70 mV, with the inside negative with respect to the outside (figure 8.4*a*). This resting potential difference is referred to as resting membrane potential (RMP) (Enoka 1994).

The arrival of a stimulus at a nerve fiber in a state of RMP alters the permeability of the fiber membrane to Na^+ and K^+ so that Na^+ flows into the cell and K^+ flows out

Sensory nerve ending: a nerve ending in contact with a specialized receptor organ such as a pain receptor

Motor nerve ending: a nerve ending in contact with a specialized effector organ such as a motor end plate in muscle

Synapse: where a nerve ending is in contact with another neuron

End organs: sensory and motor nerve endings

Association neuron: a neuron that only has synapses at its nerve endings; also called an interneuron

of the cell. The flow of Na$^+$ into the cell is initially greater than the flow of K$^+$ out of the cell so that the potential difference across the membrane decreases. If the decrease in potential difference, which depends on the strength of the stimulus, reaches a critical level of approximately 60 mV (with the inside of the membrane negative with respect to the outside), the membrane will be depolarized; the potential difference across the fiber membrane rapidly changes from 70 mV, with the inside of the membrane negative with respect to the outside, to approximately 30 mV, with the inside of the membrane positive with respect to the outside; the change in potential difference is approximately 100 mV (figure 8.4a).

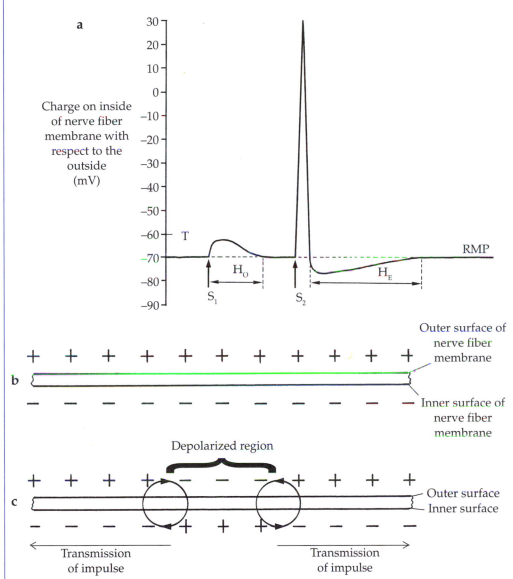

Figure 8.4   Resting membrane potential (RMP) and change in membrane potential. (a) Change in membrane potential in response to stimulation; T—level of hypopolarization necessary to trigger depolarization (T $\cong$ –60 mV); S$_1$—a stimulus that results in hypopolarization but not depolarization; H$_O$—period of hypopolarization following S$_1$; S$_2$—a stimulus that results in depolarization; H$_E$—period of hyperpolarization following repolarization. (b) Electrical charge on nerve fiber membrane during resting membrane potential. The charge on the inside of the membrane is approximately –70 mV with respect to the outside. (c) Electrical charge on depolarized region of nerve fiber membrane. The charge on the inside of the membrane is approximately +30 mV with respect to the outside.

This change in potential difference constitutes an **action potential**, which results in the flow of electrical current—called local current—between the depolarized region of the cell membrane and the adjacent unpolarized regions (both sides) (figure 8.4, *b* and *c*). The establishment of local current results in progressive (rapid wave) depolarization of the rest of the cell membrane so that the impulse is transmitted along the whole length of the fiber. After depolarization, the membrane is rapidly repolarized such that the action potential appears as a spike in a graph of the change in membrane potential with time (figure 8.4*a*).

The duration of the action potential spike (depolarization and repolarization) is less than one millisecond (Gamble 1988). Repolarization is due largely to a rapid decrease in the flow of Na^+ into the cell (due to reduced permeability of the membrane to Na^+ ions) and continued flow of K^+ out of the cell. Following repolarization there is usually a period (15–100 ms) of hyperpolarization in which the potential difference across the membrane is slightly greater than RMP as RMP is gradually restored (figure 8.4*a*). A stimulus not strong enough to cause depolarization results in a period of hypopolarization prior to restoration of RMP, that is, a period in which the potential difference across the membrane is slightly less than RMP (figure 8.4*a*).

Synapses

Every branch of an axon terminates in an end bulb (or end foot) that rests on the surface of a neighboring neuron to form a synapse—a specialized region that facilitates one-way communication between the two neurons (figure 8.5). Each neuron synapses with hundreds or thousands of other neurons (Williams et al. 1995). The most common type of synapse is axodendritic (between an axon and a dendrite), but synapses may also be axosomatic (between an axon and a cell body), and axoaxonic (between two axons). A synapse consists of a presynaptic membrane (the base of the end bulb), a postsynaptic membrane (the corresponding region of the adjacent neuron), and an intervening synaptic cleft.

The end bulb contains a number of synaptic vesicles full of neurotransmitters. Some of the vesicles contain excitatory transmitters and some contain inhibitory

Table 8.1   Classification of Peripheral Nerve Fibers

Class	Speed (m/s)	Innervation
Efferent (motor) fibers		
Aa	65 - 120	Fast twitch extrafusal fibers
Aβ	40 - 80	Slow twitch extrafusal fibers, muscle spindle intrafusal fibers
Aγ	10 - 50	Muscle spindle intrafusal fibers
B	4 - 25	Presynaptic autonomics
C	0.2 - 2.0	Postsynaptic autonomics
Afferent (sensory) fibers		
Ia	65 - 130	Muscle spindle intrafusal fibers
Ib	65 - 130	Golgi tendon organs
II	20- 90	Muscle spindle intrafusal fibers, pressure receptors
III	12 - 45	Temperature and pain receptors
IV	0.2 - 2.0	Viscera, pain receptors

Adapted, by permission, from J. Gamble, 1988, *The musculoskeletal system* (NY: Raven Press).

transmitters. The arrival of an action potential at a synapse results in neurotransmitters being released into the synaptic cleft. If excitatory neurotransmitters are released, the postsynaptic membrane will be depolarized, resulting in an action potential and transmission of the impulse. If inhibitory transmitters are released, the postsynaptic membrane will be hyperpolarized, thereby preventing the development of an action potential so that the impulse will not be transmitted.

Speed of Impulse Transmission

The speed of transmission or conduction of impulses is directly proportional to fiber diameter and the thickness of the myelin sheath. Nonmyelinated fibers do not have a myelin sheath and, thus, have much lower conduction speeds than myelinated fibers. Nerve fibers of the peripheral nervous system are classified on the basis of conduction speed and fiber diameter (table 8.1). There are five main categories of afferent fibers: Ia, Ib, II, III and IV, and five main categories of efferent fibers: Aα, Aβ, Aγ, B, and C. Skeletal muscle is innervated by the fastest afferent fibers (Ia and Ib) and the fastest efferent fibers (Aα, Aβ, and Aγ).

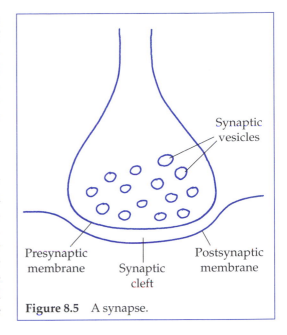

Figure 8.5   A synapse.

Nerve Tissue Organization in the Brain

The brain is the largest and most complex aggregation of neurons in the nervous system. It consists of the cerebrum and the cerebellum (see figure 8.1). The cerebrum occupies most of the cranium and is bigger than the cerebellum. The region of the cerebrum close to and including its surface is called the cerebral cortex and consists of **gray matter**—the cell bodies of neurons and their processes, which are largely nonmyelinated, together with their synapses and supporting glial cells. The cerebrum is heavily convoluted with fissures of varying depth. The largest fissure is the longitudinal central fissure that divides the cerebrum in the median plane into right and left cerebral hemispheres.

The cerebellum, which occupies the posterior inferior aspect of the cranium, is separated from the cerebrum by the transverse fissure. Like the cerebrum, the region of the cerebellum close to and including its surface consists of gray matter and is called the cerebellar cortex. The cerebellum is not convoluted but is traversed by numerous small furrows. The convolutions of the cerebrum and furrows of the cerebellum significantly increase their surface areas and, thus, the volume of gray matter. The inner parts of the cerebrum and cerebellum consist of **white matter**, largely myelinated nerve fibers organized into groups in which the fibers are parallel to each other. There are five groups of fibers resembling an intricate system of electrical wiring:

1. Fibers linking the different parts of the cerebral cortex
2. Fibers linking the different parts of the cerebellar cortex
3. Fibers linking the different parts of the cerebral cortex with the different parts of the cerebellar cortex
4. Fibers passing between the cerebral cortex and spinal cord
5. Fibers passing between the cerebellar cortex and spinal cord

The brain is the largest and most complex aggregation of neurons in the nervous system. It consists of the cerebrum and the cerebellum. The cerebral cortex and cerebellar cortex consist of gray matter. The inner parts of the cerebrum and cerebellum consist of white matter.

Gray matter: the cell bodies of neurons of the central nervous system and their fibers, which are largely nonmyelinated, together with their synapses and supporting glial cells

White matter: myelinated nerve fibers in the central nervous system

Nerve Tissue Organization in the Spinal Cord and Spinal Nerves

In transverse section the spinal cord is roughly oval shaped with an anterior median fissure and a posterior medial septum (figure 8.6). The central area is dominated by a roughly H-shaped mass of gray matter, with the rest of the cord consisting of white matter in which the groups of fibers run parallel with the spinal cord. The posterior projections (or horns) of the gray matter are continuous with afferent fibers that enter the spinal cord via the left and right dorsal (posterior) roots of the corresponding left and right spinal nerves. The cell bodies of the afferent fibers are located in the dorsal root ganglia; a ganglion is an aggregation of cell bodies outside the spinal cord.

The anterior projections of the gray matter are continuous with efferent fibers that leave the spinal cord via the left and right ventral (anterior) roots of the corresponding spinal nerves. A spinal nerve is formed by the aggregation of the afferent and efferent fibers as they pass through the corresponding intervertebral foramen. Each spinal nerve divides into an anterior and posterior branch just outside the intervertebral foramen; the anterior branch is usually much larger than the posterior branch. In each branch the individual nerve fibers are enveloped in a thin layer of connective tissue called the endoneurium, which supports a blood capillary network. The fibers are grouped together in bundles called fasciculi or funiculi and each fasciculus is sheathed within a layer of connective tissue called the perineurium. The fasciculi are grouped together and sheathed within another layer of connective tissue called the epineurium to form the complete spinal nerve (figure 8.7). Each spinal nerve consists of a mixture of myelinated and nonmyelinated nerve fibers.

> *In the spinal cord the posterior horns of the gray matter are continuous with afferent fibers of the corresponding spinal nerves, and the anterior horns are continuous with efferent fibers of the corresponding spinal nerves. Each spinal nerve consists of myelinated and nonmyelinated nerve fibers.*

Plexus: network of spinal nerves that innervate a specific region of the body

In all regions of the spinal cord apart from most of the thoracic region, the spinal nerves link up with each other on each side of the spinal cord to form networks called *plexuses*. The upper four pairs of cervical nerves form the left and right cervical plexuses that innervate the head and upper part of the neck. The other cervical spinal nerves combine with the first pair of thoracic spinal nerves to form the left and right brachial plexuses that innervate the arms. The twelfth thoracic spinal nerves and the spinal nerves of the lumbar and sacral regions combine to form the left and right lumbosacral plexuses that innervate the legs. The second to the eleventh thoracic spinal nerves, which innervate the trunk, do not form plexuses.

Voluntary and Reflex Movements

In general terms, the brain interprets sensory information from the various receptors and on the basis of this information brings about appropriate responses via effectors to ensure normal bodily functioning. Whereas the autonomic nervous system operates largely at a subconscious level, the cerebrospinal nervous system normally operates at the level of consciousness; that is, the individual consciously processes information, such as visual, auditory, and kinesthetic (sense of movement) information, and consciously brings about appropriate responses. For example, consider a child just about to touch a hot coffeepot resting on a stove (figure 8.8).

1. As his hand moves close to the pot, the heat radiating from the pot may excite the heat receptors in the skin of his hand.

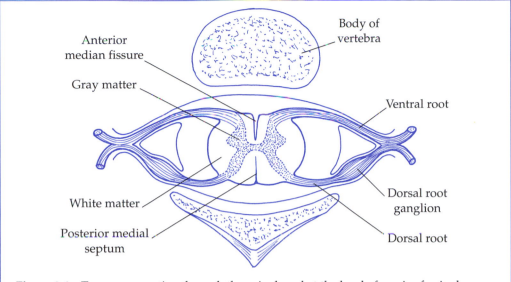

Figure 8.6   Transverse section through the spinal cord at the level of a pair of spinal nerves.

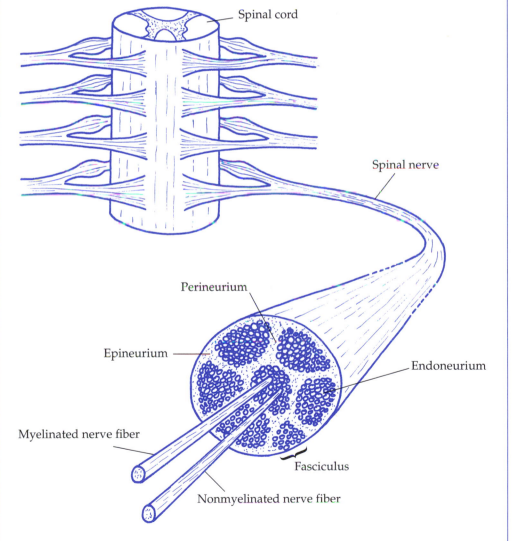

Figure 8.7   Structure of a spinal nerve and its relationship to the spinal cord.

2. This information is transmitted by sensory neurons in the peripheral nervous system to the spinal cord, where the nerve endings of the axons of the sensory neurons synapse with many other neurons.

3. Some of these neurons relay the information to the brain center responsible for heat sensation in the hand, and the child experiences the sensation of heat.

4. Having sensed the heat from the coffeepot, the child decides to move his hand away from the pot to avoid the danger.

5. Impulses are sent from the brain to motor neurons in the spinal cord that innervate the muscles of the arm.

6. The impulses are relayed to the muscles of the arm resulting in movement of the hand away from the coffeepot.

The above sequence, illustrated in figure 8.8, is typical of all voluntary movements. If the child had not sensed the danger and had actually touched the hot coffeepot, he would have jerked his hand away from the pot with lightening speed before he was even conscious of the heat. This extremely rapid involuntary reaction is called a **reflex action** and is one of a host of similar reflexes that result in instant reactions to protect the body from potentially harmful stimuli. In this particular case, the reflex action results in the child's hand being in contact with the coffeepot for only a fraction of the time it would have taken for the child to take voluntarily (and, therefore, consciously) his hand away from the pot. Consequently, the child may sustain a relatively slight burn rather than a serious burn, which would have resulted from more prolonged contact with the coffeepot.

The increased speed of a reflex action compared to a voluntary action is due to a reduction in the distance over which the impulses travel from receptor to effector. In this particular case, the heat from the coffeepot is sensed by the heat receptors in the hand, which send impulses to the spinal cord. This is basically the same as in a voluntary action—stages 1 and 2 in figure 8.8. However, in a reflex action the impulses are transmitted directly across the spinal cord to the motor neurons that

Reflex action: an involuntary movement that provides protection by rapidly removing part of the body from a potential source of danger without conscious effort

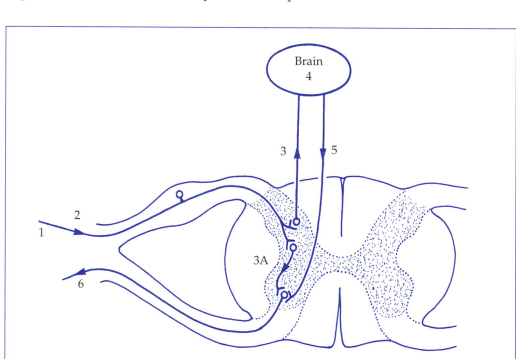

Figure 8.8   Pathway of impulses in voluntary and reflex movements (see text for description of stages).

innervate the muscles of the arm—stage 3A in figure 8.8. The impulses are relayed to the muscles of the arm and the reflex action is completed. Consequently, a reflex action is faster than a voluntary action because no time is spent in transmitting impulses up the spinal cord to the brain, making a decision, and transmitting impulses back down the spinal cord—stages 3, 4, and 5 in figure 8.8 are omitted. A reflex action is triggered when the intensity of the impulse output from the receptors is above a certain threshold indicating extreme danger to the body. In this particular case, the hotter the coffeepot, the greater the intensity of output from the heat receptors and the greater the likelihood of a reflex action being triggered.

As the impulses are transmitted across the spinal cord (stage 3A), the impulses are simultaneously transmitted to the brain as in a voluntary action (stage 3). However, stage 3 (transmission of impulses to the brain and sensation of heat) takes longer than stages 3A and 6 such that it is a few tenths of a second after the boy jerks his hand away from the coffeepot before he perceives any pain in the hand.

> *The autonomic nervous system operates at a subconscious level. The cerebrospinal nervous system normally operates at the level of consciousness, but may produce involuntary reflex actions.*

Nerve Fiber Injuries

Injuries to peripheral nerves usually occur as a result of compression or traction (stretching) or a combination of the two (Kleinrensink et al. 1994). Compression and traction may damage any or all of the main structures: connective tissue sheaths, Schwann cells, myelin sheath (when present), and nerve fibers. Compression and traction damage the nerve structures directly due to crushing and tearing, respectively. Prolonged compression also may damage the nerve structures indirectly as a result of ischemia—a disruption of the local blood supply (due to compression of the local capillary network), which results in a deficiency of oxygen and nutrients to the affected tissues.

Injuries to peripheral nerves are classified on the basis of degree of structural and functional damage into neuropraxia, axonotmesis, and neurotmesis (Seddon 1972). Neuropraxia is the lowest level of damage. It involves damage to Schwann cells and myelin sheaths, but little or no damage to nerve fibers or endoneurium. Neuropraxia is characterized by a disruption to impulse transmission, which is often associated with pain and tingling in the areas innervated by the affected nerves. Recovery of a nerve fiber from neuropraxia usually occurs within 10 to 14 days. In axonotmesis, the nerve fiber and Schwann cell covering or myelin sheath are severed, but the endoneurium remains intact. The severed part of the fiber degenerates. Recovery begins when fiber sprouts emerge from the severed end of the part of the fiber still attached to the cell body. One of the sprouts eventually dominates and gradually grows along the tube formed by the endoneurium. Growth occurs at a rate of approximately 2.5 cm per month, and function gradually returns to normal as the fiber and myelin sheath (when present) returns to normal. In neurotmesis, the nerve fiber, Schwann cell covering or myelin sheath, and endoneurium are all severed. The severed part of the fiber degenerates and without surgical repair it is unlikely that much recovery will occur. With surgical repair some recovery occurs similarly to the way in which a fiber recovers from axonotmesis, but functional outcome is often less than satisfactory.

> *Compression and traction can damage nerves directly due to crushing and tearing, respectively. Prolonged compression may damage the nerve structures indirectly as a result of ischemia.*

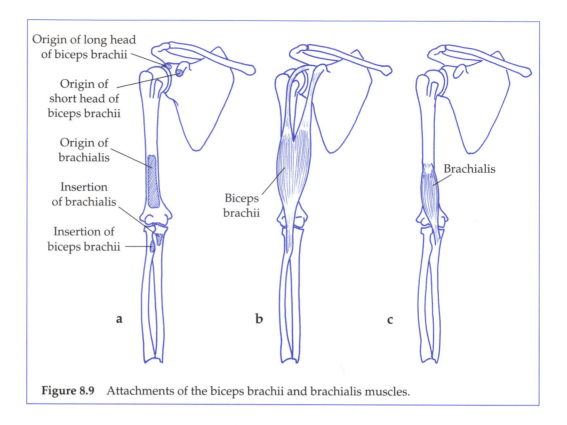

Figure 8.9   Attachments of the biceps brachii and brachialis muscles.

Skeletal Muscle Structure

The composition and basic function of the muscular system are described in chapter 2. This section considers the macrostructure of skeletal muscle.

Origins and Insertions

Most of the muscles are attached to the skeletal system by tendons or aponeuroses (as described in chapters 2 and 4). However, one or both attachments of some muscles attach directly onto bone without an intervening tendon or aponeurosis. These muscles are often located adjacent to bones. For example, the brachialis, a muscle that flexes the elbow, arises directly from a large area on the lower half of the anterior aspect of the humerus and is inserted via a thick and broad tendon onto the coronoid process of the ulna (figure 8.9, *a* and *c*). Most of the muscles of the upper and lower limbs are arranged in line with the direction of the long bones. For descriptive purposes, the proximal and distal attachments of each of these muscles are referred to as the origin and insertion of the muscles, respectively. For example, the origin of the brachialis is on the humerus and the insertion is on the ulna. Some muscles have more than one site of origin and more than one site of insertion. For example, the biceps brachii has two sites (or heads) of origin and one site of insertion (figure 8.9, *a* and *b*). The origins and insertions of the muscles of the trunk and the muscles that link the trunk to the limbs tend to be medial and lateral, respectively, but there are many exceptions.

> *The origins and insertions of the muscles of the limbs tend to be proximal and distal, respectively. The origins and insertions of the muscles of the trunk and the muscles that link the trunk to the limbs tend to be medial and lateral, respectively.*

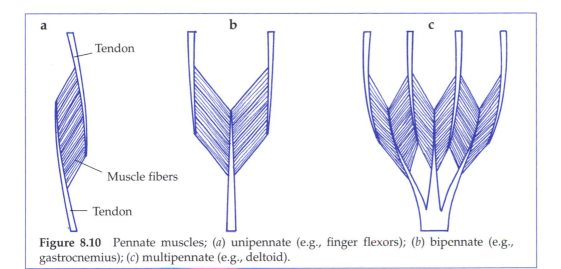

Figure 8.10   Pennate muscles; (*a*) unipennate (e.g., finger flexors); (*b*) bipennate (e.g., gastrocnemius); (*c*) multipennate (e.g., deltoid).

Pennate and Nonpennate Muscles

A skeletal muscle is made up of skeletal muscle cells bound by various layers of connective tissue that support extensive networks of nerves and blood vessels. Muscle cells are long and thin and, as such, they are usually referred to as **muscle fibers**. Each muscle fiber is approximately 50 μm (1 μm = one-millionth of a meter) wide. All the fibers in an individual muscle are about the same length. In some muscles the fibers are relatively short; for example, in the muscles that move the eyes, the fibers are 2 to 4 mm long. In contrast, the fibers in some other muscles are very long; for example, the fibers of the sartorius are approximately 30 cm long. The muscle fiber length in most muscles is between these two extreme values.

The fibers in all muscles are organized into bundles (as described later), and the fibers in each bundle run parallel with each other. However, the arrangement of the bundles of fibers with respect to the origin and insertion of the muscle is either **pennate** or **nonpennate**. In a pennate muscle, the fibers run obliquely with respect to the origin and insertion so that the line of pull of the fibers is oblique to the line of pull of the muscle (figure 8.10). Pennate muscles have a featherlike appearance and are classified according to the number of groups of fibers into unipennate, bipennate,

Muscle fiber: a muscle cell

Pennate muscle: a muscle in which the fibers run obliquely with respect to the origin and insertion so that the line of pull of the fibers is oblique to the line of pull of the muscle

Nonpennate muscle: a muscle in which the fibers run in line with the line of pull of the muscle

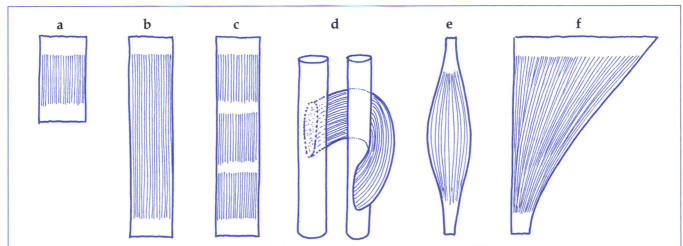

Figure 8.11   Nonpennate muscles; (*a*) quadrilateral (e.g., pronator quadratus); (*b*) strap (e.g., sartorius); (*c*) strap with tendinous intersections (e.g., rectus abdominis); (*d*) spiral (e.g., supinator); (*e*) fusiform (spindle shaped) (e.g., biceps brachii); (*f*) fan shaped (e.g., pectoralis major).

and multipennate. In a unipennate muscle there is one group of fibers that inserts onto the sides of two tendons (or one bony attachment and one tendon) (figure 8.10*a*). The flexors and extensors of the fingers are unipennate muscles. In a bipennate muscle such as the gastrocnemius there are two groups of fibers that insert onto the opposite sides of a central tendon (figure 8.10*b*). A multipennate muscle such as the deltoid is, in effect, two or more bipennate muscles combined into a single muscle (figure 8.10*c*). In a nonpennate muscle the fibers run in line with the line of pull of the muscle. There are five main types: quadrilateral, strap, spiral, fusiform, and fan shaped (figure 8.11). In a spiral muscle, the muscle curves around other muscles or bones. In a fusiform (or spindle-shaped) muscle the fibers are gathered at each end to attach onto long, relatively narrow, tendons. The biceps brachii is a fusiform muscle (see figure 8.9*b*). In a fan-shaped muscle the fibers converge from a broad origin to a relatively small insertion. The effect of pennate and nonpennate arrangements on muscle function is described later in the chapter.

Fusiform Muscle-Tendon Units

Figure 8.12 shows the structure of a fusiform muscle-tendon unit. Each muscle fiber is enveloped in a layer of areolar tissue called the endomysium, which helps to bind the muscle fibers together and provides a supporting framework for blood capillaries and the terminal branches of nerve fibers. The muscle fibers are grouped together by irregular connective tissue (a mixture of collagenous and elastic) into bundles of up to 200 fibers; each bundle is called a fasciculus (or funiculus), and the connective tissue sheath is called the perimysium. The fasciculi are bound together to form the belly of the muscle by a layer of irregular collagenous connective tissue called the epimysium. The muscle fibers gradually taper at each end. The tapering of the fibers accompanies a gradual thickening in the epimysium and perimysium layers and a change in the composition of the epimysium and perimysium layers from irregular collagenous and irregular elastic connective tissue to regular collagenous connective tissue as the epimysium and perimysium layers merge to form a tendon or aponeurosis (see figure 4.1).

Each muscle receives one or more nerves (collections of sensory and motor nerve fibers), which usually enter the muscle together with the main blood vessels (arteries enter, veins leave) at a region of the muscle that does not move a great deal during normal movement; this region is referred to as the neurovascular hilus (Williams et al. 1995). The blood vessels and nerves branch through the epimysium and perimysium layers down to the endomysium of the individual muscle fibers.

A skeletal muscle is made up of skeletal muscle fibers bound together by connective tissue that supports extensive networks of nerves and blood vessels. The muscle fibers are organized into bundles that have a pennate or nonpennate arrangement.

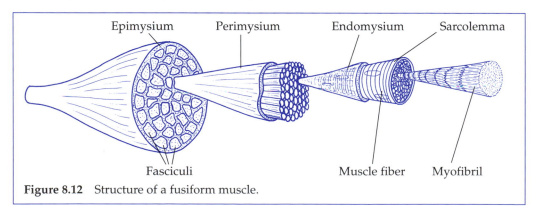

Figure 8.12   Structure of a fusiform muscle.

Fascial or Osseofascial Compartments

Any group of muscles that functions as a single unit, such as the quadriceps muscles, is usually confined within its own fascial or osseofascial **compartment**, which separates it from adjacent muscles or muscle groups. The quadriceps are confined within an osseofascial compartment formed by the anterior aspect of the femur and the merging of the epimysial layers of the vastus lateralis, rectus femoris, and vastus medialis (figure 8.13). Consequently, the quadriceps muscles are normally closely packed together, with sufficient space for the passage of blood vessels and nerves. However, any decrease in the size of the osseofascial compartment or any increase in the volume of the contents of the osseofascial compartment may increase the pressure on the blood vessels and nerves within the compartment. **Compartment pressure syndrome** occurs when the pressure within a fascial or osseofascial compartment surrounding a particular muscle or group of muscles becomes elevated to the extent that blood flow within all or part of the compartment is reduced to an abnormally low level resulting in ischemia of muscle and nerve cells (Segan et al. 1988). Ischemia results in a variety of symptoms including persistent pain, increased pain with stretching, and decreased sensitivity to light touch and pinprick. If severe compartment pressure syndrome is not treated, localized tissue necrosis (degeneration and death of cells) may occur.

Increased pressure within a compartment can occur due to increased stiffness of the fascial binding following injury. However, it is more likely to occur due to an increase in the volume of the contents of the compartment. There are three main causes of an increase in volume: muscle hypertrophy, muscle injury, and prolonged

Compartment: a fascial or osseofascial enclosure surrounding a group of muscles that normally function as a single unit

Compartment pressure syndrome: a painful condition resulting from increased pressure within a compartment

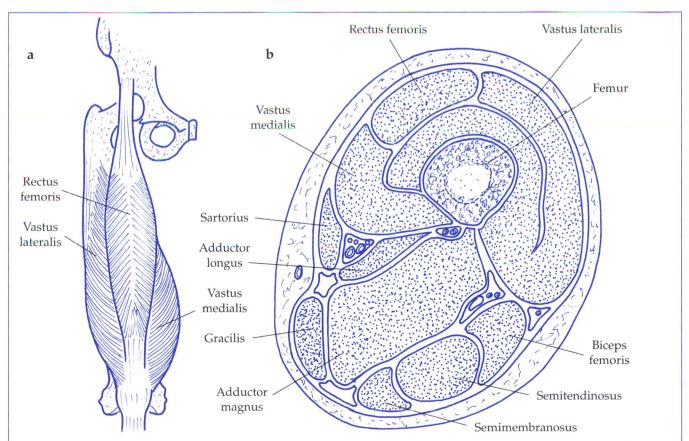

Figure 8.13 Muscle compartments of the right upper leg; (*a*) anterior aspect of the upper leg showing three of the four quadriceps muscles; (*b*) superior aspect of horizontal section through the middle of the upper leg.

Friction: the force exerted parallel to the surfaces of two objects in contact with each other, which opposes sliding, or the tendency to slide, of the surfaces on each other

high-intensity activity. Strength training can increase the cross-sectional area of the muscles that will increase the pressure inside a compartment if the fascial bindings do not adapt to the increased volume of muscle. Muscle injury results in leakage of blood from torn blood vessels, which may cause an increase in pressure within a compartment (Martinez, Steingard, and Steingard 1993). Prolonged high-intensity activity in sports such as kayaking and swimming in which certain muscle groups are more or less constantly active can result in compartment pressure syndrome during the activity if the associated fascial bindings are too tight (Ryan, Fricker, and Hannaford 1987).

Bursas and Synovial Sheaths

Most of the structures of the musculoskeletal system are closely packed together; the compartmentalization of muscle groups is an example of this. Due to the close packing there are many situations where adjacent structures move relative to each other, that is, slide over each other, during the course of normal movements. Sliding results in **friction**—a force parallel to the surfaces in contact that opposes the sliding movement (Watkins 1983). Friction generates heat and too much heat can damage or wear the surfaces. In machines, the surfaces of components that slide over each other are usually highly polished and lubricated to minimize friction. Similar mechanisms

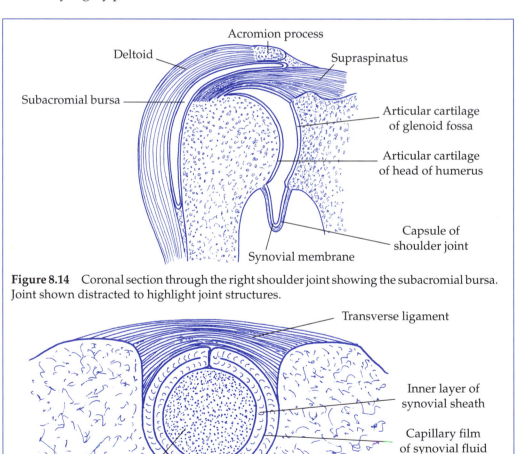

Figure 8.14   Coronal section through the right shoulder joint showing the subacromial bursa. Joint shown distracted to highlight joint structures.

Figure 8.15   Horizontal section through the bicipital groove showing the synovial sheath around the tendon of the long head of the biceps brachii.

exist within the musculoskeletal system—**bursas** and **synovial sheaths**—and the key component of both these mechanisms is synovial fluid.

A bursa is a closed sac comprised of synovial membrane containing synovial fluid. Bursas develop between structures that move relative to each other, providing complete separation (apart from the synovial fluid) and freedom of movement over a limited range. Depending on their location, bursas are described as subcutaneous, subtendinous, submuscular, and subfascial. Subcutaneous bursas are interposed between the deeper layers of the skin and underlying bone, ligament, or tendon; the prepatellar bursa is a subcutaneous bursa (see figure 7.13). A blister that develops on the skin in response to unaccustomed friction is a form of subcutaneous bursa. A blister is a short-term safety mechanism that separates the superficial and deep layers of the skin. Subtendinous bursas are interposed between tendons and bone, ligaments, or other tendons. The suprapatellar and infrapatellar bursas are subtendinous bursas. Submuscular bursas are interposed between muscles and bones, tendons, ligaments, or other muscles. The subacromial bursa between the deltoid and supraspinatus muscles is a submuscular bursa (figure 8.14). Subfascial bursas are interposed between aponeurotic areas and bone. The majority of bursas are close to joints and some bursas, such as the suprapatellar bursa and gastrocnemius bursa, are continuous with the joint capsule. Excessive repetitive or prolonged pressure on a bursa may inflame it, resulting in a painful condition referred to as **bursitis**. For example, subacromial bursitis, inflammation of the subacromial bursa, is a common injury in competitive swimmers.

Synovial sheaths cover tendons that pass under retinacula or within osseofibrous tunnels. For example, the tendon of the long head of the biceps brachii runs within an osseofibrous tunnel formed by the bicipital groove and the transverse humeral ligament (see figure 8.9b). A synovial sheath consists of a closed flattened sac comprised of synovial membrane containing a capillary film of synovial fluid wrapped around the tendon. The inner (or visceral) layer of the sheath is attached to the tendon, and the outer (or parietal) layer is attached to the osseofibrous tunnel (figure 8.15).

The median nerve, which innervates part of the hand and fingers, and the tendons of the wrist and finger flexor muscles run in an osseofibrous tunnel formed by the carpals and the flexor retinaculum of the wrist (figure 8.16). This tunnel is usually referred to as the carpal tunnel. All of the tendons are enclosed within synovial sheaths. Unaccustomed or prolonged, repetitive flexion and extension of the wrist can inflame the synovial sheaths resulting in a painful condition, called **tenosynovitis**, in which the synovial sheaths become swollen. Tenosynovitis of the wrist may

Bursa: a closed sac comprised of synovial membrane containing synovial fluid interposed between structures that move relative to each other to prevent or minimize friction between them

Synovial sheath: a closed sac comprised of synovial membrane containing a capillary film of synovial fluid interposed between a tendon and other structures such as an osseofibrous tunnel or retinaculum to prevent or minimize friction

Bursitis: a painful condition resulting from inflammation of a bursa

Tenosynovitis: a painful condition resulting from inflammation of a synovial sheath

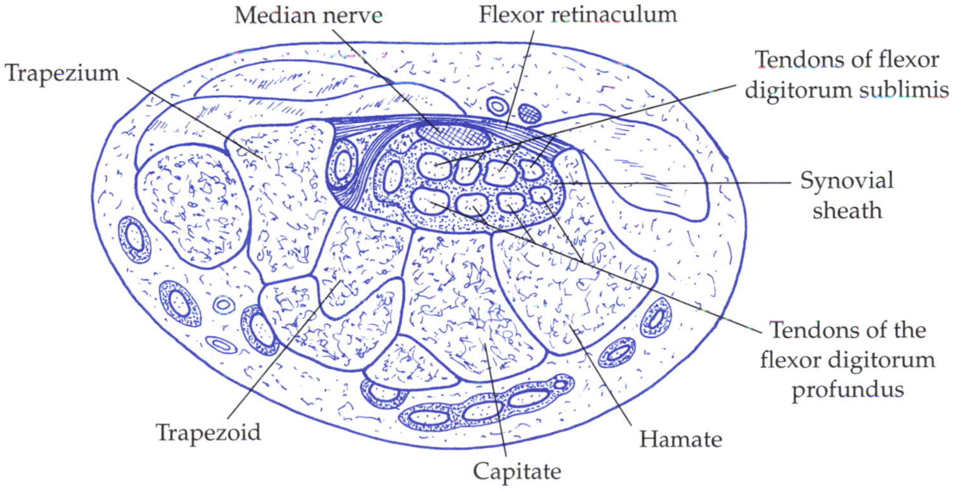

Figure 8.16   Horizontal section through the left wrist showing the osseofibrous carpal tunnel.

occur, for example, as a result of a long, strenuous game of squash or badminton in a person unaccustomed to such activity. It may also occur as a result of prolonged, repetitive tasks in the food processing industry, such as in preparing fish and poultry with a knife. The swelling increases the pressure within the carpal tunnel. The greater the pressure, the greater the possibility of neuropraxia of the median nerve, which is usually referred to as carpal tunnel syndrome (Steyers and Schelkun 1995). Carpal tunnel syndrome is characterized by pain and numbness or tingling in the sensory distribution of the median nerve—the palmer surfaces of the thumb, index finger, middle finger, and radial half of the third finger.

Muscle Fiber Structure and Function

Myofibril: a chain of sarcomeres

Myosin: a protein that largely comprises the thick filaments within a sarcomere

Actin: a protein that largely comprises the thin filaments within a sarcomere

Sarcomere: the basic structural unit of a muscle fiber that contains the contractile apparatus

A muscle fiber consists of hundreds or thousands of **myofibrils** embedded in sarcoplasm (muscle cytoplasm) and enclosed by a cell membrane called the sarcolemma (figure 8.17, *a* and *b*). Each muscle fiber has a large number of nuclei that lie just beneath the sarcolemma. Each myofibril is approximately 1 μm wide; the myofibrils are arranged parallel to each other and run the whole length of the muscle fiber. Each myofibril exhibits a characteristic pattern of alternate light and dark transverse bands due to the way the components of the myofibril reflect light (under an electron microscope). The light and dark bands are referred to as I (isotropic) and A (anisotropic) bands, respectively. Since the light and dark bands coincide in adjacent myofibrils the muscle fiber has a striped or striated appearance; thus, skeletal muscle is often referred to as striated muscle.

Each myofibril has two types of filaments arranged in a highly ordered way that gives rise to the light and dark bands. One type of filament is thicker than the other. The thicker filaments occupy the A bands and are composed of the protein **myosin**; these filaments are usually referred to as myosin filaments or A filaments (figure 8.17*c*). The thinner filaments—actin filaments or I filaments—occupy the I bands and are composed largely of the protein **actin** (Edman 1992). Each I band is divided by a transverse Z disc. The section of a myofibril between two successive Z discs is called a **sarcomere**—the basic structural unit of a muscle fiber. A myofibril consists of a chain of sarcomeres. The actin filaments project from each side of the Z discs and reach into the A bands of the corresponding sarcomeres where they interdigitate with the myosin filaments (figure 8.17*c*). The region between the ends of the two groups of actin filaments in the middle of a sarcomere is referred to as the H zone.

Each actin filament basically consists of two strands of actin molecules wound together longitudinally in a helical manner (figure 8.17*d*). Each myosin filament is composed of myosin molecules. Each myosin molecule is a clublike structure consisting of two adjacent globular heads attached by a relatively short curved neck to a long shaft (figure 8.17*e*). In a myosin filament the myosin molecules are packed such that the shafts form the main body of the filament with the heads projecting from the main body at regular intervals (figure 8.17*f*). The two halves of each myosin filament are mirror images of each other; that is, the myosin molecules in one half of the filament are oriented in the opposite direction to the myosin molecules in the other half. This arrangement, significant in the functional interaction between the actin and myosin filaments, is such that no myosin heads project from the central region of each filament; this central region is sometimes referred to as the inert zone, M band, or M line. Each myosin filament is surrounded by six actin filaments in a regular hexagonal arrangement (figure 8.17*g*). In each half of a myosin filament the myosin molecules are arranged in groups of six; a line joining the heads of the molecules in each group forms a spiral around the myosin filament. The corresponding head in each group faces the same actin filament.

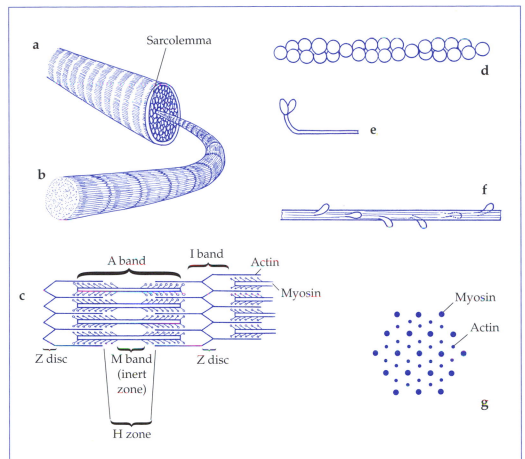

Figure 8.17 The structure of a muscle fiber; (*a*) muscle fiber; (*b*) myofibril; (*c*) arrangement of actin and myosin filaments in a sarcomere; (*d*) part of an actin filament; (*e*) myosin molecule; (*f*) part of a myosin filament; (*g*) hexagonal arrangement of actin and myosin filaments in a sarcomere.

Types of Muscular Contraction

When a muscle fiber contracts, the heads of the myosin molecules attach to special sites on the actin filaments in the regions where the actin and myosin filaments overlap. The attachments are referred to as cross bridges and each cross bridge generates a certain amount of tension. When a muscle fiber relaxes, the myosin heads detach from the actin and no tension is exerted between them.

When a muscle contracts, the tension exerted by the muscle is directly proportional to the number of cross bridges (the larger the number of cross bridges, the greater the tension). When a muscle contracts it tends to shorten, that is, it tends to pull its origin and insertion closer together. However, the muscle may shorten, lengthen, or stay the same length depending on the external load on the muscle (the load tending to lengthen the muscle). If the muscle force is greater than the external load, the muscle will shorten; this type of contraction is called a **concentric contraction**. If the muscle force is less than the external load, the muscle will lengthen; this type of contraction is called an **eccentric contraction**. Concentric and eccentric contractions are often referred to as **isotonic contractions**—contractions that involve a change in the length of the muscle. If the muscle force is equal to the external load, the length of the muscle will not change; this type of contraction is called an **isometric contraction**.

Concentric contraction: a contraction during which the muscle shortens

Eccentric contraction: a contraction during which the muscle lengthens

Isotonic contraction: a contraction involving a change in the length of the muscle, that is, a concentric contraction or an eccentric contraction

Isometric contraction: a contraction during which the length of the muscle does not change

Sliding Filament Theory of Muscular Contraction

During all types of muscular contraction the actin and myosin filaments stay the same length, but in isotonic contractions the degree of interdigitation between the two sets of filaments changes as the length of the muscle fibers changes; the width of the A bands stays the same, but the width of the I bands and H zones varies. As the muscle fibers shorten, the region of interdigitation increases and the width of the I bands and H zones decreases. As the muscle fibers lengthen, the region of interdigitation decreases and the width of the I bands and H zones increases. These observations led to the formulation of the **sliding filament theory** of muscular contraction (Huxley and Hanson 1954). The essential features of the theory are as follows:

1. When a muscle contracts, force is generated by the formation of cross bridges.

2. Flexion of the myosin molecule necks while the heads are in contact with the actin filaments exerts a pulling action on the actin filaments, which causes the actin filaments to slide relative to the myosin filaments, so that the muscle shortens.

3. After exerting their pulling action, the heads of the myosin molecules detach (or decouple) from the actin filaments and swing back to reattach (or recouple) onto the actin filaments farther along the actin filaments.

4. The coupling and decoupling of different cross bridges occurs at different times so that tension can be maintained while the muscle fiber shortens.

Whereas details of the processes involved have still to be discovered, the sliding filament theory has now gained general acceptance (Edman 1992).

Sliding filament theory: the generally accepted mechanism of muscle contraction involving coupling and decoupling and sliding between actin and myosin filaments

Motor unit: the functional unit of skeletal muscle consisting of a motor neuron with an Aα axon, together with all the terminal branches of the axon and the muscle fibers that they innervate

Motor Units

The functional unit of skeletal muscle is the **motor unit**. A motor unit consists of a motor neuron with an Aα axon (sometimes referred to as an alpha motoneuron), together with all the terminal branches of the axon and the muscle fibers that they innervate (Gamble 1988) (figure 8.18). The number of muscle fibers innervated by a single alpha motoneuron is referred to as the innervation ratio. The innervation ratio of different motor units varies considerably, from approximately 1:4 in the muscles that move the eyes, to approximately 1:2000 in the large back extensor and leg extensor muscles. When an alpha motoneuron transmits an action potential, all of the fibers in the motor unit contract. Consequently, muscles associated with very fine motor control, such as the muscles that move the eyes, have motor units with low innervation ratios so that the amount of force produced can be finely controlled. Muscles associated with forceful movements are made up of motor units with relatively high innervation ratios. The fibers of a single motor unit are usually mixed with the fibers of other motor units, but are grouped within a relatively small area of the muscle.

The innervation ratio is the number of muscle fibers in a motor unit. Muscles associated with fine motor control have motor units with low innervation ratios, whereas muscles associated with forceful movements are made up of motor units with relatively high innervation ratios.

A muscle fiber contracts when an action potential is generated in the sarcolemma at the junction with a motor end plate. The action potential is transmitted along and across the muscle fiber (by a system of tubules continuous with the sarcolemma) resulting in contraction. A single action potential produces a muscle twitch—

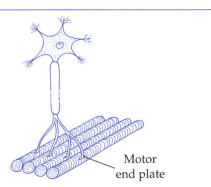
Motor end plate

Figure 8.18   The composition of a motor unit; an alpha motoneuron with all the terminal branches of the axon and the muscle fibers that they innervate.

a force that rapidly peaks and then equally rapidly dies away. If a series of action potentials is generated at a high enough frequency, the twitches fuse to produce tetanus, a sustained level of force.

Slow and Fast Twitch Fibers

Whereas the basic structure and function of all muscle fibers is the same, muscle fibers and, consequently, motor units vary in relation to the following:

1. Activation threshold—the level of stimulus required to generate an action potential
2. Contraction time—the time from force onset to peak force
3. Resistance to fatigue

Muscle fibers are classified into **slow twitch fibers** and **fast twitch fibers** on the basis of their contraction times. Slow twitch fibers, also referred to as type I and red fibers, have contraction times of 100 to 120 ms. Fast twitch fibers, also referred to as type II fibers and white fibers, have contraction times of 40 to 45 ms (Gregor 1993; Gamble 1988). The metabolism of slow twitch fibers is basically aerobic and, therefore, they are resistant to fatigue. Fast twitch fibers are subdivided on the basis of their metabolic characteristics into type IIa (aerobic and fatigue resistant) and type IIb (anaerobic and fatigue sensitive). The muscle fibers in a particular motor unit have the same functional characteristics and, as such, motor units can be classified into three categories:

- Slow contracting, fatigue resistant (S)
- Fast contracting, fatigue resistant (FR)
- Fast contracting, fatigable (FF)

Individuals differ in the proportions of the different types of muscle fibers in their muscles. The average person has approximately 50% type I and 25% type IIa and 25% type IIb fibers in her calf muscles. In comparison, elite distance runners have a much higher proportion of type I fibers, and elite sprinters have a much higher proportion of type IIb fibers (Gamble 1988).

Slow twitch fibers: muscle fibers with contraction times of 100 to 120 ms; also referred to as type I fibers and red fibers

Fast twitch fibers: muscle fibers with contraction times of 40 to 45 ms; also referred to as type II fibers and white fibers

Table 8.2   Characteristics of Slow and Fast Twitch Motor Units

	Slow twitch, fatigue resistant (S)	Fast twitch, fatigue resistant (FR)	Fast twitch, fatigable (FF)
Activation threshold of muscle fibers	Low	Moderate	High
Contraction time of fibers (ms)	100–120	40–45	40–45
Innervation ratio of motor unit	Low	Moderate	High
Type of muscle fibers	I	IIa	IIb
Type of axon	Aβ	Aα	Aα
Diameter of axon (μm)	7–14	12–20	12–20
Speed (m/s)	40–80	65–120	65–120
Duration and size of force	Prolonged low force	Prolonged relatively high force	Intermittent high force
Types of activities	Long distance running and swimming	Kayaking and rowing	Sprinting, throwing, jumping, and weight lifting

Whereas the classification of muscle fibers into types I, IIa, and IIb is widely used, the categories represent ranges of metabolic and functional characteristics rather than discrete categories (Sargeant 1994). The metabolic and functional characteristics of muscle fibers appear to be influenced considerably by the type of innervation the fibers receive. In experimental animals, it has been shown that altering the type of innervation results, over time, in a change in the metabolic and functional characteristics of muscle fibers (Noth 1992). Table 8.2 lists the characteristics of slow and fast twitch motor units.

> *Muscle fibers are classified into slow twitch and fast twitch fibers on the basis of their contraction times. The muscle fibers in a particular motor unit have similar functional characteristics and, as such, motor units are classified into three categories: slow contracting, fatigue resistant; fast contracting, fatigue resistant; and fast contracting, fatigable.*

Kinesthetic Sense and Proprioception

The central nervous system constantly receives sensory information from a wide variety of sources concerning the different aspects of physiological functioning. Awareness of body position and body movement is provided by a range of sensory organs, in particular, those concerned with the sensations of effort and heaviness, timing of the movement of individual body parts, the position of the body in space, joint positions, and joint movements.

Kinesthetic sense: awareness of body position and body movement

The input from these sources contributes to what is referred to as **kinesthetic sense** (or **kinesthesia**) (figure 8.19). Some aspects of kinesthetic sensitivity, such as a sense of effort and heaviness and a sense of timing of actions, are generated by sensory centers that monitor the motor commands sent to muscles. Other aspects of kinesthetic sensitivity are generated largely by input from peripheral receptors that monitor the execution of motor commands—the actual movements. For example, input from the eyes and ears is responsible for generating a sense of the position of the body in space. The sense of the position of a joint and movement of a joint is

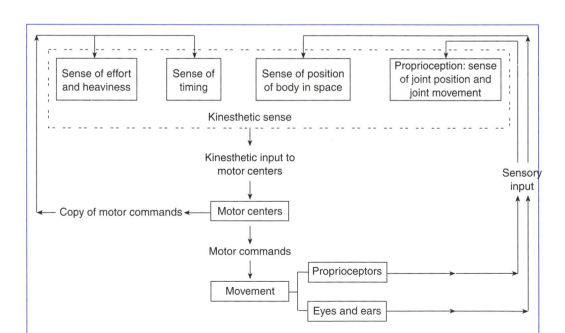

Figure 8.19   Relationship between kinesthetic sense and proprioception.

generated by a group of receptors located in the skin and musculoskeletal tissues. These receptors are called proprioceptors, and the sensation they provide is called **proprioception**. Proprioceptors are mechanoreceptors; they are activated by physical distortion.

Proprioceptors

The existence of proprioceptors in the skin, muscles, tendons, joint capsules, and ligaments is well established, but the precise role of the different proprioceptors and the interrelationships between them is less clear (Grigg 1994). However, it appears that proprioceptors in joint capsules and ligaments are largely responsible for generating a sense of joint position and joint movement in the end ranges of joint movements—where joint capsules and ligaments may become taut. Between the end ranges it seems unlikely that the proprioceptors in joint capsules and ligaments provide much sensory information since they are unlikely to be under sufficient tension to excite them.

Joint and Ligament Proprioceptors

There are two main types of proprioceptors in joint capsules: Ruffini end organs (or Ruffini corpuscles) and Pacinian corpuscles (figure 8.20). In a Ruffini end organ there are a small number of main terminal branches with a profuse system of small branches on the end of each main branch. The end of each main branch and its smaller branches are enclosed within a connective tissue covering and located between the collagenous fibers of the capsule. Ruffini end organs appear to be mainly responsive to tension. A Pacinian corpuscle is a more complex structure than a Ruffini end organ. It consists of a single terminal branch surrounded by several concentric layers of Schwann cells, all enclosed in a connective tissue cover. Pacinian corpuscles, which appear to be responsive to compression, are widely distributed between the collagenous fibers of the joint capsule and surrounding fascia.

The proprioceptors in ligaments and skin are similar to those found in joint capsules. However, there are few proprioceptors in ligaments and skin compared to the number in joint capsules. In addition, since tension in ligaments and skin may be caused by movement in a number of different directions, it is unlikely that the proprioceptors can provide information on movement in specific directions. For these reasons, it is thought that the contribution of the proprioceptors in ligaments and skin to proprioception is relatively small (Grigg 1994). However, they may make a significant contribution to joint stabilization in reflexive muscular activity.

Muscle and Tendon Proprioceptors

Whereas proprioceptors in joint capsules and, to a lesser extent, in ligaments, appear to generate information on joint position and joint movement in the end ranges of joint movements, proprioceptors in muscles and tendons seem to be responsible for generating this type of information for movement between end ranges. In tendons, there are proprioceptors called Golgi tendon organs that are responsive to tension. Golgi tendon organs are similar in structure to Ruffini end organs and are located between the collagenous fibers of the tendon. In muscle, there are specialized proprioceptors called **muscle spindles,** which, like Golgi tendon organs, are responsive to tension.

Proprioception: the sense of the position of a joint and movement of a joint; proprioception is part of kinesthetic sense

Muscle spindle: a tension receptor in muscle

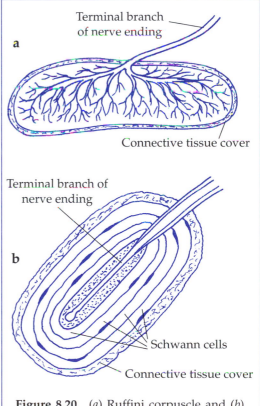

Figure 8.20   (*a*) Ruffini corpuscle and (*b*) Pacinian corpuscle.

Awareness of body position and body movement is provided by a range of sense organs, in particular, those concerned with the sensations of effort and heaviness, timing of the movement of individual body parts, the position of the body in space, joint positions, and joint movements. The input from all of these sources provides what is referred to as kinesthetic sense.

Each muscle spindle consists of a number of tiny muscle fibers enclosed within a spindle-shaped connective tissue capsule (figure 8.21). The muscle spindle fibers are referred to as intrafusal (inside the spindle) to distinguish them from the larger and more numerous extrafusal fibers that make up the vast majority of the muscle. Muscle spindles are embedded among extrafusal fibers. Intrafusal fibers are classified on the arrangement of their nuclei into nuclear bag fibers and nuclear chain fibers. In both types of fibers the nuclei are located in the central region of the fiber, which is devoid of myofibrils. In nuclear bag fibers the nuclei cluster in a group, whereas in nuclear chain fibers the nuclei are arranged in a line parallel to the long axis of the fiber. The regions of the fibers on each side of the central region contain a large number of myofibrils and are referred to as the polar regions (Enoka 1994).

Nuclear bag fibers and nuclear chain fibers are usually about 8 mm and 4 mm long, respectively, and in a typical muscle spindle there are two bag fibers and four chain fibers (Roberts 1995). The central regions of both types of fibers are supplied with type Ia and type II sensory nerve endings. The endings of type Ia nerve fibers spiral around the intrafusal fibers and the endings of type II nerve fibers consist of a number of branches with end bulbs; the spirals and end bulbs adhere closely to the sarcolemma of each fiber.

The polar regions of the intrafusal fibers are supplied with motor nerve endings from Aβ and Aγ motor nerve fibers. Stimulation of the intrafusal fibers via these nerves results in contraction of the polar regions, which stretches the central regions and excites the spiral and end bulb nerve endings. This results in sensory discharge via the type Ia and II sensory nerve fibers. The type Ia and II fibers synapse in the spinal cord directly with the Aα motor neurons that supply the extrafusal muscle fibers of the same muscle, resulting in contraction of the extrafusal fibers. The level of contraction of the extrafusal fibers depends on the level of activation, which, in turn, depends on the degree of tension in the intrafusal fibers. Under resting conditions there is always a certain amount of tension in the intrafusal muscle fibers (due to activation by the sensory centers of the brain via the Aβ and Aγ fibers), which, in turn, results in a certain amount of tension in the extrafusal muscle fibers. The resting level of tension in the intrafusal and extrafusal muscle fibers is called **muscle tone**; the level of muscle tone in the extrafusal fibers is determined by the level of muscle tone in the intrafusal fibers.

Muscle tone: the resting level of tension in muscle; the level of muscle tone in the extrafusal fibers is determined by the level of muscle tone in the intrafusal fibers

Sensory output from muscle spindles can be generated by stimulation via the Aβ and Aγ motor nerve fibers (sometimes referred to as the fusimotor nerves) that innervate the intrafusal fibers. However, sensory output from muscle spindles can also be generated by stretching the muscle as a whole, since this will stretch the muscle spindles and excite the spiral and end bulb nerve endings. Low-velocity stretching of active muscles—eccentric muscle contraction—is an essential feature of normal movements; as a muscle group shortens, its antagonist partner experiences eccentric contraction. It is thought that the sensory information provided by muscle spindles as a result of stretching during eccentric contraction provides a sense of joint position and joint movement, especially during midrange movements (Gandevia, McClosky, and Burke 1992). The amount of sensory information generated depends on the muscle tone of the intrafusal fibers; increasing the muscle tone of the intrafusal fibers increases the sensitivity of the spindles to stretch, and vice versa.

Whereas low-level stretch of a muscle seems to be important in generating proprioceptive information concerning joint movement and joint position, rapid

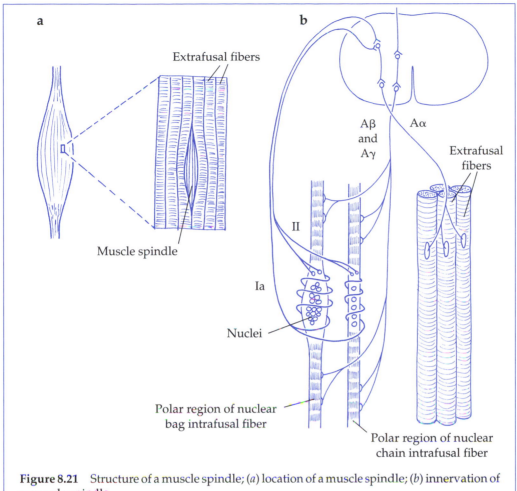

Figure 8.21   Structure of a muscle spindle; (*a*) location of a muscle spindle; (*b*) innervation of a muscle spindle.

stretch results in reflex contraction of the muscle (via the spindle afferent to muscle efferent loop) to prevent subluxation of the associated joints. These **stretch reflexes** are important in protecting joints from injury. For example, recurrent inversion sprains of the ankle are associated with deficient stretch reflex of the everters and dorsiflexors (Garn and Newton 1988). Excitation of muscle spindles appears to be mainly responsible for initiating reflex muscle contractions. However, there is evidence that proprioceptors in joint capsules and ligaments may also contribute to the initiation of such reflex contractions, especially at the ends of joint ranges of movement (Matthews 1988; Hall et al. 1994). In these situations joint capsules and ligaments can be taut and their proprioceptors most active.

> *Proprioceptors in joint capsules and ligaments (Ruffini end organs and Pacinian corpuscles) appear to generate information on joint position and joint movement in the end ranges of joint movements. Proprioceptors in muscles (muscle spindles) and tendons (Golgi tendon organs) appear to be responsible for generating this type of information for movement in between end ranges of joint movements.*

Role of Proprioceptors

The precise roles of the various types of proprioceptors are not yet clear. However, there is general agreement that proprioceptive information aids coordination and

Stretch reflex: the involuntary contraction of a muscle in response to a sudden unexpected increase in its length; the reflex is brought about by the spindle afferent to muscle efferent loop

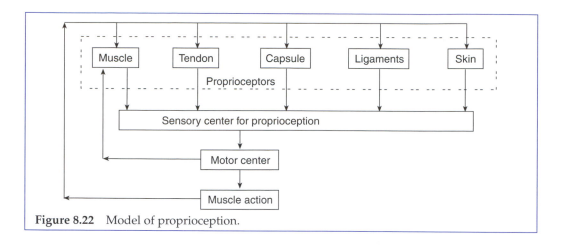

Figure 8.22   Model of proprioception.

balance and, in particular, maintains joint congruence (Grigg 1994; Wilkerson and Nitz 1994) (figure 8.22). It appears that injury to muscles, ligaments, and capsules may damage proprioceptors resulting in long-term proprioceptive deficits, which, in turn, contribute to the development of degenerative joint diseases such as osteoarthritis (Freeman and Wyke 1967; Garn and Newton 1988; Hall et al. 1995). There is evidence that proprioceptive information from muscle-tendon units can be enhanced by specific exercises that emphasize activation of muscle spindles and, thereby, improve muscle tone. Such enhancement may compensate for proprioceptive deficits in other structures such as joint capsules and ligaments (Beard et al. 1994; Skinner et al. 1986; Steiner et al. 1986).

Mechanical Characteristics of Muscle-Tendon Units

The amount of force generated by a muscle-tendon unit depends on the length of the unit at the time of stimulation and the speed with which it changes length in the ensuing contraction. In this regard, an isometric contraction is simply one point on the continuum between maximum velocity of shortening (concentric contraction) and maximum velocity of lengthening (eccentric contraction).

Length-Tension Relationship in a Sarcomere

The amount of tension generated in a sarcomere depends on the number of cross bridges between the actin and myosin filaments: the greater the number of cross bridges, the greater the force. The actual number of cross bridges depends on the following:

1. The degree of interdigitation between the actin and myosin filaments. Too much and too little interdigitation decreases the number of myosin heads that are in a position to attach to form cross bridges.
2. The level of stimulation (activation) applied to the sarcomere. The higher the stimulation, the greater the number of cross bridges that are formed and, therefore, the greater the force.

Figure 8.23 shows the isometric length-tension relationship for a sarcomere; the sarcomere was maximally stimulated, but not allowed to shorten, at a number of different sarcomere lengths and the force was recorded at each length (Gordon, Huxley, and Julian 1966). When the sarcomere is extended to the point where there

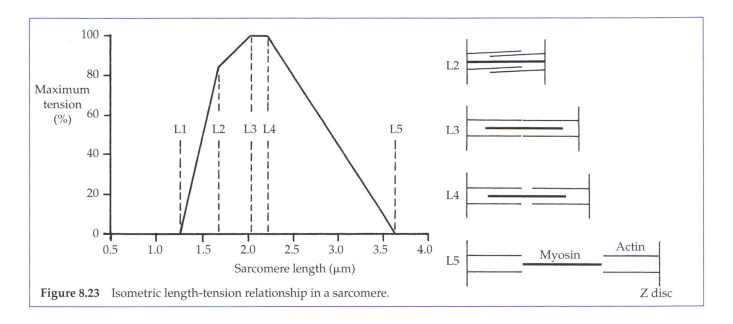

Figure 8.23   Isometric length-tension relationship in a sarcomere.

is no interdigitation between the actin and myosin filaments, no tension is generated since there are no myosin heads in a position to attach to form cross bridges. This situation is represented by the point L5 in figure 8.23. As the sarcomere shortens and the degree of interdigitation gradually increases, there is a linear increase in the amount of force generated up to the point L4 where the length-tension relationship levels off. At L4 the maximum number of cross bridges have been formed and, as such, tension is maximum.

The region of the length-tension relationship between L4 and L5 is referred to as the descending limb. At L4 the H zone corresponds to the inert zone of the myosin filaments (see figure 8.17c). As the sarcomere shortens between L4 and L3 the tension stays the same since no more cross bridges can be formed (the inert zones of the myosin filaments do not have myosin heads to form cross bridges). The region of the length-tension relationship between L3 and L4 is referred to as the plateau region. At L3 the ends of the actin filaments in each half of the sarcomere come together—the H zone is zero. As the sarcomere shortens between L3 and L2 the actin filaments progressively overlap each other, resulting in a progressive drop in tension since the overlapping actin filaments interfere with cross bridge formation in the region of overlap. The region of the length-tension relationship between L2 and L3 is referred to as the shallow ascending limb. At L2 the ends of the myosin filaments abut the Z discs. As the sarcomere shortens between L2 and L1 there is a rapid and progressive drop in tension due to progressive overlap of the actin filaments and progressive longitudinal compression of the myosin filaments, both of which reduce the number of myosin heads available to form cross bridges. At L1 no cross bridges can be formed and, consequently, no tension is generated. The region of the length-tension relationship between L1 and L2 is referred to as the steep ascending limb.

Length-Tension Relationship in a Muscle-Tendon Unit

The **length-tension relationship** of a muscle-tendon unit is different from that of a sarcomere due to the muscle-tendon unit's connective tissue, which exerts passive tension when stretched. Consequently, the tension produced by a muscle-tendon unit will be the sum of the tension produced by the contractile (muscle) component and the passive tension exerted by the connective tissue components. Figure 8.24 shows the isometric length-tension relationship (or curve) for a muscle-tendon unit. The

Length-tension relationship: the relationship between length and tension in a muscle

contributions of the contractile and connective tissue components to total tension at any particular length are shown in the separate curves. Some of the connective tissue components are parallel with the muscle fibers and some are arranged in series; this has given rise to the terms parallel elastic component and series elastic component (Huijing 1992) (figure 8.25). The parallel elastic component consists of sarcolemma, endomysia, perimysia, and epimysium. The series elastic component consists of tendons and aponeuroses, and strands of the protein titin, which connect the ends of the myosin filaments to the Z discs in each sarcomere. In addition to contributing to passive tension when stretched, the titin strands stabilize the hexagonal arrangement of the actin and myosin filaments (Lieber and Bodine-Fowler 1993).

In the absence of stimulation and any external load, the muscle-tendon unit assumes a rest length at which the tension in the unit is zero, with no tension in the contractile component, and no tension and, therefore, no stretch in the connective tissue component. Figure 8.24 shows that rest length is associated with that part of the isometric length-tension curve at which tension in the contractile component is maximum.

As in a sarcomere, the shorter the muscle-tendon unit, the lower the tension generated. Tension tends to reduce to zero when the unit shortens to approximately 60% of its rest length (see figure 8.24). However, this is unlikely to occur in practice due to the arrangement of the muscle-tendon units on the skeleton. When a muscle-tendon unit contracts in a very shortened state and, as such, generates low force, it is said to be in a state of **active insufficiency** (Elftman 1966). This is more likely to

Active insufficiency: the lower limit, minimum length, of the working range of a muscle

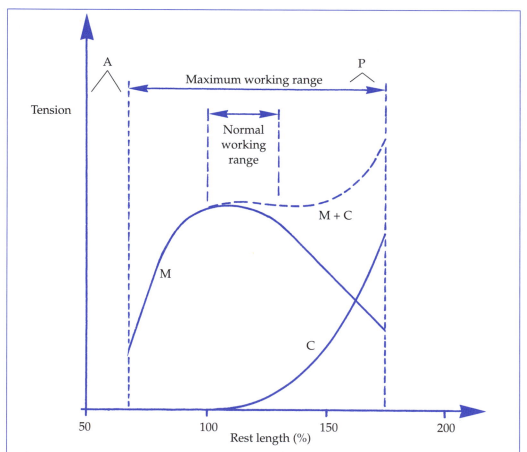

Figure 8.24 Isometric length-tension relationship in a muscle-tendon unit: M—tension exerted by contractile component; C—tension exerted by connective tissue component; M + C—total tension; P—passive insuffieciency; A—active insufficiency.

occur with muscle-tendon units that cross more than one joint. For example, the hamstrings extend the hip and flex the knee. However, the muscle fibers in the hamstrings are not long enough to fully extend the hip and flex the knee simultaneously. If a person stands on one leg and attempts to fully extend the hip and fully flex the knee of the other leg at the same time, she will be unable to flex the knee much more than 90°; that is, the hamstrings will be in a state of active insufficiency. In contrast, if the person flexes the hip and then tries to fully flex the knee she will find that the knee flexes to approximately 140°. This is possible because the hamstrings operate closer to rest length and are therefore able to exert more force (Elftman 1966).

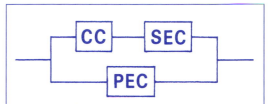

Figure 8.25   Model of the relationship between the contractile and connective tissue components in a muscle-tendon unit: CC—contractile component; SEC—series elastic component; PEC—parallel elastic component.

Figure 8.24 shows that as a muscle-tendon unit lengthens beyond rest length, the isometric tension generated is fairly constant between 100% and 150% of rest length and then increases to a maximum at approximately 175% of rest length. The change in isometric tension between rest length and maximum tension is associated with a gradual decrease in the amount of tension produced by the contractile component and a gradual increase in the amount of passive tension. Tension in the contractile component would be reduced to zero if the muscle-tendon unit were lengthened to approximately 210% of rest length. However, the parallel elastic connective tissue components ensure that this situation does not arise by limiting the maximum length of the muscle-tendon unit to approximately 175% of rest length. In this situation the muscle-tendon unit is said to be in a state of **passive insufficiency** (Elftman 1966).

When a muscle-tendon unit lengthens, all of the sarcomeres do not lengthen to the same extent. Theoretically a situation could occur where some of the sarcomeres in a muscle fiber were fully extended, that is, no interdigitation between the actin and myosin filaments, while other sarcomeres were not fully extended. If the muscle were stimulated to contract in this state, the fully stretched sarcomeres would not be able to contract, whereas the other sarcomeres would be able to contract. This would result in further stretching and, consequently, damage to the already fully stretched sarcomeres. This theoretical condition has been referred to as muscle instability, and may be prevented, at least in part, by passive insufficiency (Alexander 1989). Consequently, the maximum working length range of a muscle-tendon unit is determined by the lengths at which active insufficiency and passive insufficiency occur, that is, between approximately 75% and 175% of rest length. However, it is likely that the normal working range is between approximately 100% and 130% of rest length. This range incorporates the region of the length-tension curve where contractile tension is maximum and, as such, allows maximum flexibility in tension generation (see figure 8.24).

Passive insufficiency: the upper limit, maximum length, of the working range of a muscle

Force-Velocity Relationship in a Muscle-Tendon Unit

The everyday physical tasks individuals perform are usually well within the strength capability of the muscle-tendon units used. In such movements, the muscle-tendon units generate just enough tension to overcome the external load acting on them so they can move the external load. The external load may simply be the weight of a limb segment such as the forearm in a movement involving elbow flexion. At other times the external load consists of the weight of the limb segments together with any additional load that is being moved such as something held in the hand.

When the amount of force produced by a muscle (muscle-tendon unit) just matches the external load, the muscle contracts isometrically. The maximum load the muscle can sustain isometrically is called the isometric strength of the muscle. When the external load is less than isometric strength, the muscle is able to contract

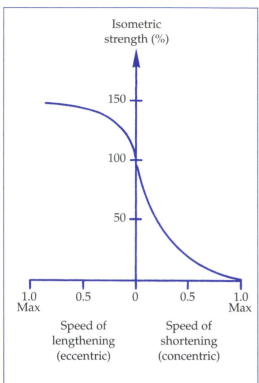

Figure 8.26   The force-velocity relationship in skeletal muscle.

concentrically. The speed of shortening in a concentric contraction depends on how much force the muscle needs to produce to move the external load. The greater the external load, the greater the muscle force needs to be, and the greater the muscle force (as a proportion of isometric strength), the slower the speed of shortening. A muscle can shorten at maximum speed when the external load on the muscle is zero. When the external load on a muscle is greater than the isometric strength of the muscle, it is forced to lengthen (contract eccentrically).

In an eccentric contraction a muscle resists the stretching load. In so doing, the cross bridges are themselves stretched, adding to the overall tension such that the force produced by the muscle is greater than the isometric strength of the muscle. The force produced by a muscle during eccentric contraction depends on the speed of lengthening, which depends on the size of the external load. The greater the external load (in relation to the isometric strength of the muscle), the greater the speed of lengthening. The greater the speed of lengthening, the greater the effect of the stretch reflex, and, therefore, the greater the force produced by the muscle. When the external force exceeds the maximum strength of the muscle the muscle and its tendon are damaged. The relationship between speed of shortening or speed of lengthening and muscle force is referred to as the **force-velocity relationship** (figure 8.26). Figure 8.27 shows the effect of the force-velocity relationship on the length-tension relationship of a muscle-tendon unit. The figure shows that at any particular length, the greater the speed of

Force-velocity relationship: the relationship between speed of shortening or speed of lengthening and tension in a muscle

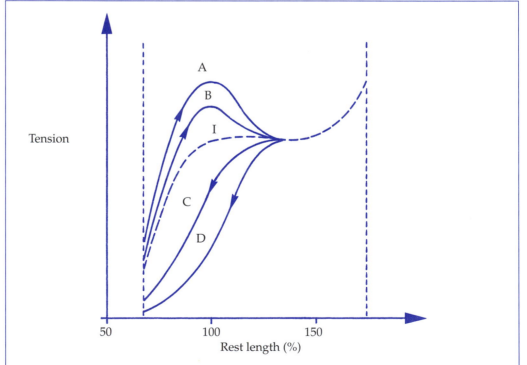

Figure 8.27   The effect of speed of shortening and speed of lengthening on the length-tension relationship in skeletal muscle. A and B show eccentric contractions: the speed of lengthening in A > B; C and D show concentric contractions: the speed of shortening in C < D; I shows the isometric tension curve.

shortening, the lower the tension, and the greater the speed of lengthening, the higher the tension.

> *The amount of force generated by a muscle-tendon unit depends on the length of the muscle-tendon unit at the time of stimulation (length-tension relationship) and the speed with which it changes length in the ensuing contraction (force-velocity relationship).*

Muscle Architecture and Function

All muscles are made up of muscle fibers. However, the length and the orientation of the fibers (pennate or nonpennate) have a considerable effect on the function of the muscles. The fundamental relationships between muscle architecture and muscle function are that excursion (the distance that the muscle can shorten) and velocity of shortening are proportional to fiber length, and force is proportional to total physiological cross-sectional area of the muscle fibers (Lieber and Bodine-Fowler 1993).

All muscle fibers are comprised of similar sarcomeres, and the number of sarcomeres determines the length of a muscle fiber. Each sarcomere in a muscle fiber is capable of shortening to the same extent as all the other sarcomeres in the muscle fiber. Consequently, the excursion of the muscle fiber is equal to the sum of the excursions of all the individual sarcomeres; the greater the number of sarcomeres, the longer the muscle fiber, the greater the excursion. Excursion and velocity of shortening are directly related since velocity of shortening is the rate of change of excursion—the rate of change in length of the muscle. The longer the muscle fiber (in terms of number of sarcomeres), the greater its excursion and velocity of shortening.

Theoretically, the ideal muscle (in terms of force and excursion capabilities) has a large cross-sectional area and very long fibers. However, such a muscle would be bulky and create considerable packing problems due to its girth and areas of attachment to the skeletal system. Since there are no muscles with both of these characteristics, it is reasonable to assume that the architecture of the muscular system has evolved to provide the best compromise between structure and function. The muscles of the body represent a broad range of combinations of force and excursion capability (Lieber 1992), and it is, perhaps, not surprising that most movements of the body involve simultaneous activity in a number of muscles with each muscle performing a particular role.

> *The fundamental relationships between muscle architecture and muscle function are that excursion (the distance that the muscle can shorten) and velocity of shortening are proportional to fiber length, and force is proportional to total cross-sectional area of the muscle fibers.*

Roles of Muscles

With regard to control of joint movements, muscles perform a number of different roles including stabilizer, agonist, prime mover, assistant mover, antagonist, synergist, and neutralizer. Each of the muscles that contributes to a particular movement may have more than one role and the relative importance of each role may change during the movement.

Joint stabilization—the maintenance of joint congruence—is a major function of the muscular system. The extent to which a muscle contributes to joint stabilization depends largely on the line of pull of the muscle in relation to the center of the joint the muscle crosses. Generally, the closer the line of pull of the muscle to the center of the joint, the greater the stabilizing effect of the muscle (see chapter 9).

An agonist is a muscle that moves a body segment in the intended direction. For example, the deltoid and supraspinatus are both agonists in abduction of the shoulder joint, with the deltoid as prime mover and the supraspinatus as the assistant mover (figure 8.28, *a*, *b*, and *c*). An antagonist is a muscle that acts in the direction opposite to that of an agonist. For example, in abduction of the shoulder joint, the latissimus dorsi and pectoralis major are antagonists (figure 8.28, *d* and *e*).

A synergist assists the action of the prime mover. For example, in abduction of the shoulder, the trapezius and serratus anterior rotate and abduct the scapula (scapulothoracic gliding mechanism, chapter 7) so that complete abduction of the arm (180° with respect to the anatomical position) can be achieved (figure 8.28, *f* and *g*).

A neutralizer prevents unwanted action of a muscle. For example, the flexors of the fingers tend to simultaneously flex the wrist (figure 8.29). They do not usually perform both movements at the same time since, in doing so, the muscles would experience active insufficiency. The length of the muscle fibers of the finger flexors

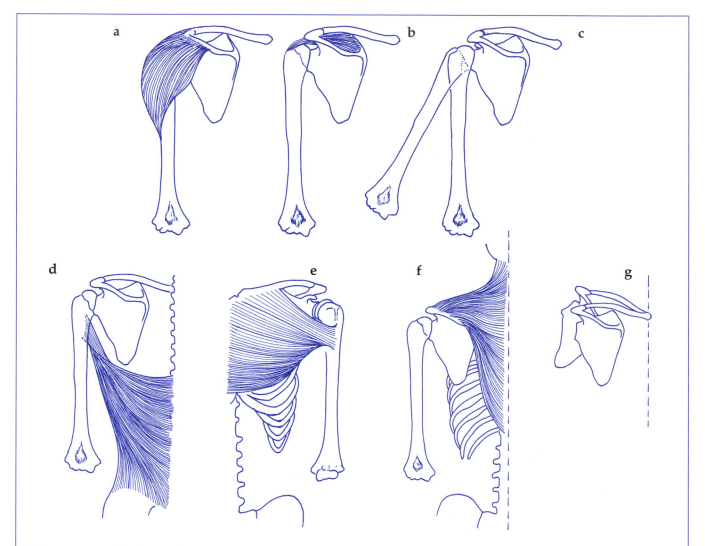

Figure 8.28   Roles of muscles; (*a*) posterior aspect of left shoulder girdle showing the deltoid muscle; (*b*) posterior aspect of left shoulder girdle showing the supraspinatus muscle; (*c*) action of the deltoid and supraspinatus, (i.e., abduction of the shoulder joint); (*d*) posterior aspect of the left shoulder girdle showing the latissimus dorsi muscle; (*e*) anterior aspect of the left shoulder girdle showing the pectoralis major muscle; (*f*) posterior aspect of the left shoulder girdle showing the trapezius muscle; (*g*) posterior aspect of the left shoulder girdle showing the action of the trapezius and serratus anterior muscles, i.e, rotation of the scapula about the acromioclavicular joint and rotation of the clavicle about the sternoclavicular joint.

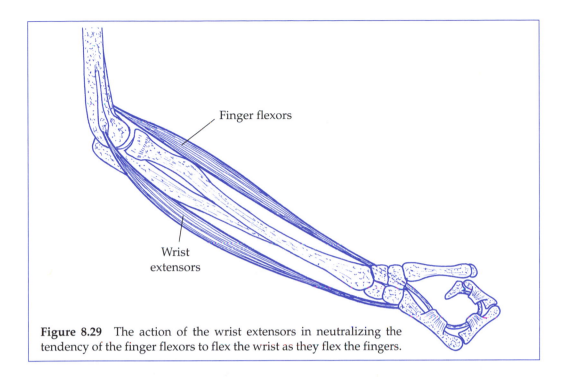

Figure 8.29 The action of the wrist extensors in neutralizing the tendency of the finger flexors to flex the wrist as they flex the fingers.

is approximately 100 mm. To flex the wrist and hold a tight position, the finger flexors need to shorten by about 27 mm. To flex the fingers into a tight fist, the finger flexors need to shorten by about 37 mm. The finger flexors can perform both movements separately since 27 mm to 37 mm is well within the normal working range of the muscles. However, the muscles cannot perform both movements simultaneously because, to do so, they would need to shorten by about 64 mm (27 mm plus 37 mm), which would produce active insufficiency (Alexander 1992). The finger flexors can only produce a tight fist when the wrist is held straight so that the tendency of the finger flexors to flex the wrist is neutralized. This is achieved by the wrist extensors (figure 8.29).

Muscle Fiber Arrangement and Force and Excursion

Figure 8.30, *a* and *c*, shows two muscle-tendon units with the same rest length and the same muscle mass (same volume of myofibrils). One muscle is a nonpennate parallel-fibered muscle and the other is a unipennate muscle. The physiological cross-sectional area of each muscle is the cross-sectional area of all of the muscle fibers perpendicular to their line of pull. Assuming the length of the muscle fibers in the pennate muscle is only half that of the fibers in the nonpennate muscle, it follows that the physiological cross-sectional area of the pennate muscle is double that of the nonpennate muscle (since the muscle mass is the same in both muscles). Consequently, the pennate muscle can exert double the force of the nonpennate muscle.

However, since the fibers in the pennate muscle are oblique to the line of pull of the muscle-tendon unit, not all of the force is in the line of pull of the muscle-tendon unit. In a pennate muscle the angle of the fibers with respect to the line of pull of the muscle-tendon unit is usually 30° or less (Alexander 1968), such that 90% or more of the force exerted by the muscle is directed in the line of pull of the muscle. Thus, pennate muscles are normally capable of exerting far more force in the line of pull of the muscle-tendon unit than nonpennate muscles of the same muscle mass.

In contrast to their capacity to generate force, the excursion range of nonpennate muscle-tendon units is usually much greater than that of pennate muscle-tendon units of similar muscle mass. The increased excursion range of nonpennate muscles is

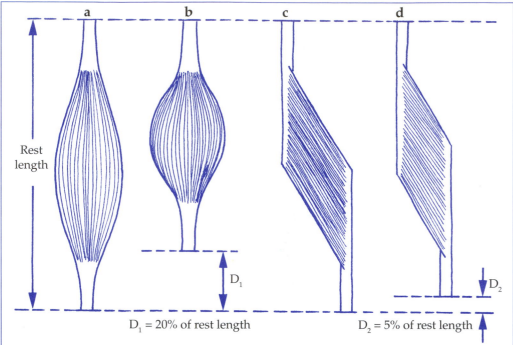

Figure 8.30   Effects of the pennate and nonpennate muscle structure on force and excursion; (*a*) nonpennate muscle at rest length; (*b*) nonpennate muscle in *a* with muscle fibers shortened to 70% of rest length; (*c*) pennate muscle at rest length; (*d*) pennate muscle in *c* with muscle fibers shortened to 70% of rest length.

due to increased fiber length in nonpennate muscles and the obliquity of the fibers in pennate muscles, which reduces excursion in the line of pull of the muscle. In figure 8.30, *b* and *d*, the muscles are shown with their fibers shortened to 70% of their rest length. The corresponding shortening of the muscle-tendon units is 20% and 5%, respectively, in the nonpennate and pennate muscles.

> *The muscles of the body represent a broad range of combinations of force and excursion capability, and most body movements involve simultaneous activity in a number of muscles with each muscle performing a particular role including stabilizer, agonist, prime mover, assistant mover, antagonist, synergist, and neutralizer.*

Biarticular Muscles

Many of the muscles of the body, especially those in the upper and lower limbs, pass over more than one joint (see figure 8.29). These muscles are usually referred to as **biarticular muscles**, since muscles that pass over more than two joints function in the same way as muscles that pass over just two joints (Lieber 1992). Biarticular muscles are too short to fully flex or fully extend simultaneously all of the joints over which they cross. For example, the hamstrings are two-joint muscles that can extend the hip and flex the knee, but these actions (as explained earlier in the chapter) cannot be performed maximally at the same time. Indeed, hip extension is normally associated with knee extension, and hip flexion is normally associated with knee flexion, as in walking and running. In this way the length of the muscle stays within the normal working range of approximately 100% to 130% of rest length. The main advantages of biarticular muscles are that tension is produced in one muscle rather than two (or more), which conserves energy, and working within the 100% to 130% range allows maximum flexibility in tension generation (Lieber 1992; Van Ingen Schenau, Pratt, and Macpherson 1994).

Biarticular muscle: a muscle that passes over more than one joint

Stretch-Shorten Cycle

When a concentric contraction occurs without prestretch, the initial phase of contraction takes up the slack in the series elastic connective tissue (SEC) components; it is only when the slack in the SEC components has been taken up, that is, when the SEC components have become taut, that the force produced by the muscle is transmitted to the skeleton. The need for the muscle to expend energy in taking up the slack in the SEC components before force can be transmitted to the skeleton is not efficient. It is not surprising, therefore, that in most, if not all, whole-body movements, many of the muscles involved in controlling the various joints are initially stretched before being allowed to shorten. For example, jumping movements generally begin with a downward movement followed by upward movement (figure 8.31). Similarly, in throwing actions, the arm is usually swung backward before being accelerated forward (see figure 2.13).

In such actions the initial movement in the opposite direction to that of the final movement is usually referred to as a countermovement. A countermovement involves two phases. In the first phase the body (as in a vertical jump) or body segments (such as the arm in a throwing action) develop speed of movement in the opposite direction to that of the final movement. Before the final movement can be initiated, the movement of the body or body segments in the opposite direction must be arrested. Consequently, in the second phase of a countermovement the muscles contract to arrest the movement of the body or body segments; in doing so they are forcibly stretched and, as such, contract eccentrically. The eccentric phase is usually immediately followed by a concentric phase to produce the final movement. The pattern of eccentric contraction followed without a pause by concentric contraction

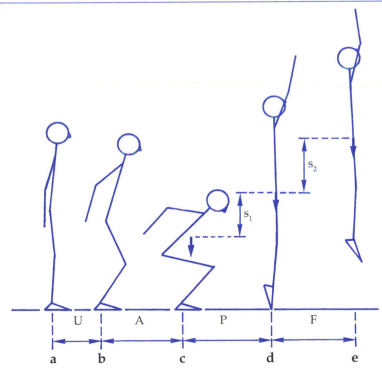

U: unweighting phase

A: absorption phase (or weighting phase)

P: propulsion phase

F: flight phase from takeoff to the top of the flight

s_1: vertical displacement of center of gravity during propulsion phase

s_2: vertical displacement of center of gravity from takeoff to top of flight

Figure 8.31   Sequence of movement in a countermovement vertical jump (*a* to *e*) and a squat jump from a stationary squat position (*c* to *e*); (*a*) standing upright at rest; (*b*) instant in countermovement when downward velocity of center of gravity is at its maximum; (*c*) instant at which vertical velocity of center of gravity is zero (i.e., maximum displacement downward); (*d*) take-off; (*e*) top of flight (i.e., vertical velocity of center of gravity is zero).

Stretch-shorten cycle: a sequence of muscle activity involving an eccentric contraction followed immediately by a concentric contraction

is referred to as the **stretch-shorten cycle**. Most whole-body actions, including relatively slow walking, utilize the stretch-shorten cycle. However, the utilization of the stretch-shorten cycle is more obvious in movements involving a backswing. For example, in a golf swing, the backswing is arrested by eccentric contraction of the same muscles that then contract concentrically to produce the downswing. The eccentric phase of the backswing is usually accentuated by starting to turn the trunk to face the flag before the backswing has been completed.

> For any given task, movements using the stretch-shorten cycle may be more effective (in terms of performance) and more efficient (in terms of utilization of energy) than movements that rely on concentric contractions.

Storage and Utilization of Elastic Strain Energy

When a muscle contracts isometrically it expends energy in creating tension, but does no work, since the length of the muscle-tendon unit does not change. When a muscle contracts concentrically it expends energy in creating tension and does work by pulling its skeletal attachments closer together. The amount of work done by a muscle in a concentric contraction is the product of the average net muscle force—the difference between the muscle force and the external load, and the distance over which the muscle-tendon unit shortens. When a muscle contracts eccentrically it expends energy in creating tension, but in contrast to a concentric contraction, work is done on the muscle-tendon unit; it is lengthened by the external load. The amount of work done on the muscle-tendon unit by the external load is the product of the average net external load—the difference between the external load and the muscle force—and the distance over which the muscle-tendon unit is lengthened. When work is done on a muscle-tendon unit—when energy is expended in stretching a muscle-tendon unit—the energy is absorbed by the muscle-tendon unit in the form of strain energy. The extent to which the strain energy can be used to move the body, rather than being dissipated as heat within the muscle, will depend on the speed of the changeover from eccentric to concentric activity within the stretch-shorten cycle.

For any given task, movements utilizing the stretch-shorten cycle may be more effective (in terms of performance) and more efficient (in terms of utilization of energy) than movements relying on concentric contractions (Anderson and Pandy 1993). For example, most individuals are able to jump higher using a countermovement jump (CMJ) (see figure 8.31) than a squat jump (SJ) (see figure 8.31, c, d, and e). In a CMJ the individual starts from a standing position and then performs a downward countermovement immediately followed by a maximum effort upward jump. The leg extensor muscles contract eccentrically to arrest the countermovement and then contract concentrically to drive the body upward into the jump. In an SJ the individual performs a maximum effort jump from a stationary squat position. The leg extensor muscles contract isometrically in the stationary squat position and then contract concentrically to drive the body upward into the jump. Assuming the same level of squat is achieved in both the CMJ and the SJ, the difference in performance—height jumped—will be largely due to differences in the storage and utilization of elastic strain energy in the connective tissues of the leg extensor muscles and the amount of force produced by the muscle components of the leg extensor muscles.

During the eccentric phase of the countermovement in a CMJ, the connective tissue components (series elastic and parallel elastic) are forcibly stretched and, therefore, elastic strain energy builds up in them like the energy stored in a stretched spring. The amount of elastic strain energy stored in the connective tissues depends on the velocity of stretch—the rate at which stretching occurs—and the magnitude of stretch—the actual amount by which the muscle-tendon units lengthen during the stretching phase.

The higher the velocity of stretch, the greater the amount of elastic strain energy stored. However, a high velocity of stretch of the connective tissues can only be achieved if the force produced by the muscle components is high; a low to moderate force results in lengthening of the muscle components with little or no stretch on the connective tissues. With regard to magnitude of stretch, the muscle components can only produce a high force within the central region of the length-tension range; consequently, the magnitude of stretch should be low to moderate. In general, the amount of elastic strain energy stored in the connective tissues of the leg extensor muscles during the countermovement of a CMJ will be maximized by a low to moderate magnitude of stretch combined with a high velocity of stretch.

In the squat position of an SJ, the leg extensor muscles will contract isometrically and their connective tissue components will be in a stretched position. Consequently, the connective tissues will have a certain amount of elastic strain energy (dependent on the magnitude of stretch), but it is likely to be less than that stored during the countermovement of a CMJ (dependent upon the velocity of stretch and magnitude of stretch). In both the SJ and CMJ (and any other movement involving stored elastic energy in connective tissue components), the elastic strain energy can contribute to movement in the form of force (additional to that produced by the muscle components) exerted during the recoil of the connective tissue components from their stretched position.

Due to the velocity of stretch it should be possible to store more elastic strain energy in the countermovement phase of a CMJ than in the static squat phase of a SJ. However, the ability to utilize the additional strain energy depends upon the delay between the eccentric and concentric phases of the movement. Some of the additional strain energy is, inevitably, converted to heat and, therefore, is not available to contribute to movement. The amount of elastic strain energy converted to heat depends on the speed of the changeover from eccentric to concentric contraction. Generally, the faster the changeover, the smaller the proportion of strain energy converted to heat, and, consequently, the greater the proportion available to contribute to movement (Gregor 1993).

Sometimes the elastic strain energy that is converted to heat is referred to as lost energy. However, this is misleading since energy can never be completely lost, it can only be converted from one form of energy to another. For example, the muscles convert energy in the form of chemical substances (such as adenosine triphosphate) into mechanical energy in the form of movement (kinetic energy), which itself may be converted into another form of mechanical energy (strain energy). With regard to elastic strain energy in muscle-tendon units, the proportion of this energy converted to heat is only lost in the sense that it is not available to contribute to movement.

The concentric phase of the SJ is preceded by isometric contraction of the leg extensor muscles, whereas the concentric phase of the CMJ is preceded by eccentric contraction. The elicitation of the stretch reflex during the eccentric phase of the CMJ is likely to enhance performance over the SJ in two ways:

1. The high force during the eccentric phase (greater than isometric force) is likely to increase the possibility of a high velocity of stretching of the connective tissues and, consequently, increase the amount of elastic strain energy stored in the connective tissues compared to the SJ.

2. The force generated during the concentric phase of the CMJ is likely to be greater than that generated in the concentric phase of the SJ.

Power Output of the Leg Extensors in a Countermovement Jump

In the first part of the countermovement, the unweighting phase (see figure 8.31, *a* and *b*), the leg extensor muscles are temporarily relaxed and the center of gravity

(CG) accelerates downward. During this phase the body loses a certain amount of gravitational potential energy, but gains an equivalent amount of kinetic energy. During the second part of the countermovement, the weighting or absorption phase (see figure 8.31, b and c), the downward kinetic energy generated in the unweighting phase is absorbed in the form of strain energy by eccentric activity of the leg extensor muscles such that the downward velocity of the CG is reduced to zero. The shorter the duration of the absorption phase, the greater the average power input to the leg extensor muscles (the greater the average rate at which work is done on the leg extensor muscles). During the propulsion phase (see figure 8.31, c and d) the leg extensors contract concentrically to generate as much upward velocity as possible in the CG prior to takeoff. If the speed of the changeover from eccentric to concentric activity, from absorption to propulsion, is rapid, some of the strain energy in the leg extensor muscles may be used in the generation of upward velocity. The work done by the leg extensors during the propulsion phase is equivalent to the change in gravitational potential energy of the body between positions c and e in figure 8.31, when the body is instantaneously stationary; that is, work done by leg extensors = $f(s_1 + s_2)$, where f = body weight, s_1 = vertical displacement of CG during the propulsion phase, and s_2 = vertical displacement of CG from takeoff to top of flight.

Consequently, the average power output of the leg extensor muscles during the propulsion phase is given by $f(s_1 + s_2)/t$, where t is the duration of the propulsion phase. For example, if f = 70 kg wt, s_1 = 0.3 m, s_2 = 0.5 m, and t = 0.2 s, then average power output P is given by

$$P = \frac{70 \text{ kg wt} \times 0.8 \text{ m}}{0.2 \text{ s}}$$

$$P = \frac{686.7 \text{ N} \times 0.8 \text{ m}}{0.2 \text{ s}}$$

$$P = 2747 \text{ W}$$

$$P = 39 \text{ W/kg}.$$

The average level of power output the human body can sustain depends on the duration of the effort; the shorter the duration, the greater the power output (Wilkie 1960). Consequently, maximum levels of power output are achieved in short explosive-type activities such as rebound jumping; in such activities, highly trained athletes may achieve power outputs in the region of 50 to 60 W/kg. For activities lasting between 15 seconds and one minute, power output is limited to about 20 to 25 W/kg, and for activities lasting longer than five minutes, power output is limited to about 4 to 5 W/kg (Wilkie 1960, Bosco, Luhtanen, and Komi 1983).

Stretch Load and Jump Performance in Drop Jumping

An individual's ability to make use of the stretch-shorten cycle in jumping activities depends on her ability to tolerate stretch loading. As stretch load and, therefore, velocity of stretch increases, performance increases up to a particular optimum velocity of stretch, and then decreases with further increases in velocity of stretch. For example, it has been shown that in drop jumping—when subjects are asked to drop onto the floor from various heights and then immediately perform a maximum vertical jump—performance (jump height) increases as drop height increases up to approximately 50 to 60 cm, and then decreases as drop height increases further (Komi and Bosco 1978). As drop height increases, stretch load (reflected in the magnitude of the ground reaction force) during the eccentric phase of ground contact tends to increase. However, for each individual there is a drop height D that corresponds to a maximum tolerable stretch load (a maximum stretch load at which maximal effort in the drop jump can be maintained). This drop height may be associated with maximal performance due to the beneficial effects of the stretch-shorten cycle. At drop heights greater than D, the individual is unable to tolerate the stretch load associated with maximal effort and may reduce the stretch load by increasing the amount of hip, knee, and ankle flexion during the eccentric phase of landing. This increases the magnitude of stretch of the leg extensors. The increased magnitude of stretch reduces the velocity of stretch of the leg extensors and, as such, reduces the potential benefits of increased muscle force and elastic strain energy resulting in decreased performance. Not surprisingly, jump training based on drop jumping has been shown to increase performance and the size of the stretch load that can be tolerated (Komi 1992).

Using the Stretch-Shorten Cycle in Movements Without a Countermovement

In movements such as the CMJ, it is only possible to use the stretch-shorten cycle by incorporating a countermovement. In some whole-body movements it may be possible to use the stretch-shorten cycle to accelerate the relevant body part(s) in the intended direction during the eccentric phase as well as during the concentric phase. In such actions there is no countermovement, but there is a premovement that results in eccentric contraction of the relevant muscles. For example, in the final stage of throwing a javelin, the javelin is pulled by the throwing arm from a position behind the trunk to a release point in front of the trunk (figure 8.32). If, at the start of this pulling action, the (right-handed) thrower can thrust his right shoulder forward fast enough, the inertia of the arm and javelin results in eccentric contraction of the shoulder flexors and abductors and elbow flexors, which are (in association with muscles in the legs and trunk) responsible for the pulling action. The arm and javelin are accelerated forward though the shoulder and elbow muscles contract

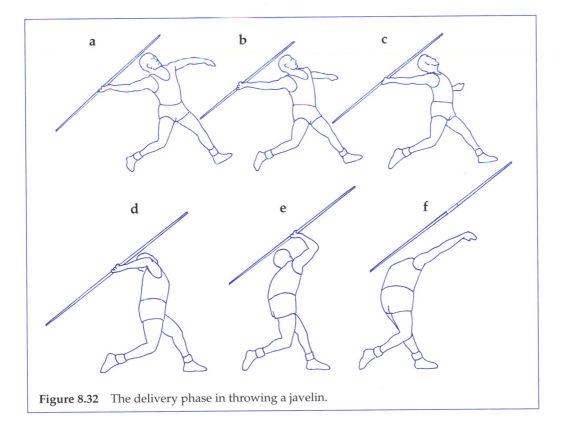

Figure 8.32   The delivery phase in throwing a javelin.

eccentrically. The eccentric phase of the elbow flexors is relatively short, and the elbow flexes and then extends prior to release of the javelin. In contrast, the eccentric phase of the shoulder flexors and abductors may last for most of the pulling action with only a short concentric phase just prior to release (see figure 8.32, *e* and *f*). Ideally, the thrower should thrust the right shoulder forward for as long as possible during the pulling action, since this will maximize the force exerted on the javelin by

1. prolonging the eccentric (high-force) phases of the shoulder and elbow muscles and

2. increasing the force exerted in the concentric phases of the muscles by decreasing their speed of shortening.

Summary

This chapter described the structure and function of the neuromuscular system. Considerable variation exists among muscles in the type, arrangement, and innervation ratio of muscle fibers, and these variations are reflected in the force and excursion capabilities of the individual muscles. Most body movements involve simultaneous activity in a number of muscles with each muscle performing a particular role such that the functional characteristics of the muscles complement each other and maximize the effectiveness and efficiency of the movement. The chapter also described the role of the neuromuscular system in kinesthetic sense. One of the main functions of muscles is to stabilize joints, and the next chapter considers the effects of the open-chain arrangement of the skeleton on forces exerted in muscles and joints.

Review Questions

1. Differentiate between the cerebrospinal nervous system and the central nervous system.

2. Differentiate between resting membrane potential and action potential in a nerve fiber.

3. Describe the general organization of nerve tissue in the spinal cord and spinal nerves.

4. Describe the sequence of events within the nervous system in a typical voluntary movement.

5. Describe the various categories of pennate and nonpennate muscles.

6. Describe the possible causes and effects of compartment pressure syndrome.

7. Describe the sliding filament theory of muscular contraction.

8. Differentiate between kinesthetic sense and proprioception.

9. Differentiate between active and passive insufficiency in skeletal muscle.

10. Describe the fundamental relationships between muscle architecture and muscle function.

11. Describe the roles of prime mover, synergist, and neutralizer in skeletal muscle.

12. Describe the stretch-shorten cycle of muscular contraction in relation to a countermovement jump.

13. Explain how the stretch-shorten cycle can be utilized in a movement that does not have a countermovement.

Chapter 9

Forces in Muscles and Joints

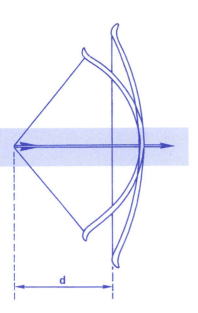

The musculoskeletal system exerts internal forces to counteract external forces acting on the body. In general, the mechanical advantage of muscles is low such that muscle forces and joint reaction forces are high relative to the external loads. However, muscles tend to work together, which spreads the load on the muscles and reduces bending stress on bones. This chapter examines the effect of changes in the leverage of external loads on the magnitude of internal forces. Before reading this chapter you may want to review the sections in chapter 1 concerning resolution of a vector and angular kinetics.

Objectives

After reading this chapter you should be able to do the following:

1. Define or describe the key terms.
2. Identify different classes of levers within the musculoskeletal system.
3. Describe the effects of leverage of external loads on muscle forces and joint reaction forces.
4. Describe the effect of joint angle on the swing and stabilization components of a muscle.

Force Vectors

In the analysis of human movement, force vectors are used to represent the internal and external forces that act on the body. The representation of muscle forces by force vectors is particularly useful in the analysis of muscle actions.

Figure 9.1a shows the right biceps brachii muscle, which has two heads of origin. The short head arises from the coracoid process and the long head arises from the supraglenoid tubercle (tubercle on the superior border of the glenoid fossa). The tendon of the long head passes through the bicipital groove of the humerus. In the upper half of the muscle, the muscle fibers in the long head portion are separate from those in the short head portion, but the two groups of fibers merge in the lower

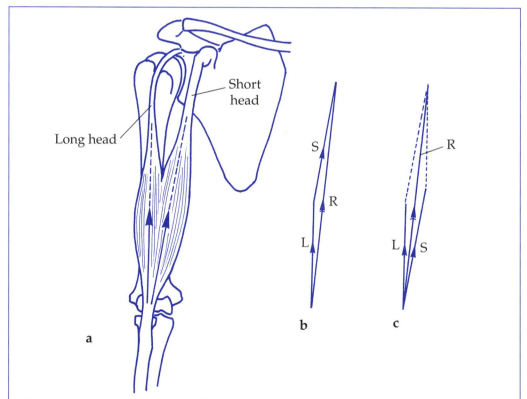

Figure 9.1 Resultant force exerted by the biceps brachii; (a) anterior view of the right biceps brachii; (b) vector chain representation of the resultant R of the force L exerted by the long head and the forces exerted by the short head; (c) parallelogram of vectors representation of the resultant R of L and S.

half of the muscle. The muscle is attached by a single tendon to the radial tuberosity. When the whole muscle is stimulated to contract, the line of action of the force exerted by the fibers in the long head portion is slightly different from that in the short head portion due to the separation in the upper half of the muscle (figure 9.1*a*). However, the net effect of the two forces is a single resultant force as shown by the vector chain method in figure 9.1*b*. The actual magnitude and direction of the resultant force produced by the biceps brachii in a particular movement depends on the component forces produced by the short head and long head portions of the muscle. By varying the component forces (by selective recruitment of appropriate motor units) and, therefore, the resultant force, the action of the biceps brachii can be matched (in association with other muscles) to the specific requirements of each particular movement that involves the biceps brachii.

The capacity to vary the magnitude and direction of force is characteristic of most muscles and reflects the tendency of muscles to work together to produce movements. The extent of this variation depends considerably on the size, shape, and number of attachments of the muscle or muscle-tendon unit to the skeletal system. In general, the greater the size, the broader the shape, and the greater the number of attachments, the greater the variation in magnitude and direction of force produced (in muscles of similar muscle mass).

> *The magnitude and direction of the resultant force produced by a muscle in a particular movement depends on the component forces produced by the various portions of the muscle. By selective recruitment of appropriate motor units the component forces and, therefore, the resultant force, can be matched to the specific requirements of each particular movement.*

Lever Systems in the Musculoskeletal System

The segments of the body are essentially levers, and each joint constitutes a fulcrum between adjacent segments. The muscles pull on the bones of the segments to control their movement in the same way that effort forces counteract resistance forces in inanimate lever systems. The resistance to movement exerted by a body segment is in the form of the segment's weight and any other external loads attached to the segment. Most, if not all, of the muscles of the body operate in first or third class lever systems. Like the third class lever systems, most of the first class lever systems have mechanical advantages less than 1.0 because the muscles or muscle-tendon units, which operate within them, are inserted close to the joints they control, and, therefore, they have shorter moment arms than the resistance forces they counteract.

> *Most if not all of the muscles of the body operate in first or third class lever systems. Like the third class lever systems, most of the first class lever systems have mechanical advantages less than 1.0.*

External Versus Internal Forces

Figure 9.2*a* shows the position of the head in normal upright standing. In this position the line of action of the weight of the head passes in front of the vertebral column and, as such, exerts a clockwise moment that rotates the head forward and downward about a transverse axis through the fulcrum, the joint between the occipital bone and the atlas. The tendency of the weight W of the head to rotate the head forward and downward is counteracted by the neck extensor muscles. Figure 9.2*b* shows a free body diagram of the head where F is the force exerted by the neck extensor muscles and J is the force exerted at the fulcrum, that is, the **joint reaction force** exerted by the

Joint reaction force: the equal and opposite force exerted on articular surfaces in contact with each other in a joint

atlas on the occipital bone. Figure 9.2c shows the corresponding force-moment arm diagram. Force F and weight W constitute a first class lever system about the fulcrum. In this example, it is assumed that F acts vertically downward. The weight of the head of an adult is in the region of 7.9% of total body weight (see p. 281).

Consequently, if total body weight is 70 kg wt the weight of the head is approximately 5.6 kg wt. The moment arms of W (M_W) and F (M_F) are in the region of 2 cm and 7 cm, respectively. Since the head is in equilibrium, the resultant moment acting on the head is zero and the resultant force acting on the head is zero. Consequently, the forces F and J can be determined by equating moments and then equating forces as follows:

1. Equating moments:

$$(W \times M_W) - (F \times M_F) = 0.$$

That is,

$$W \times M_W = F \times M_F$$

$$F = \frac{W \times M_W}{M_F},$$

where W = 5.6 kg wt, M_W = 2 cm, and M_F = 7 cm. Thus,

$$F = \frac{5.6 \ \text{kg wt} \times 2 \ \text{cm}}{7 \ \text{cm}}$$

$$F = 1.6 \ \text{kg wt}.$$

2. Equating forces: Since F and W are vertical forces, J must also be vertical, that is,

$$J = F + W$$

$$J = 1.6 \ \text{kg wt} + 5.6 \ \text{kg wt}$$

$$J = 7.2 \ \text{kg wt}.$$

Thus, to counteract W, an external force, the musculoskeletal system has to exert two internal forces, F and J. Force F is an active force, a muscle force, and J is a passive force, a joint reaction force. This illustrates the relationship between the internal and external forces that act on the body; the musculoskeletal system exerts internal forces to counteract or overcome the external forces.

> *In counteracting or overcoming external forces the musculoskeletal system exerts two kinds of internal forces: active forces (muscle forces) and passive forces (joint reaction forces and forces in passive joint support structures).*

In the previous example, all of the forces acting on the head are vertical forces and, as such, the vector chain of the forces is a straight line. The downward and upward forces have been separated in figure 9.2d to demonstrate that when the resultant force acting on an object is zero, the vector chain of the component forces starts and ends at the same point. In most musculoskeletal lever systems, the internal and external forces are not usually parallel. For example, figure 9.3a shows the head position of a person writing at a desk. In this situation the head and trunk are tilted forward such that the moment arm of the weight of the head about the atlanto-occipital joint is greater than when the head is in the upright position (see figure 9.2), that is, about 4.0 cm compared to 2.0 cm. The moment arm of the force F exerted by the neck extensor muscles is about the same as when the head is in the upright position, that is, about 7.0 cm, but its line of action will be at approximately 50° to the horizontal. Figure 9.3b shows a free body diagram of the head, and figure 9.3c shows the corresponding force-moment arm diagram. By taking moments about the fulcrum

$$(W \times M_W) - (F \times M_F) = 0.$$

Thus,

$$W \times M_W = F \times M_F.$$

That is,

$$F = \frac{W \times M_W}{M_F},$$

where $W = 5.6$ kg wt, $M_W = 4$ cm, and $M_F = 7$ cm. Thus,

$$F = \frac{5.6 \text{ kg wt} \times 4 \text{ cm}}{7 \text{ cm}}$$

$$F = 3.2 \text{ kg wt}.$$

The joint reaction force J can be found two ways: constructing the vector chain and equating the forces. Figure 9.3d shows the vector chain solution; J has a magnitude of approximately 8.3 kg wt and acts at an angle of approximately 75.5° to the horizontal. The vector chain method of determining the resultant of a number of component forces is useful and fairly quick. However, the accuracy with which the resultant is determined depends on the accuracy of the scale diagram; the greater the number of component vectors, the more difficult it becomes to determine the resultant accurately.

The resultant force can be determined accurately by equating the forces acting on the object. Figure 9.4a shows a free body diagram of the head (same as figure 9.3b). Figure 9.4b shows the force F resolved into its horizontal and vertical components, and figure 9.4c shows the force J resolved into its horizontal and vertical components. Figure 9.4d shows the forces acting on the head in terms of their horizontal and vertical components. Since the head is in equilibrium, the resultant force on the head is zero. Consequently, the resultant of the horizontal forces is zero and the resultant of the vertical forces is zero. Using the convention that forces acting to the right or upward are positive and that forces acting to the left or downward (with respect to figure 9.4, a through d) are negative, it follows that:

Horizontal forces: $J_H - F_H = 0,$

where J_H = horizontal component of J and F_H = horizontal component of F. Therefore,

$$J_H = F_H$$

$$J_H = F \cos 50°$$

$$J_H = 3.2 \text{ kg wt} \times 0.6428$$

$$J_H = 2.06 \text{ kg wt}.$$

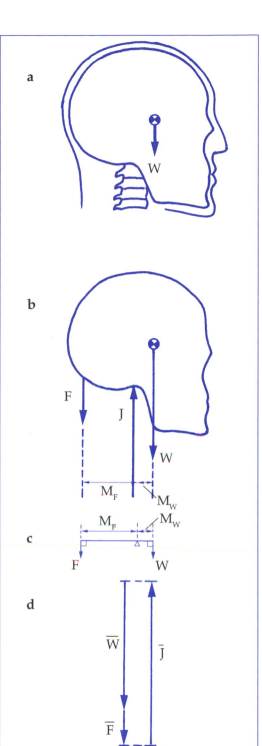

Figure 9.2   Forces acting on the head during normal upright standing; W—weight of the head; F—force exerted by neck extensors; M_W—moment arm of W; M_F—moment arm of F; J—joint reaction force exerted by atlas on occipital bone. (a) Location of the center of gravity of the head. (b) Free body diagram of the head. (c) Force-moment arm diagram corresponding to b. (d) Vector chain of forces acting on the head.

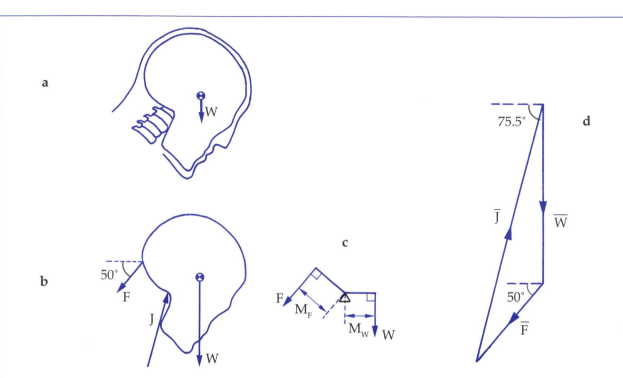

Figure 9.3 Forces acting on the head while writing at a desk; W—weight of the head; F—force exerted by neck extensors; M_W—moment arm of W; M_F—moment arm of F; J—joint reaction force exerted by atlas on occipital bone.

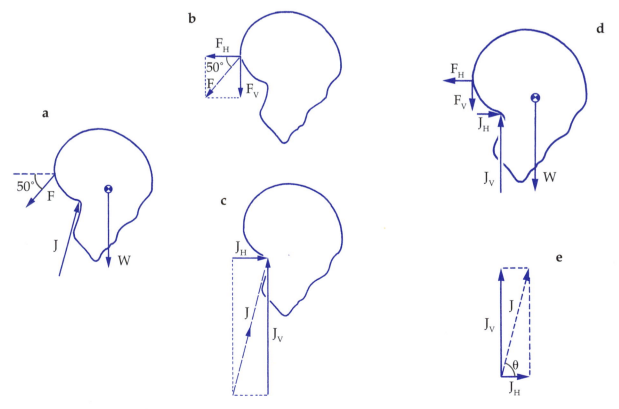

Figure 9.4 Determination of joint reaction force by equating forces; W—weight of the head; F—force exerted by neck extensors; F_H—horizontal component of F; F_V—vertical component of F; J—joint reaction force exerted by atlas on occipital bone; J_H—horizontal component of J; J_V—vertical component of J.

Vertical forces: $J_V - F_V - W = 0$, where J_V = vertical component of J, F_V = vertical component of F, and W = weight of head. Therefore,

$$J_V = F_V + W$$

$$J_V = F \sin 50° + W$$

$$J_V = 3.2 \text{ kg wt} \times 0.7664 + 5.6 \text{ kg wt}$$

$$J_V = 2.45 \text{ kg wt} + 5.6 \text{ kg wt}$$

$$J_V = 8.05 \text{ kg wt}.$$

Since J_H and J_V are at right angles to each other, the magnitude of their resultant J can be found by applying Pythagoras's theorem, and the direction of J can be found by calculating the cosine of the angle θ, which it makes to the horizontal from J_H and J (figure 9.4e).

Magnitude of J, by Pythagoras's theorem:

$$J^2 = J_V^2 + J_H^2$$

$$J^2 = (8.05^2 + 2.06^2) \text{ (kg wt)}^2$$

$$J^2 = (64.8 + 4.24) \text{ (kg wt)}^2$$

$$J^2 = 69.04 \text{ kg wt}^2$$

$$J = 8.3 \text{ kg wt}.$$

Direction of J:

$$\cos \theta = \frac{J_H}{J}$$

$$\cos \theta = \frac{2.06 \text{ kg wt}}{8.30 \text{ kg wt}}$$

$$\cos \theta = 0.2482$$

$$\theta = 75.5°.$$

In this example the vector chain method and the calculation method (equating the forces) produced the same result, indicating, in this case, the accuracy of the vector chain method.

The Moment of External Forces Versus the Magnitude of Internal Forces

The examples of the forces acting on the head in the upright (see figure 9.2) and writing positions (see figure 9.3) show the general effect of increases in the moment of external forces on the magnitude of the internal forces that counteract them; increasing the moment of external forces results in an increase in muscle forces, which in turn results in an increase in joint reaction forces. This is illustrated in figure 9.5, which shows a 6 m long beam of weight W balanced in three positions on a knife-edge support. In figure 9.5a, the beam is balanced with the line of action of its weight passing through the knife-edge support; that is, the moment of the weight of the beam about the fulcrum is zero. In this situation the reaction force R is sufficient to counteract W and maintain equilibrium (figure 9.5b). In figure 9.5c, the beam has been displaced 1 m to the right such that W exerts a clockwise turning moment on the beam of W × 1 m. The beam is held in equilibrium by the reaction force R_1 and a force F_1 exerted by a tie that attaches the left end of the beam to the base of support. If it is assumed that F_1 acts vertically, R_1 will also act vertically. In this situation F_1 and R_1 are

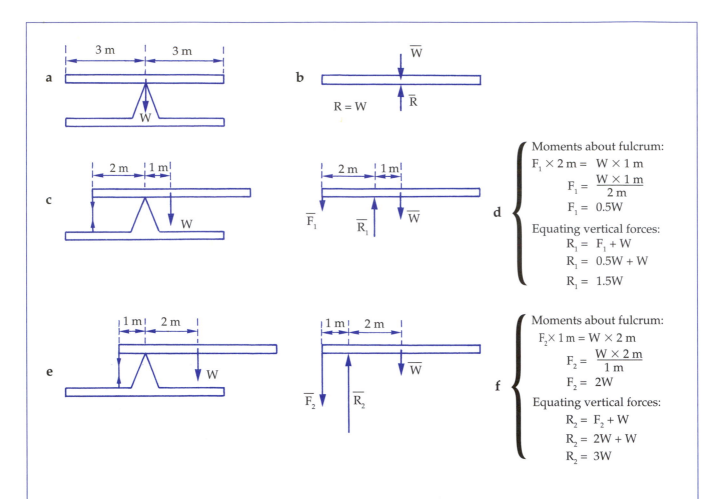

Figure 9.5 Effect of increasing the moment arm of the weight W of a beam balanced on a knife edge on the restraining forces needed to maintain equilibrium; (*a*) beam balanced with line of action of W through the base of support; (*b*) free body diagram of *a*; (*c*) beam balanced with the aid of a restraining force F₁; (*d*) free body diagram of *c*; (*e*)beam balanced with the aid of a restraining force F₂; (*f*) free body diagram of *e*.

found to be 0.5W and 1.5W, respectively (figure 9.5*d*). In figure 9.5*e*, the beam has been displaced 2 m to the right with respect to its original position such that W exerts a clockwise turning moment on the beam of W × 2 m. The beam is held in equilibrium by the reaction force R_2 and a force F_2 exerted by the tie; it is assumed that F_2 and, consequently, R_2 act vertically. In this situation F_2 and R_2 are found to be 2W and 3W, respectively (figure 9.5*f*).

> *In general, increasing the moment of external forces results in an increase in the magnitude of muscle forces, which results in an increase in the magnitude of joint reaction forces.*

Forces About the Hip Joint in Single-Leg Stance

The situation depicted in figure 9.5*f* is similar to that of single-leg support during standing or walking (figure 9.6, *a*, *b*, and *c*). In this position the pelvis acts as a first class lever and rotates about the hip joint under the action of the weight of the body and the hip abductor muscles. Figure 9.6*c* shows a free body diagram of the pelvis in this situation, and figure 9.6*d* shows the corresponding force-moment arm diagram. W is the weight of the body less the weight of the grounded leg. For a man of total body weight of 70 kg wt, W is approximately 59 kg wt (84.4% of total body weight; see p. 281). Weight W exerts a clockwise moment on the pelvis. The hip abductor muscles maintain the pelvis in a level position by exerting an equal and opposite moment on the pelvis. The line of action of the force A exerted by the hip abductors is approximately 80° with respect to the horizontal. Force J is the joint reaction force, that is, the force exerted by the head of the femur on the pelvis via the acetabulum. When the pelvis is in equilibrium, the resultant of W, A, and J is zero. The moment arms of A and W with respect to the hip joint axis of rotation are approximately 6 cm and 11 cm, respectively. By taking moments about the fulcrum

$$(W \times M_W) - (A \times M_A) = 0.$$

That is,

$$W \times M_W = A \times M_A$$

$$A = \frac{W \times M_W}{M_A},$$

where $W = 59$ kg wt, $M_W = 11$ cm, and $M_A = 6$ cm. That is,

$$A = \frac{59 \text{ kg wt} \times 11 \text{ cm}}{6 \text{ cm}}$$

$$A = 108 \text{ kg wt.}$$

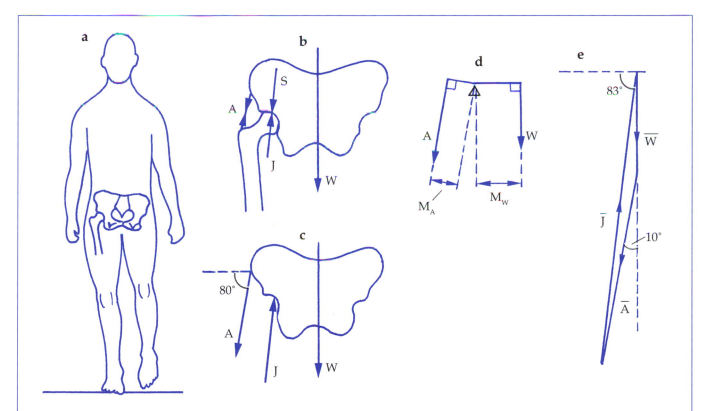

Figure 9.6 Effect of one-legged stance on hip abductor muscle force and hip joint reaction force; (*a*) one-legged stance; (*b*) forces on pelvis and hip joint during one-legged stance (A—force exerted by hip abductors, J—joint reaction force exerted by head of femur on acetabulum, S—joint reaction force exerted by acetabulum on head of femur: J is equal and opposite to S, W—weight of body less the weight of the grounded leg); (*c*) free body diagram of pelvis during one-legged stance; (*d*) force-moment arm diagram corresponding to *c*; (*e*) vector chain determination of joint reaction force.

(continued)

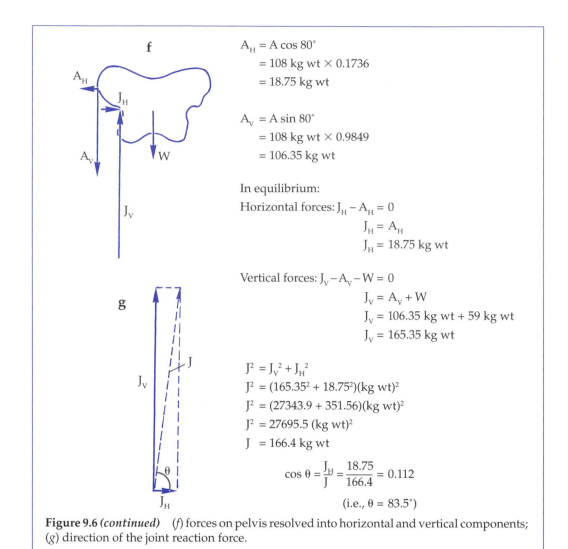

f

$A_H = A \cos 80°$
$= 108 \text{ kg wt} \times 0.1736$
$= 18.75 \text{ kg wt}$

$A_V = A \sin 80°$
$= 108 \text{ kg wt} \times 0.9849$
$= 106.35 \text{ kg wt}$

In equilibrium:

Horizontal forces: $J_H - A_H = 0$
$J_H = A_H$
$J_H = 18.75 \text{ kg wt}$

g

Vertical forces: $J_V - A_V - W = 0$
$J_V = A_V + W$
$J_V = 106.35 \text{ kg wt} + 59 \text{ kg wt}$
$J_V = 165.35 \text{ kg wt}$

$J^2 = J_V^2 + J_H^2$
$J^2 = (165.35^2 + 18.75^2)(\text{kg wt})^2$
$J^2 = (27343.9 + 351.56)(\text{kg wt})^2$
$J^2 = 27695.5 \text{ (kg wt)}^2$
$J = 166.4 \text{ kg wt}$

$$\cos \theta = \frac{J_H}{J} = \frac{18.75}{166.4} = 0.112$$

(i.e., $\theta = 83.5°$)

Figure 9.6 *(continued)* *(f)* forces on pelvis resolved into horizontal and vertical components; *(g)* direction of the joint reaction force.

The vector chain determination of J is shown in figure 9.6e; J = 166 kg wt at an angle of 83° to the horizontal. The determination of J by equating the forces is shown in figure 9.6, f and g; J = 166.4 kg wt at an angle of 83.5° to the horizontal. In this example the vector chain and calculation methods of determining J give almost identical results. The results show that the joint reaction force in one-legged stance is in the region of 2.4 times body weight. The force S—the force exerted by the pelvis on the head of the femur—is equal and opposite to that of J (see figure 9.6b).

When someone is recovering from a serious leg injury such as a broken femur, it is desirable to reduce the load on the leg during weight-bearing activities such as standing and walking by using crutches and walking canes. Figure 9.7a shows a man walking with the aid of a cane in his left hand. During the right leg single-support phase of the walking cycle the cane helps to support the weight of the body and, therefore, to reduce the load on the right leg. The cane is, in effect, an extension of the left arm, enabling the left arm to help support the weight of the body by pushing against the floor. In this example the left arm can be considered to be a lateral extension of the pelvis such that the free body diagram of the pelvis can be represented as in figure 9.7b (Watkins 1983). The corresponding force-moment arm diagram is shown in figure 9.7c.

When walking without a cane, as in figure 9.6a, the moment of W is counteracted by the moment of A on its own. However, when walking with a cane the moment of

W is counteracted by the combined moment of A and the force C exerted by the cane on the man's left hand. When the pelvis is in equilibrium, the resultant of W, A, C, and J will be zero. For a man of total body weight 70 kg wt, C is about 16 kg wt with a moment arm of approximately 35 cm with respect to the hip joint axis of rotation; it is assumed that C acts vertically. By taking moments about the fulcrum

$$(W \times M_W) - (A \times M_A) - (C \times M_C) = 0.$$

That is

$$(W \times M_W) = (A \times M_A) + (C \times M_C)$$

$$(A \times M_A) = (W \times M_W) - (C \times M_C)$$

$$A = \frac{(W \times M_W) - (C \times M_C)}{M_A},$$

where W = 59 kg wt, C = 16 kg wt, M_A = 6 cm, M_W = 11 cm, and M_C = 35 cm. Thus,

$$A = \frac{(59 \text{ kg wt} \times 11 \text{ cm}) - (16 \text{ kg wt} \times 35 \text{ cm})}{6 \text{ cm}}$$

$$A = 14.8 \text{ kg wt.}$$

The vector chain determination of J is shown in figure 9.7d; J = 57.5 kg wt at an angle of 87° to the horizontal. The determination of J by equating the forces is shown in figure 9.7, e and f; J = 57.6 kg wt at an angle of 87.4° to the horizontal. As in the previous example, the vector chain and calculation methods of determining J give almost identical results. The results indicate that using a cane to aid one-legged stance considerably reduces the force exerted in the hip abductor muscles (108 kg wt to 14.8 kg wt) and, consequently, reduces the hip joint reaction force (166 kg wt to 57.5 kg wt).

Effect of Squat and Stoop Postures on Forces in the Lumbar Region

When the trunk is inclined forward from the upright position, the weight of the upper body (head, neck, arms, and trunk) exerts a moment tending to flex the trunk; the greater the degree of inclination of the trunk, the greater the trunk flexor moment exerted by upper body weight. When the trunk is inclined forward, but in equilibrium, the trunk flexor moment exerted by upper body weight is counteracted by the trunk extensor muscles (figure 9.8, a and b). Upper body weight and the force exerted by the trunk extensor muscles constitute a first class lever system with respect to the intervertebral joints (figure 9.8c).

The moment arm of the trunk extensor muscles is similar throughout the vertebral column. However, the moment arm of upper body weight (the proportion of upper body weight superior to a particular intervertebral joint) increases as the degree of forward inclination of the trunk increases. Consequently, the greater the degree of forward inclination of the trunk, the greater the force exerted by the trunk extensors to maintain equilibrium. Since the moment arm of the trunk extensors is usually much shorter than the moment arm of upper body weight, the force exerted by the trunk extensors and, therefore, the joint reaction forces exerted on the intervertebral joints, are relatively large when maintaining the trunk in a forward inclined position. Since the proportion of upper body weight above a particular intervertebral joint and its moment arm about the joint tend to increase from above downward, the force exerted by the trunk extensor muscles and the intervertebral joint reaction forces also tend to increase from above downward.

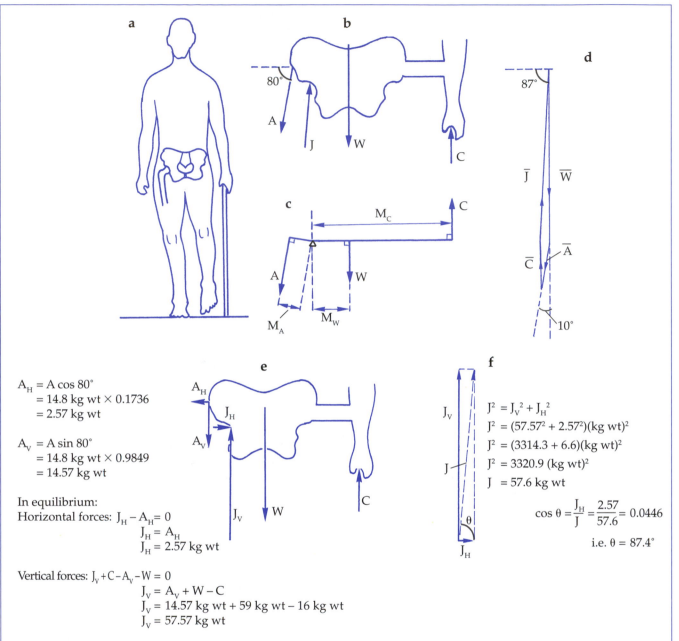

$A_H = A \cos 80°$
$= 14.8 \text{ kg wt} \times 0.1736$
$= 2.57 \text{ kg wt}$

$A_V = A \sin 80°$
$= 14.8 \text{ kg wt} \times 0.9849$
$= 14.57 \text{ kg wt}$

In equilibrium:
Horizontal forces: $J_H - A_H = 0$
$J_H = A_H$
$J_H = 2.57 \text{ kg wt}$

Vertical forces: $J_V + C - A_V - W = 0$
$J_V = A_V + W - C$
$J_V = 14.57 \text{ kg wt} + 59 \text{ kg wt} - 16 \text{ kg wt}$
$J_V = 57.57 \text{ kg wt}$

$J^2 = J_V^2 + J_H^2$
$J^2 = (57.57^2 + 2.57^2)(\text{kg wt})^2$
$J^2 = (3314.3 + 6.6)(\text{kg wt})^2$
$J^2 = 3320.9 \, (\text{kg wt})^2$
$J = 57.6 \text{ kg wt}$

$\cos \theta = \dfrac{J_H}{J} = \dfrac{2.57}{57.6} = 0.0446$

i.e. $\theta = 87.4°$

Figure 9.7 Effect of using a cane on hip abductor muscle force and hip joint reaction force in one-legged stance; (*a*) one-legged stance with the aid of a cane; (*b*) free body diagram of pelvis during one-legged stance with a cane (A—force exerted by hip abductors, J—joint reaction force exerted by head of femur on acetabulum, W—weight of body less the grounded leg, C—force exerted by the cane on the left hand); (*c*) force-moment arm diagram corresponding to *b*; (*d*) vector chain determination of joint reaction force; (*e*) forces on pelvis resolved into horizontal and vertical components; (*f*) direction of the joint reaction force.

Squat Posture

Any load held or lifted in front of the body exerts a flexor moment on the trunk and, therefore, increases trunk extensor muscle forces and intervertebral joint reaction forces. Figure 9.8*a* shows a person in a squat posture holding a load of 10 kg wt just above the floor. Figure 9.8*b* shows a free body diagram of the upper body and load, where W is the combined weight of the upper body (W_1) and load (W_2), F is the force exerted by the trunk extensor muscles across the L5/S1 intervertebral joint, and J is

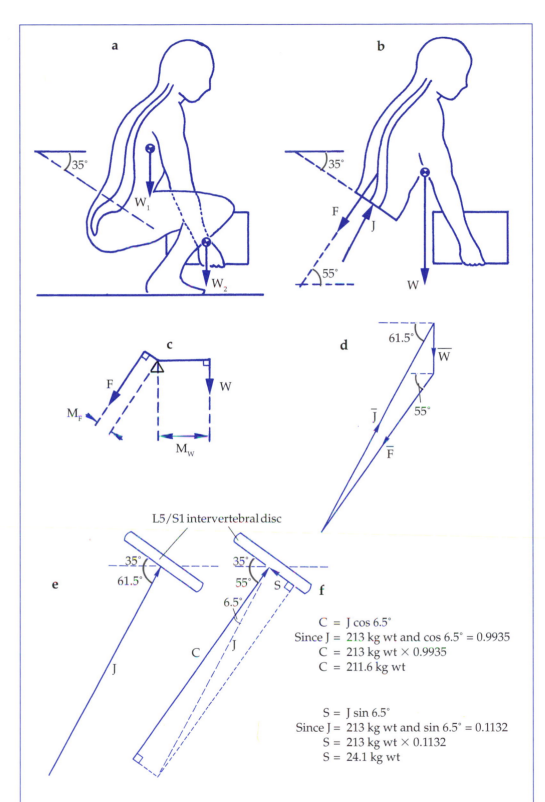

Figure 9.8 Force exerted by the trunk extensor muscles and joint reaction force in the L5/S1 intervertebral joint when lifting a load of 10 kg wt from a squat position; (*a*) orientation of trunk and L5/S1 joint; (*b*) free body diagram of upper body weight and load relative to L5/S1 joint; (*c*) force-moment arm diagram corresponding to *b*; (*d*) vector chain determination of L5/S1 joint reaction force; (*e*) orientation of joint reaction force to plane of L5/S1 joint; (*f*) compression and shear components of joint reaction force.

the L5/S1 intervertebral joint reaction force. Figure 9.8c shows the corresponding force-moment arm diagram. The weight W_1 of the upper body above the L5/S1 joint, that is, the head, neck, arms, and the proportion of the trunk above the L5/S1 joint, is approximately 48% of total body weight (Morris, Lucas, and Bresler 1961). Consequently, for a man of total body weight 70 kg wt, W_1 is approximately 34 kg wt. The moment arm of W (M_W) is approximately 20 cm. The plane of the L5/S1 joint in the squat position is approximately 35° with respect to the horizontal. The line of action of F is approximately perpendicular to the plane of the L5/S1 joint, and the moment arm of F (M_F) about the fulcrum, that is, the middle of the L5/S1 joint, is approximately 5 cm. By taking moments about the fulcrum

$$(W \times M_W) - (F \times M_F) = 0$$

$$(W \times M_W) = (F \times M_F).$$

That is,

$$F = \frac{(W \times M_W)}{M_F},$$

where W = 44 kg wt, M_W = 20 cm, and M_F = 5 cm. Thus,

$$F = \frac{44 \text{ kg wt} \times 20 \text{ cm}}{5 \text{ cm}}$$

$$F = 176 \text{ kg wt.}$$

The vector chain determination of J is shown in figure 9.8d; J = 213 kg wt at an angle of 61.5° to the horizontal. Consequently, to hold a load of 10 kg wt just above the floor in a squat position, the force exerted by the trunk extensors adjacent to the L5/S1 intervertebral joint is in the region of 2.5 times body weight, and the L5/S1 joint reaction force is about 3 times body weight.

As shown in figure 9.8e, the line of action of J is oblique to the plane of the L5/S1 joint. Consequently, J has a compression component C (perpendicular to the plane of the joint) and a shear component S (parallel to the plane of the joint) (figure 9.8f). Since J (213 kg wt) makes an angle of 83.5° with the plane of the joint, C is relatively large (212 kg wt) and S is relatively small (24 kg wt).

Stoop Posture

In figure 9.9a, the person is shown holding the 10 kg wt just above the floor in a stoop posture. Figure 9.9b shows a free body diagram of the upper body and load relative to the L5/S1 joint. Figure 9.9c shows the corresponding force-moment arm diagram. In this situation the moment arm of W is approximately 30 cm (compared to 20 cm in the squat posture). The plane of the L5/S1 joint in the stoop posture is approximately 70° with respect to the horizontal (compared to 35° in the squat posture). If the line of action of F is assumed to be perpendicular to the plane of the L5/S1 joint with a moment arm of 5 cm (as in the squat position), then by taking moments about the fulcrum

$$(W \times M_W) - (F \times M_F) = 0$$

$$(W \times M_W) = (F \times M_F).$$

That is,

$$F = \frac{(W \times M_W)}{M_F},$$

where W = 44 kg wt, M_W = 30 cm, and M_F = 5 cm. Then

$$F = \frac{44 \text{ kg wt} \times 30 \text{ cm}}{5 \text{ cm}}$$

$$F = 264 \text{ kg wt.}$$

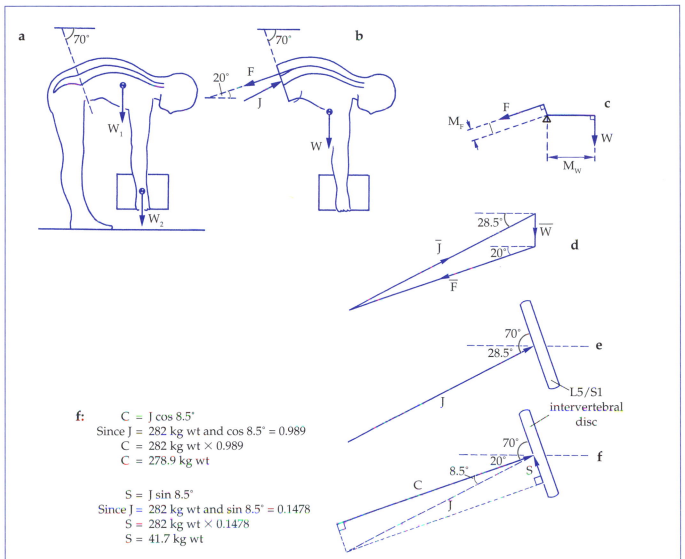

f:
$$C = J \cos 8.5°$$
Since J = 282 kg wt and cos 8.5° = 0.989
$$C = 282 \text{ kg wt} \times 0.989$$
$$C = 278.9 \text{ kg wt}$$

$$S = J \sin 8.5°$$
Since J = 282 kg wt and sin 8.5° = 0.1478
$$S = 282 \text{ kg wt} \times 0.1478$$
$$S = 41.7 \text{ kg wt}$$

Figure 9.9   Force exerted by the trunk extensor muscles and joint reaction force in the L5/S1 intervertebral joint when lifting a load of 10 kg wt from a stoop position; (*a*) orientation of trunk and L5/S1 joint; (*b*) free body diagram of upper body weight and load relative to L5/S1 joint; (*c*) force-moment arm diagram corresponding to *b*; (*d*) vector chain determination of L5/S1 joint reaction force; (*e*) orientation of joint reaction force to plane of L5/S1 joint; (*f*) compression and shear components of joint reaction force.

The vector chain determination of J is shown in figure 9.9*d*; J = 282 kg wt at an angle of 28.5° to the horizontal. Consequently, to hold a load of 10 kg wt just above the floor in a stoop position, the force exerted by the trunk extensor muscles adjacent to the L5/S1 intervertebral joint will be about 3.8 times body weight (compared to 2.5 times body weight in the squat posture), and the L5/S1 joint reaction force will be about 4 times body weight (compared to 3 times body weight in the squat posture). As in the squat posture, the line of action of J is oblique to the plane of the L5/S1 joint (figure 9.9*e*). In this case the compression component is approximately 279 kg wt (compared to 212 kg wt in the squat posture) and the shear component is about 42 kg wt (compared to 24 kg wt in the squat posture) (figure 9.9*f* and table 9.1).

Effect of Intratruncal Pressure

In the above estimates of trunk extensor muscle forces and L5/S1 joint reaction forces, no account was taken of the effect of intratruncal pressure. As described in chapter 6,

Table 9.1 Effect of Type of Posture (Squat, Stoop) and Intratruncal Pressure (P) on Trunk Extensor Muscle Force and L5/S1 Joint Reaction Force (J)

| | Force (kg wt) | | Squat as percentage of | (2) as a percentage of (1) (%) | |
	Squat	Stoop	Stoop (%)	Squat	Stoop
Trunk extensor muscle force					
(1) Without P	176	264	67		
(2) With P	135	203	66	77	77
Joint reaction force J					
(1) Without P	213	282	75		
(2) With P	155	192	81	73	68
Compression component of J					
(1) Without P	212	279	76		
(2) With P	153	188	81	72	67
Shear component of J					
(1) Without P	24	42	60		
(2) With P	24	40	60	100	95

intratruncal pressure may reduce trunk extensor muscle forces and intervertebral joint reaction forces in postures involving forward inclination of the trunk by exerting a trunk extensor moment. It has been estimated that the trunk extensor moment exerted by intratruncal pressure is approximately 30% of the moment exerted by the trunk extensor muscles and that the moment arm of intratruncal pressure is approximately 11 cm (Morris, Lucas, and Bresler 1961). Figure 9.10a shows the orientation of the trunk and the thoracic and abdominal cavities in the squat posture. Figure 9.10b shows a free body diagram of the upper body and load relative to the L5/S1 joint, including the force P exerted by intratruncal pressure. Figure 9.10c shows the corresponding force-moment arm diagram. In this situation the trunk flexor moment exerted by W will be counteracted by F and P. By taking moments about the fulcrum

$$(W \times M_W) - (F \times M_F) - (P \times M_P) = 0.$$

That is,

(1)
$$(W \times M_W) = (F \times M_F) + (P \times M_P)$$

If the moment exerted by P is approximately 30% of the moment exerted by F, then

(2)
$$(P \times M_P) = 0.3(F \times M_F).$$

Replacing $(P \times M_P)$ in (1) gives

$$(W \times M_W) = 1.3(F \times M_F).$$

Thus,

$$F = \frac{W \times M_W}{1.3 \times M_F},$$

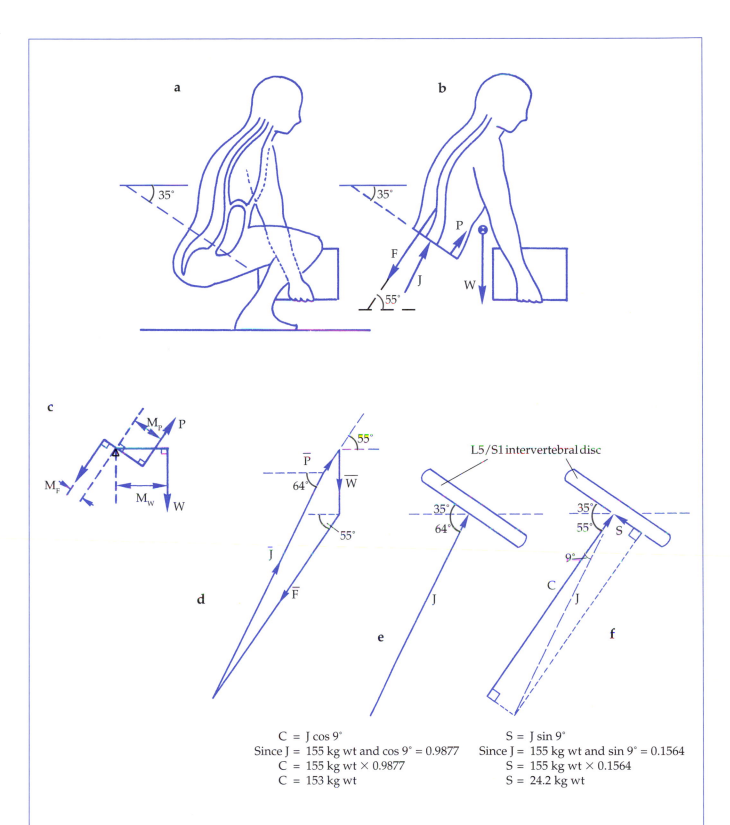

$$C = J \cos 9°$$
Since J = 155 kg wt and cos 9° = 0.9877
$$C = 155 \text{ kg wt} \times 0.9877$$
$$C = 153 \text{ kg wt}$$

$$S = J \sin 9°$$
Since J = 155 kg wt and sin 9° = 0.1564
$$S = 155 \text{ kg wt} \times 0.1564$$
$$S = 24.2 \text{ kg wt}$$

Figure 9.10 Effect of intratruncal pressure on the force exerted by the trunk extensor muscles and joint reaction force in the L5/S1 intervertebral joint when lifting a load of 10 kg wt from a squat position; (*a*) orientation of trunk and L5/S1 joint; (*b*) free body diagram of upper body weight and load relative to L5/S1 joint; (*c*) force-moment arm diagram corresponding to *b*; (*d*) vector chain determination of L5/S1 joint reaction force; (*e*) orientation of joint reaction force to plane of L5/S1 joint; (*f*) compression and shear components of joint reaction force.

where W = 44 kg wt, M_W = 20 cm, and M_F = 5 cm. Then

$$F = \frac{44 \text{ kg wt} \times 20 \text{ cm}}{1.3 \times 5 \text{ cm}}$$

$$F = 135.4 \text{ kg wt}$$

From (2),

$$P = \frac{0.3(F \times M_F)}{M_P},$$

where F = 135.4 kg wt, M_F = 5 cm, and M_P =11 cm. Then

$$P = \frac{0.3 \times (135.4 \text{ kg wt} \times 5 \text{ cm})}{11 \text{ cm}}$$

$$P = 18.4 \text{ kg wt.}$$

The vector chain determination of J is shown in figure 9.10d; J = 155 kg wt at an angle of 64° to the horizontal. As in the previous examples (figures 9.8 and 9.9), J is oblique to the plane of the L5/S1 joint (figure 9.10e). Figure 9.10f shows the compression and shear components of J. In this example, intratruncal pressure reduces the trunk extensor muscle force, the L5/S1 joint reaction force, and the compression component of the joint reaction force by 23%, 27%, and 28%, respectively, in the squat posture (see table 9.1).

Figure 9.11a shows the orientation of the trunk and thoracic and abdominal cavities in the stoop posture. Figure 9.11b shows a free body diagram of the upper body and load relative to the L5/S1 joint, including the force P exerted by intratruncal pressure. Figure 9.11c shows the corresponding force-moment arm diagram. By taking moments about the fulcrum

$$(W \times M_W) - (F \times M_F) - (P \times M_P) = 0.$$

That is,

$$(W \times M_W) = (F \times M_F) + (P \times M_P).$$

As previously discussed, if the moment exerted by P is about 30% of the moment exerted by F, then

$$F = \frac{W \times M_W}{1.3 \times M_F},$$

where W = 44 kg wt, M_W = 30 cm, and M_F = 5 cm. Then

$$F = \frac{44 \text{ kg wt} \times 30 \text{ cm}}{1.3 \times 5 \text{ cm}}$$

$$F = 203 \text{ kg wt.}$$

Also,

$$P = \frac{0.3(F \times M_F)}{M_P},$$

where F = 203 kg wt, M_F = 5 cm, and M_P =11 cm, such that

$$P = \frac{0.3 \times 203 \text{ kg wt} \times 5 \text{ cm}}{11 \text{ cm}}$$

$$P = 27.7 \text{ kg wt.}$$

The vector chain determination of J is shown in figure 9.11d; J = 192 kg wt at an angle of 32° to the horizontal. As in the previous examples (figures 9.8 through 9.10), J is oblique to the plane of the L5/S1 joint (figure 9.11e). The compression and shear components are shown in figure 9.11f. In this example, intratruncal pressure reduces the trunk extensor muscle force, the L5/S1 joint reaction force, and the compression component of the joint reaction force by 23%, 32%, and 33%, respectively, in the stoop posture (see table 9.1).

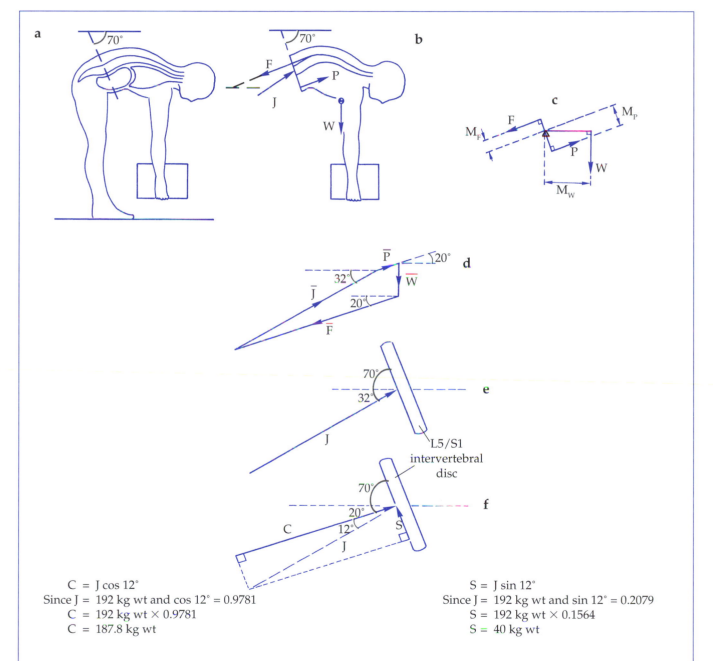

C = J cos 12°
Since J = 192 kg wt and cos 12° = 0.9781
C = 192 kg wt × 0.9781
C = 187.8 kg wt

S = J sin 12°
Since J = 192 kg wt and sin 12° = 0.2079
S = 192 kg wt × 0.1564
S = 40 kg wt

Figure 9.11   Effect of intratruncal pressure on the force exerted by the trunk extensor muscles and joint reaction force in the L5/S1 intervertebral joint when lifting a load of 10 kg wt from a stoop position; (a) orientation of trunk and L5/S1 joint; (b) free body diagram of upper body weight and load relative to L5/S1 joint; (c) force-moment arm diagram corresponding to b; (d) vector chain determination of L5/S1 joint reaction force; (e) orientation of joint reaction force to plane of L5/S1 joint; (f) compression and shear components of joint reaction force.

Swing and Stabilization Components of Muscle Force

In postures other than very relaxed postures such as lying down or sitting in an arm chair, most of the muscles of the body are active to control the movements of the joints. In controlling joint movements muscles exert two effects on joints: stabilization and linear/angular displacement. Since joints need to be stabilized—joint congruence needs to be maintained—whether or not the joints are moving, it follows that the stabilization of joints is a major function of the muscles. The contribution of a muscle to stabilization and angular movement of a joint is determined by the **stabilization component** and the **swing component**, respectively, of the muscle force. The stabilization component is directed at the axis of rotation to maintain joint congruence. The swing component is at right angles to the stabilization component and, as such, exerts a turning moment about the joint. In movement of a particular joint, it is likely that all of the muscles involved contribute to both stabilization and swing, but the contributions of each muscle depend on the angle of the joint.

Figure 9.12 shows the orientation of the line of action of the biceps brachii at three different elbow angles. In figure 9.12*a* the elbow is close to full extension. In this position the stabilization component is much larger than the swing component. As the elbow flexes from the extended position, the stabilization component progressively decreases and the swing component progressively increases such that a point is reached, when the elbow angle is about 90°, where the stabilization component is zero (figure 9.12*b*). Further elbow flexion results in the component of muscle force in line with the axis of rotation being directed away from the axis of rotation, that is, tending to sublux the joint and reduce joint congruence (figure 9.12*c*). Under normal circumstances this is not likely a problem since the stabilization component is relatively small due to active insufficiency of the muscles concerned, and the other muscles involved in controlling the movement of the joint are likely to counteract the subluxation component.

Prime Movers and Synergists in Elbow Flexion

The biceps brachii and brachialis are prime movers in elbow flexion and, as such, they exert relatively large swing components and small stabilization components

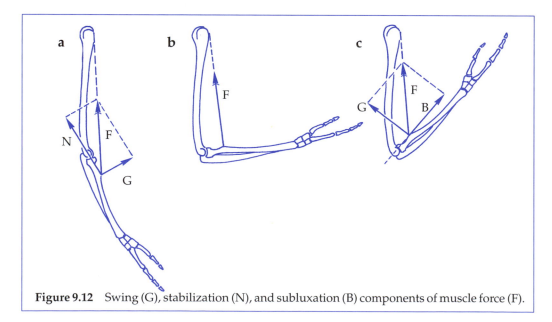

Figure 9.12   Swing (G), stabilization (N), and subluxation (B) components of muscle force (F).

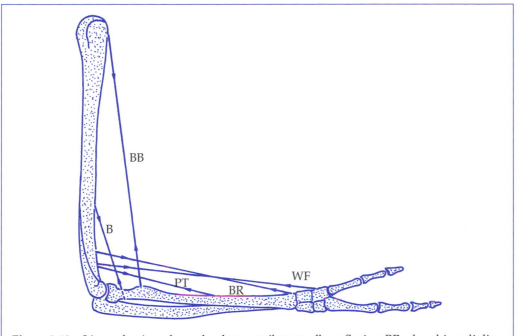

Figure 9.13   Lines of action of muscles that contribute to elbow flexion: BR—brachioradialis; PT—pronator teres; BB—biceps brachii; B—brachialis; WF—wrist and finger flexors.

during most of the elbow flexion and extension range of motion; they also exert subluxation components close to full flexion. Figure 9.13 shows the lines of action of the biceps brachii (BB) and brachialis (B) when the forearm is held horizontal with the upper arm vertical. The biceps brachii and brachialis are assisted in elbow flexion by a number of other muscles in the role of synergist. These muscles, which include the pronator teres (PT), brachioradialis (BR), and wrist and finger flexors (WF)—flexor carpi ulnaris, flexor carpi radialis, and flexor digitorum sublimis—exert relatively small swing components and large stabilization components during most of the elbow flexion and extension range. Consequently, during elbow flexion, these muscles mainly function to stabilize the elbow joint while providing some assistance to the biceps brachii and brachialis in terms of swing. Figure 9.13 shows these muscles' lines of action.

Calculation of Muscle Forces

Figure 9.14, *a* and *b*, shows a free body diagram of the forearm of an adult held in a horizontal position with the upper arm vertical and a load (W_L) of 2 kg wt in the palm of the hand. For a person weighing 70 kg wt, the weight of the lower arm and hand (W_{AH}) is approximately 1.5 kg wt (2.26% of total body weight; see p. 281). In this position W_{AH} and W_L exert clockwise moments on the forearm about the elbow. In equilibrium these clockwise moments are counteracted by the combined counterclockwise moment exerted by the five muscles or muscle groups shown in figure 9.14*a*: brachialis (B), biceps brachii (BB), pronator teres (PT), brachioradialis (BR), and the wrist and finger flexors (WF). The wrist and finger flexors comprise the flexor carpi ulnaris, flexor carpi radialis, and that portion of the flexor digitorum sublimis that crosses the elbow joint. The pronator teres represents that portion of the muscle that crosses the elbow joint. Figure 9.14*b* shows a simplified version of the free body diagram in figure 9.14*a*. The simplified version assumes that the points of application of the muscle forces are in the same horizontal plane as the axis of elbow flexion and extension, that is, the fulcrum.

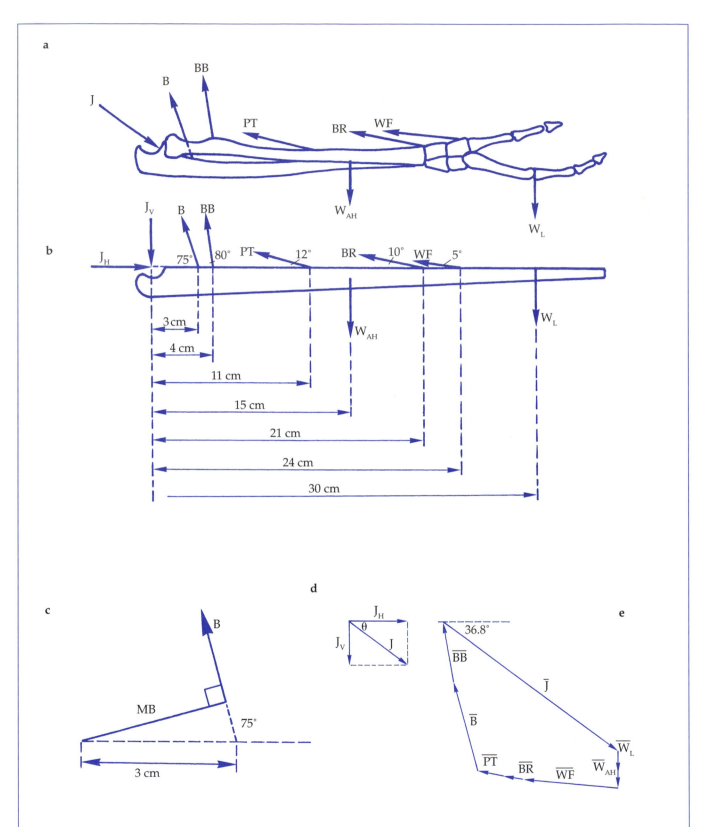

Figure 9.14   Forces acting on the forearm when held horizontally with the upper arm vertical and a load of 2 kg wt in the palm of the hand; (*a*) free body diagram of forearm and hand; (*b*) simplified free body diagram of forearm and hand; (*c*) determination of the moment arm of the force exerted by the brachialis; (*d*) determination of elbow joint reaction force from horizontal and vertical components; (*e*) vector chain determination of elbow joint reaction force.

The calculation of the moment arm of each muscle is illustrated with reference to the brachialis. Figure 9.14c shows the line of action of the brachialis and its moment arm M_B. Since the line of action of the brachialis makes an angle of 75° with the horizontal plane through the fulcrum, it follows that

$$\frac{M_B}{3 \text{ cm}} = \sin 75°$$

$$M_B = \sin 75° \times 3 \text{ cm}$$

$$M_B = 0.9659 \times 3 \text{ cm}$$

$$M_B = 2.9 \text{ cm.}$$

The moment arms of the other muscles can be calculated in the same way. By taking moments about the fulcrum

$$(W_{AH} \times M_{AH}) + (W_L \times M_L) = (B \times M_B) + (BB \times M_{BB}) + (PT \times M_{PT})$$
$$+ (BR \times M_{BR}) + (WF \times M_{WF}),$$

where W_{AH} = 1.5 kg wt; W_L = 2 kg wt; B = force exerted by the brachialis; BB = force exerted by the biceps brachii; PT = force exerted by the pronator teres; BR = force exerted by the brachioradialis; WF = force exerted by the wrist and finger flexors; M_{AH} = moment arm of W_{AH} = 15 cm; M_L = moment arm of W_L = 30 cm; M_B = moment arm of B = 2.9 cm; M_{BB} = moment arm of BB = 3.94 cm; M_{PT} = moment arm of P_T = 2.29 cm; M_{BR} = moment arm of BR = 3.65 cm; M_{WF} = moment arm of WF = 2.09 cm. Thus,

(3) (1.5 kg wt × 15 cm) + (2 kg wt × 30 cm) = (B × 2.9 cm) + (BB × 3.94 cm)
+ (PT × 2.29 cm) + (BR × 3.65 cm)
+ (WF × 2.09 cm)

22.5 kg wt cm + 60 kg wt cm = (B × 2.9 cm) + (BB × 3.94 cm)
+ (PT × 2.29 cm) + (BR × 3.65 cm)
+ (WF × 2.09 cm)

82.5 kg wt cm = (B × 2.9 cm) + (BB × 3.94 cm)
+ (PT × 2.29 cm) + (BR × 3.65 cm)
+ (WF × 2.09 cm)

To calculate the muscle forces, it is necessary to estimate the force produced by each muscle relative to the other muscles. These estimates of relative force are made on the basis of the physiological cross-sectional areas of the muscles (An et al. 1981; Lieber, Fazeli, and Botte 1990). Consequently, in relation to the force BR exerted by the brachioradialis, the relative magnitudes of the muscle forces are as shown in table 9.2. By substitution of the relative muscle forces in (3),

82.5 kg wt cm = (4.7BR × 2.9 cm) + (3.1BR × 3.94 cm)
+ (1.3BR × 2.29 cm) + (BR × 3.65 cm)
+ (4.9BR × 2.09 cm)

82.5 kg wt cm = (13.63BR cm) + (12.21BR cm) +
(2.98BR cm) + (3.65BR cm)
+ (10.24BR cm)

82.5 kg wt cm = 42.71BR cm.

That is,

$$BR = \frac{82.5 \text{ kg wt cm}}{42.71 \text{ cm}}$$

$$BR = 1.93 \text{ kg wt.}$$

Table 9.2   Relative Magnitudes of Muscle Forces

Muscle	Relative force exerted by muscle
Brachioradialis (BR)	BR
Pronator teres (PT)	1.3 BR
Biceps brachii (BB)	3.1 BR
Brachialis (B)	4.7 BR
Wrist and finger flexors (WF)	4.9 BR

Table 9.3   Muscle Forces

Muscle	Force exerted (kg wt)		
Brachioradialis (BR)	=	BR	= 1.93
Pronator teres (PT)		= 1.3 BR	= 2.51
Biceps brachii (BB)		= 3.1 BR	= 5.98
Brachialis (B)		= 4.7 BR	= 9.07
Wrist and finger flexors (WF)		= 4.9 BR	= 9.46

Consequently, table 9.3 shows the forces exerted by the muscles.

Calculation of Elbow Joint Reaction Force

Since the forearm and hand are in equilibrium, the resultant force acting on the forearm and hand is zero. Consequently, the sum of the horizontal forces is zero, and the sum of the vertical forces is zero. With respect to figure 9.14b, the horizontal and vertical forces can be calculated as follows:

Horizontal forces:

$$J_H - B \cos 75° - BB \cos 80° - PT \cos 12° - BR \cos 10° - WF \cos 5° = 0.$$

That is,

$$J_H = B \cos 75° + BB \cos 80° + PT \cos 12° + BR \cos 10° + WF \cos 5°$$

$$J_H = 2.35 \text{ kg wt} + 1.04 \text{ kg wt} + 2.45 \text{ kg wt} + 1.90 \text{ kg wt} + 9.42 \text{ kg wt}$$

$$J_H = 17.16 \text{ kg wt.}$$

Vertical forces:

$$B \sin 75° + BB \sin 80° + PT \sin 12° + BR \sin 10° + WF \sin 5° - W_{AH} - W_L - J_V = 0.$$

That is,

$$J_V = B \sin 75° + BB \sin 80° + PT \cos 12° + BR \sin 10° + WF \sin 5° - W_{AH} - W_L$$

$$J_V = 8.76 \text{ kg wt} + 5.89 \text{ kg wt} + 0.52 \text{ kg wt} + 0.33 \text{ kg wt} + 0.82 \text{ kg wt} - 1.5 \text{ kg wt} - 2 \text{ kg wt}$$

$$J_V = 12.82 \text{ kg wt.}$$

Table 9.4 Contribution of Prime Movers and Synergists to the Total Elbow Flexor Swing Moment

Group	Muscle	Moment (kg wt cm)	Group total (kg wt cm)	(%)
Prime mover	Brachialis	26.28	49.84	60.5
Prime mover	Biceps brachii	23.56		
Synergist	Pronator teres	5.74		
Synergist	Brachioradialis	7.04	32.57	39.5
Synergist	Wrist and finger flexors	19.79		
	Total swing moment =	82.41	82.41	100

Magnitude of elbow joint reaction force J:

By Pythagoras's theorem

$$J^2 = J_V^2 + J_H^2$$

$$J^2 = (12.82 \text{ kg wt})^2 + (17.16 \text{ kg wt})^2$$

$$J^2 = (164.35 + 294.45) \text{ (kg wt)}^2$$

$$J^2 = 458.81 \text{ (kg wt)}^2$$

$$J = 21.42 \text{ kg wt.}$$

Direction of J (figure 9.14d):

$$\text{Cos } \theta = \frac{J_H}{J} = \frac{17.16 \text{ kg wt}}{21.42 \text{ kg wt}} = 0.8011$$

$$\theta = 36.8° \text{ with respect to the horizontal.}$$

Figure 9.14e shows the vector chain determination of J.

Contribution of Prime Movers and Synergists to Swing and Stabilization

The contribution of the prime movers and synergists to swing is given by their contributions to the total swing moment. As shown in table 9.4, the prime movers and synergists contribute approximately 60% and 40%, respectively, to the total swing moment. The contributions of the prime movers and synergists to joint stabilization is given by their contributions to the total stabilization component, which, in this example, is equal and opposite to J_H. As shown in table 9.5, the prime movers and synergists contribute approximately 20% and 80%, respectively, to the total stabilization component.

Contribution of Synergists to Reducing Bending Stress on the Forearm and Hand

Whereas an object may be in equilibrium, it may also be subject to bending or torsional stress depending on the orientation of the forces acting on it. In the elbow example,

Table 9.5   **Contribution of Prime Movers and Synergists to the Total Elbow Flexor Joint Stabilization Component**

Group	Muscle	Stabilization component (kg wt)	Group total (kg wt)	(%)
Prime mover	Brachialis	2.35	3.39	19.7
Prime mover	Biceps brachii	1.04		
Synergist	Pronator teres	2.45		
Synergist	Brachioradialis	1.90	13.77	80.3
Synergist	Wrist and finger flexors	9.42		

Total stabilization component =   17.16

the synergists not only contribute to swing and joint stabilization, but they also reduce bending stress on the forearm and hand. In the absence of the synergists, the clockwise moment CM exerted by W_{AH} and W_L about the radial tuberosity (site of attachment of the biceps brachii) would bend the arm downward. With respect to figure 9.14b,

$$CM = (W_{AH} \times 11 \text{ cm}) + (W_L \times 26 \text{ cm})$$

$$CM = 16.5 \text{ kg wt cm} + 52 \text{ kg wt cm}$$

$$CM = 68.5 \text{ kg wt cm.}$$

However, the synergists counteract the bending moment by exerting a counterclockwise moment AM (about the radial tuberosity); that is,

$$AM = (PT \sin 12° \times 7 \text{ cm}) + (BR \sin 10° \times 17 \text{ cm}) + (WF \sin 5° \times 20 \text{ cm})$$

$$AM = 3.65 \text{ kg wt cm} + 5.69 \text{ kg wt cm} + 16.49 \text{ kg wt cm}$$

$$AM = 25.83 \text{ kg wt cm.}$$

Consequently, in this example, the synergists reduce the bending moment on the forearm and hand by approximately 38%. It is likely that the inclusion of more muscles in the free body diagram and more accurate data regarding muscle forces and their moment arms would reduce the bending moment even further.

Summary

This chapter explored the effects of changes in the leverage of external loads on the magnitude of internal forces required to counteract the external loads. Increasing the leverage of external loads results in an increase in the magnitude of muscle forces, which, in turn, results in an increase in the magnitude of joint reaction forces. Furthermore, due to the low mechanical advantage of most muscles, the magnitude of muscle forces and joint reaction forces is much larger than the external loads. Muscles work together to counteract external loads; this distributes the load over many muscles and, therefore, reduces the force exerted by each muscle and reduces bending stress on bones. Under normal circumstances the musculoskeletal components adapt their size, shape, and structure to more readily withstand the loads exerted on them. The response and adaptation of the musculoskeletal system to loading is covered in part III, starting with consideration of the mechanical characteristics of musculoskeletal components in the next chapter.

Segmental Masses and Mass Center Loci

Part	% of total weight	Mass center locus as a proportion of segment length
Upper arm	2.700	0.436 (with respect to proximal joint)
Forearm	1.600	0.430 (with respect to proximal joint)
Hand	0.665	0.506 (wrist to knuckle of middle digit)
Forearm and hand	2.260	0.677 (elbow to ulnar styloid process)
Whole upper limb	5.010	0.512 (shoulder to ulnar styloid process)
Upper leg	9.905	0.433 (with respect to proximal joint)
Lower leg	4.685	0.433 (with respect to proximal joint)
Foot	1.455	0.429 (with respect to length of foot, heel to toe)
Lower leg and foot	6.185	0.434 (knee to ankle with ankle in neutral position)
Whole lower limb	16.190	0.434 (hip axis to medial malleolus)
Trunk, head, and neck	57.61	0.604 (top of head to hip joint)
Head and neck	7.90	0.433 (top of head to seventh cervical vertebra)

Adapted from W.T. Dempster, 1955, "Space requirements for the seated operator," *WADC Technical Report 55159* (Wright-Patterson Air Force Base, Ohio: Wright Air Development Center).

Review Questions

1. Explain the relationship between the moment of external forces and the magnitude of muscle forces and joint reaction forces.
2. With regard to muscle forces, differentiate between stabilization component and subluxation component.
3. By equating the horizontal and vertical components of force, determine the magnitude and direction of the joint reaction force in each of the examples given in figures 9.8, *a* through *f*; 9.9, *a* through *f*; 9.10, *a* through *f*; and 9.11, *a* through *f*.

Part III

MUSCULOSKELETAL RESPONSE AND ADAPTATION TO LOADING

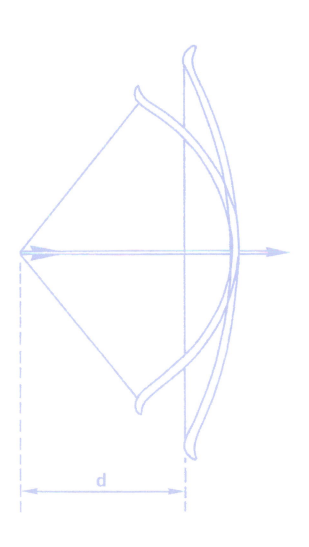

Part I described the basic structure and function of the musculoskeletal system and part II described the functional anatomy of the musculoskeletal components. Part III describes the immediate and long-term effects of loading on the external form and internal architecture of the musculoskeletal system. Chapter 10—Mechanical Characteristics of Musculoskeletal Components—details the musculoskeletal system's mechanical response to loading. Chapter 11—Structural Adaptation of the Musculoskeletal System—describes how the structure of the musculoskeletal system adapts to the time-averaged loads exerted on it. And finally, chapter 12—Etiology of Musculoskeletal Disorders and Injuries—explains the main groups of risk factors that influence the level of loading on the musculoskeletal system.

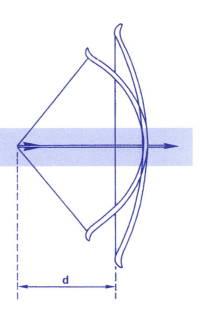

Chapter 10

Mechanical Characteristics of Musculoskeletal Components

All materials deform in response to loading in a manner that reflects their mechanical characteristics. The components of the musculoskeletal system have different mechanical characteristics, but in combination they provide considerable protection against potentially harmful loads. This chapter describes the mechanical characteristics of the musculoskeletal components and the response of the musculoskeletal system to loading. Before reading this chapter consider reviewing the sections in chapter 1 concerning work, energy, and power.

Objectives

After reading this chapter you should be able to do the following:

1. Define or describe the key terms.
2. Differentiate between a load-deformation curve and a stress-strain curve.
3. Describe the general stress-strain behavior of the musculoskeletal components.
4. Differentiate between passive and active loading.
5. Describe the causes of shock and vibration.
6. Describe the shock absorption mechanisms that operate within the human body.
7. Describe the effects of impact loading on synovial joints.

Stress-Strain Relationships in Solids

Load-deformation relationship: the relationship between load and deformation for a material

Stress-strain relationship: the relationship between stress and strain for a material

Yield point: the upper limit of the elastic range; also referred to as elastic limit

Elastic range: the strain range within which a material remains perfectly elastic

Failure point: the point at which a material fails

Plastic range: the strain range between the yield point and the failure point

Ultimate stress: the stress experienced by a material immediately prior to failure; also referred to as the strength of the material

All materials deform to a certain extent in response to loading; the greater the load, the greater the deformation. A material's **load-deformation relationship** reflects its mechanical characteristics. In order to compare the mechanical characteristics of different materials, it is usual to express load and deformation in terms of stress (load per unit cross-sectional area) and strain (deformation as a proportion of the dimensions of the material when unloaded), respectively. Stress is denoted by the Greek letter sigma (σ), and strain is denoted by the Greek letter epsilon (ϵ). Figure 10.1 shows a generalized **stress-strain curve.**

In response to tension loading, a material tends to lengthen in the direction of the tension load. Consequently, tension strain refers to increase in length of the material as a proportion of the length of the material when it is not loaded. Similarly, in response to compression loading, a material tends to shorten in the direction of the compression load—compression strain refers to decrease in length of the material as a proportion of the length of the material when it is not loaded. Note that stress is measured in units of force per unit cross-sectional area, such as N/cm^2, but strain has no units since it is the ratio of two lengths.

All materials deform in response to loading. The load-deformation relationship of a material reflects its mechanical characteristics. To compare the mechanical characteristics of different materials, load is expressed in terms of stress and deformation in terms of strain.

Most materials are elastic to a certain extent; that is, they deform in response to loading and then restore their original dimensions when the load is removed. In figure 10.1 the point B represents the **yield point** or **elastic limit**—the point at which the material starts to tear or fracture. The strain range between zero strain and the strain at B (ϵ_B) is the **elastic range**; that is, provided that the strain on the material is within the elastic range, the material will not be damaged and will remain perfectly elastic. However, if the strain on the material exceeds the elastic range the material will be deformed plastically, that is, in the case of an inanimate object, permanently damaged to a degree corresponding to the amount of strain. As such it will lose some of its elasticity and will not return to its original dimensions when the load is removed. If the amount of strain is allowed to increase progressively beyond the yield point the material will eventually fail completely; this is the **failure point** and is represented by the point C in figure 10.1. The strain range between ϵ_B and ϵ_C is the **plastic range**. The stress at C (σ_C) is the **ultimate stress** or **strength** of the material.

Stiffness and Compliance

In most materials all or part of the stress-strain curve within the elastic range is linear—the increase in strain is directly proportional to the increase in stress. Any material that behaves this way is said to obey **Hooke's Law** (after Robert Hooke, 1635–1703, who first described this characteristic of elastic behavior). Consequently, the linear region of the stress-strain curve is referred to as the **Hookean region**. The upper limit of the Hookean region is the **proportional limit**. In some materials the proportional limit and the yield point coincide, but in most materials the proportional limit occurs before the yield point The gradient of the stress-strain curve in the Hookean region reflects the **stiffness** of the material, that is, the resistance of the material to deformation. The greater the stress required to produce a given amount of strain, the stiffer the material.

The gradient of the stress-strain curve in the Hookean region is referred to as **Young's modulus of elasticity** (or **Young's modulus** or **elastic modulus**) for the material. Young's modulus (after Thomas Young, 1773–1829) indicates the amount of stress needed to produce 100% strain. Most materials fail long before 100% strain, but Young's modulus provides a standard measure of stiffness for comparing different materials. Young's modulus is denoted by the Greek epsilon E. With reference to figure 10.1, if $\sigma_A = 6000$ N/cm^2, and $\epsilon_A = 0.015$ (1.5% strain), then

$$E = \frac{\sigma_A}{\epsilon_A}$$

$$E = \frac{6000 \text{ N/cm}^2}{0.015} = 400000 \text{ N/cm}^2 = 400 \text{ kN/cm}^2$$

$$(1 \text{ kN} = 1 \text{ kilonewton} = 1000 \text{ N}).$$

> **Hooke's Law:** elastic behavior in which strain is directly proportional to stress
>
> **Hookean region:** the linear region of a stress-strain curve in the elastic range
>
> **Proportional limit:** the upper limit of the Hookean region
>
> **Stiffness:** the resistance of a material to deformation; stress per unit strain
>
> **Young's modulus:** the gradient of the stress-strain curve in the Hookean region

The larger the Young's modulus, the stiffer the material. Slight changes in the composition of a material may affect its stiffness (and other mechanical characteristics). For example, different kinds of steel have different levels of stiffness. Similarly, the stiffness of musculoskeletal components depends considerably on the age, nutrition, and physical activity level of the individual. Furthermore, the mechanical characteristics of similar musculoskeletal components vary with location in the body. For example, the stiffness of compact bone in the femur is different from that in the tibia of the same individual (Burstein and Wright 1994). Due to the variation in the magnitude of the mechanical properties of different types of the same material there are no standard reference data. Table 10.1 shows some data reported in the literature concerning the stiffness of certain materials in tension.

> *Slight changes in the composition of a material may affect its stiffness (and other mechanical characteristics). The stiffness of musculoskeletal components depends considerably on the age, nutrition, and physical activity level of the individual.*

Figure 10.2 shows generalized stress-strain curves for bone and ligament. It is clear that bone is stiffer and stronger than ligament. However, bone tends to fail suddenly, whereas ligament exhibits progressive, albeit rapid, failure following the elastic limit. Furthermore, in contrast to bone, ligament has a characteristic toe region in which a relatively small increase in stress results in a relatively large increase in strain during the first 1% of strain. The

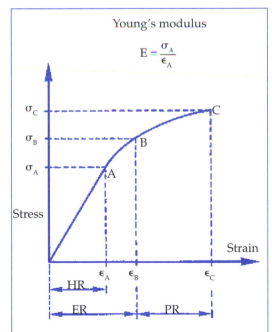

Figure 10.1 Generalized stress-strain curve: ER—elastic range; PR—plastic range; HR—Hookean range; A—proportional limit; B—yield point (elastic limit); and C—failure point (complete rupture).

Table 10.1 Young's Modulus for Some Materials Under Tension (kN/cm²)

Material		Mean	Range
1. Normal compact bone	Femur	1700	
(6th decade of life)	Tibia	2000	
2. Normal compact bone	Femur	1569	1269–1943
3. Osteoporotic compact bone	Femur	1155	397–1834
4. Patellar ligament		40	
5. Elastin		0.06	
6. Tool steel		19000	
7. Stainless steel		17000	
8. Glass		7000	
9. Oak		1000	
10. Lightly vulcanized rubber		0.14	

1 and 4 reported by Burstein and Wright (1994); 2 and 3 reported by Dickenson, Hutton,and Stott (1981); 5, 9, and 10 reported by Alexander (1968); 6, 7, and 8 reported by Frost (1967).

Strain isotropy: when the Young's modulus of a material is the same in tension, compression, and shear

Strain anisotropy: when the Young's modulus of a material is different in tension, compression, and shear

Compliance: the reciprocal of stiffness; strain per unit stress

Toughness: the amount of strain energy a material can absorb prior to failure; the greater the energy absorption, the tougher the material

Toughness modulus: strain energy absorbed to failure per unit volume of material

toe region reflects the straightening out of the collagen fibers, which, under resting conditions, have a wavy arrangement. Tendons, muscle-tendon units, and cartilage respond to loading in a similar manner to ligament.

In real-life situations many materials, like bone, are loaded simultaneously in tension, compression, and shear. Young's modulus can be calculated for each type of loading. When the Young's modulus of a particular material is the same in all three forms of loading, the material is said to exhibit **strain isotropy**. When the Young's modulus is different for the three forms of loading the material is said to exhibit **strain anisotropy**. Most materials with a physical grain—like wood, bone, tendon, ligament, and cartilage—exhibit strain anisotrophy (Frost 1967). For example, bone is stiffer in compression than in tension, and stiffer in tension than in shear.

Compliance is the reciprocal of stiffness; that is, increasing the stiffness of a material decreases its compliance and vice versa. The greater the strain produced by a given amount of stress, the more compliant the material. The yield point of the load-deformation (or stress-strain) curve of a material is usually signified by a marked increase in compliance (decrease in stiffness).

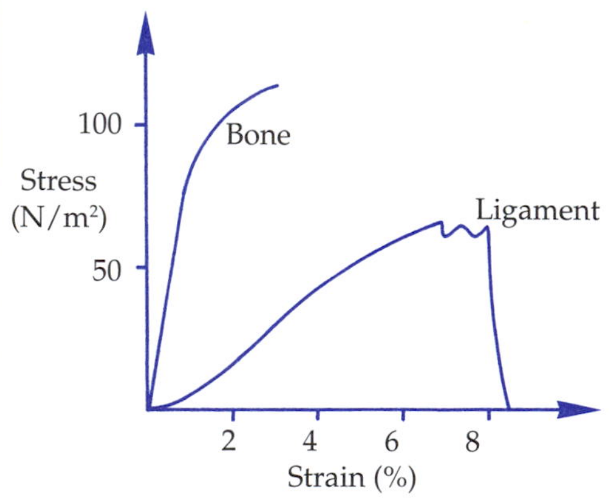

Toughness, Fragility, and Brittleness

For any given piece of material the area under the load-deformation curve represents the strain energy absorbed by the material—the work done on the material. The **toughness** of a material is defined as the amount of strain energy the material absorbs prior to failure; the greater the amount of energy absorbed, the tougher the material. For the purpose of comparing the toughness of different materials the **toughness modulus** is defined as the strain energy absorbed per unit volume of the material prior to failure. For example, if the unit volume is the cubic centimeter (cm³), then

Figure 10.2 Generalized stress-strain curves for bone and ligament in tension.

the toughness of the material is expressed in joules per cubic centimeter (J/cm^3). Similarly, if the unit volume is the cubic meter (m^3), then the toughness of the material is expressed in joules per cubic meter (J/m^3). The toughness modulus of a material is represented by the area under the stress-strain curve of the material.

Figure 10.3 shows typical stress-strain curves for normal compact bone and osteoporotic compact bone. The area under the osteoporotic curve is much smaller than that under the normal curve; that is, normal bone is much tougher than osteoporotic bone. **Fragility** refers to a low level of toughness—a fragile material absorbs relatively little strain energy prior to failure. It follows that osteoporotic bone is more fragile than normal bone. Figure 10.3 shows that the greater toughness of normal bone is due to greater strength (ultimate stress) and greater strain to failure compared to osteoporotic bone. A **brittle** material fails after relatively little strain; the lower the strain to failure, the more brittle the material. It follows that osteoporotic bone is more brittle than normal bone.

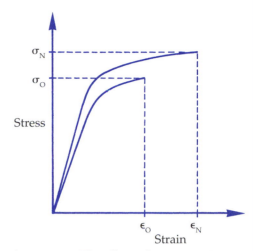

Figure 10.3   The effects of osteoporosis on the stress-strain characteristics of compact bone; N—normal bone; O—osteoporotic bone.

Hysteresis, Resilience, and Damping

If a rubber ball is dropped from a certain height onto the floor, it will strike the floor with kinetic energy equivalent to the gravitational potential energy it possessed at release. During contact with the floor the ball will undergo a loading phase in which it is compressed, and the kinetic energy of the ball is transformed into strain energy in the compressed ball. Following the loading phase the ball undergoes an unloading phase in which it recoils and the strain energy is released as kinetic energy in the form of the upward bounce of the ball. However, the ball does not bounce as high as the point from which it is dropped. This situation is shown in figure 10.4a, where h_1 is drop height and h_2 is the bounce height. Since the ball is at rest at A and B, some of the energy of the ball was dissipated during contact with the floor in the form of, for example, heat and sound. The amount of energy dissipated is reflected in the load deformation curves of the ball during loading and unloading (figure 10.4b).

The amount of strain energy absorbed by the ball during loading—the area under the loading curve—is greater than the amount of energy returned during unload-

Fragility: a low level of toughness; the lower the toughness, the greater the fragility

Brittleness: the amount of strain to failure; the lower the strain to failure, the more brittle the material

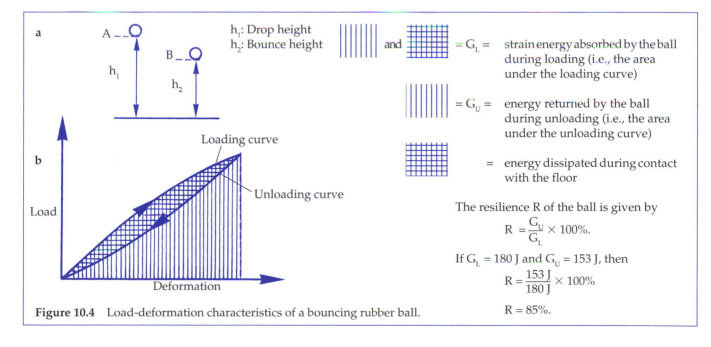

Figure 10.4   Load-deformation characteristics of a bouncing rubber ball.

a

A h_1: Drop height
B h_2: Bounce height

||||||| and ▦ = G_L =   strain energy absorbed by the ball during loading (i.e., the area under the loading curve)

||||||| = G_U =   energy returned by the ball during unloading (i.e., the area under the unloading curve)

▦ =   energy dissipated during contact with the floor

b

Load

Loading curve
Unloading curve
Deformation

The resilience R of the ball is given by

$$R = \frac{G_U}{G_L} \times 100\%.$$

If $G_L = 180$ J and $G_U = 153$ J, then

$$R = \frac{153\,J}{180\,J} \times 100\%$$

$$R = 85\%.$$

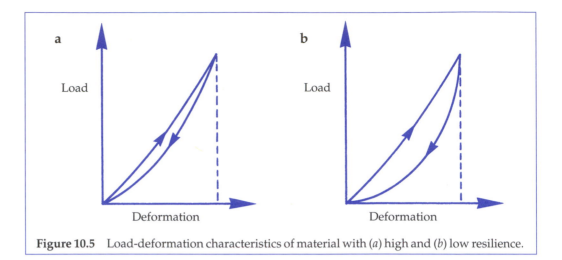

Figure 10.5   Load-deformation characteristics of material with (*a*) high and (*b*) low resilience.

ing—the area under the unloading curve. The loop described by the loading and unloading curves is the **hysteresis loop** (from the Greek word *husteros* meaning later or delayed). The area of the hysteresis loop represents the energy dissipated. All materials exhibit hysteresis to a certain extent. The extent of hysteresis in a material is reflected in the **resilience** of the material, which is defined as the amount of energy returned during unloading as a percentage of the amount of energy absorbed during loading. Resilience can range, theoretically, from 100% to zero, but no 100% resilient materials have been found (Frost 1967).

Figure 10.5*a* shows the load-deformation characteristics of highly resilient material such as ligament and tendon, and figure 10.5*b* shows the load-deformation characteristics of low-resilience material, such as some forms of vinyl acetate foam. **Damping** refers to a low level of resilience; a damping material returns very little energy during unloading compared to the amount of energy that it absorbs during loading. Damping materials are employed in situations where energy needs to be dissipated rather than returned; for example, the matting used in high jump and pole vault landing areas have low resilience—good damping properties. Similarly, for protection during transportation, fragile goods are usually packed in materials with good damping properties.

Viscosity and Viscoelasticity

As described previously, stiffness is a measure of the resistance of a solid material to deformation by a load. The viscosity of a liquid or semiliquid substance is a measure of the resistance of the substance to shear deformation in response to a shear load. Figure 10.6 shows a flat, square piece of wood separated from a larger flat surface by a layer of liquid or semiliquid substance. If the area of the piece of wood is A and a horizontal force F is applied to the piece of wood, then the shear stress on the

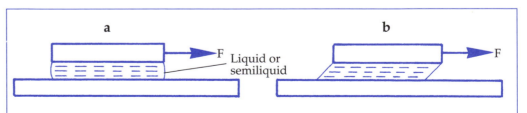

Figure 10.6   Response of a liquid or semiliquid to shear stress; (*a*) at the start of the application of F; (*b*) some time after the start of the application of F.

substance is given by F/A. The movement of the wood is resisted by the viscosity of the substance; the greater the viscosity, the greater the resistance. The viscosity V of the substance is defined as follows:

$$V = \frac{\text{shear stress}}{\text{shear strain rate}}.$$

Consequently, for a given level of shear stress the lower the shear strain rate, the higher the viscosity of the substance and vice versa (Alexander 1968).

In contrast to an elastic material in which the response to loading (and unloading) is immediate, the response of a viscous substance to loading is time dependent; deformation occurs gradually. Most biological materials, including all the musculoskeletal components, have elastic and viscous properties—they behave viscoelastically. As described in chapter 4, a viscoelastic material deforms gradually in response to loading and gradually restores its original dimensions when unloaded. In mechanical models of viscoelastic materials, the elastic elements are usually represented by springs and the viscous elements by Newton's model of a hydraulic piston (a "dashpot") (Taylor et al. 1990; figure 10.7). Figure 10.7c shows a mechanical model of a viscoelastic material based on a piston and three springs; one of the springs is in series with the piston and the other two springs are in parallel with the piston. In response to tension loading, the model gradually lengthens in the direction of T at a rate that depends on the stiffness of the springs and the viscosity of the piston. When the load is removed the model gradually restores its original length at a rate depending on the force exerted by the springs and the viscosity of the piston.

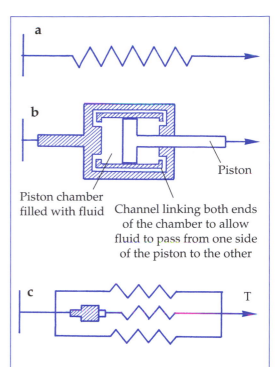

Piston chamber filled with fluid   **Channel linking both ends of the chamber to allow fluid to pass from one side of the piston to the other**   **Piston**

Figure 10.7   A mechanical model of viscoelastic material; (*a*) a spring represents elastic behavior; (*b*) a piston represents viscous behavior; (*c*) a model of a viscoelastic material that incorporates a piston and three springs with one of the springs in series with the piston and the other springs in parallel with the piston.

Properties of Viscoelastic Materials

The response of a viscoelastic material to loading is always time dependent. However, the actual response depends on the type of load. When a viscoelastic material is subjected to a constant load (lower than yield stress), the material deforms asymptotically with time—it gradually deforms at a progressively decreasing rate until a point at which further deformation ceases. This property of viscoelastic materials is called **creep**. If the load is then removed, the material gradually restores its original dimensions. For example, if a viscoelastic material is subjected to a constant tension load (lower than yield stress) the material gradually lengthens until a point is reached where lengthening ceases (figure 10.8*a*). If the load is then removed the material gradually restores its original dimensions.

If a viscoelastic material is deformed (within its elastic range) and then held in the deformed position, the stress experienced by the material decreases asymptotically with time until a point at which no further decrease in stress occurs. This property of viscoelastic materials is called **stress relaxation**. If the load is then removed the material gradually restores its original dimensions. For example, if a viscoelastic material is stretched within its elastic range and then held at a constant length, the tension stress experienced by the material decreases with time until a point where no further decrease in tension stress occurs (figure 10.8*b*). If the load is then removed the material gradually restores its original dimensions.

Continued

Creep: the property of viscoelastic materials to deform asymptotically in response to a constant load

Stress relaxation: the property of viscoelastic materials to reduce asymptotically the level of stress experienced in response to a constant level of deformation

Strain rate dependency: the change in the mechanical characteristics of a viscoelastic material in response to changes in strain rate

In addition to creep and stress relaxation, viscoelastic materials also exhibit **strain rate dependency**—the mechanical characteristics of a viscoelastic material depend on the rate of strain. The higher the rate of strain, the smaller the degree to which creep and stress relaxation can occur during deformation, which, in turn, increases the stiffness, strength (ultimate stress), and toughness of the material (Garrett et al. 1987; Taylor et al. 1990) (figure 10.8c). In general, the higher the strain rate, the greater the stiffness, strength, and toughness of the material. The increased toughness of a viscoelastic material in response to an increased strain rate enables it to absorb more energy. With regard to the musculoskeletal system, the increased toughness of the components, especially muscle-tendon units and bone, in response to increased strain rates may enable the body to dissipate large amounts of energy as heat. However, if the large amounts of stored energy associated with high strain rates are dissipated in the form of failure, then the damage to the musculoskeletal system may be considerable. For example, when a bone fails in response to a low rate of strain, the amount of energy dissipated is relatively low and the bone may fracture in the form of a clean break with relatively little damage to the surrounding soft tissues. However, if the bone fails in response to a high rate of strain, a large amount of energy is dissipated and the bone may shatter into many small pieces resulting in considerable damage to the bone and surrounding soft tissues. Bones may be subjected to very high rates of strain, for example, in high-speed collisions.

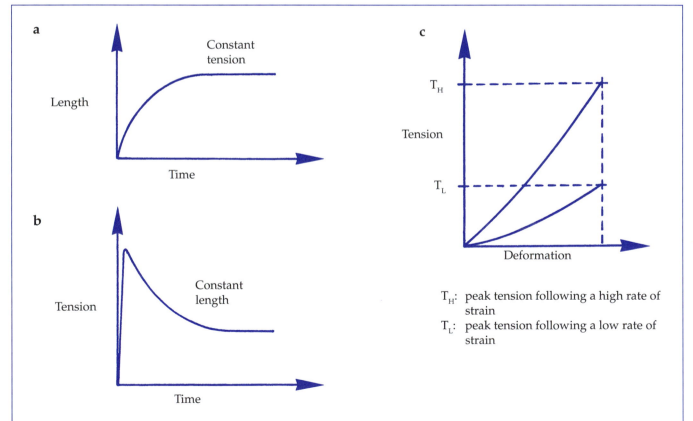

Figure 10.8 Properties of viscoelastic materials. (*a*) Creep: length-time curve for a viscoelastic material subjected to a constant tension load. (*b*) Stress relaxation: tension-time curve for a viscoelastic material stretched and then held at a constant length. (*c*) Strain rate dependency: effect of strain rate on stiffness, peak tension, and energy absorption in a viscoelastic material (at the same degree of deformation).

Biological materials, including all the musculoskeletal components, behave viscoelastically. Viscoelastic materials exhibit creep, stress relaxation, and strain rate dependency.

Active and Passive Loading

Figure 10.9*a* shows a woman standing upright. Her weight W is transmitted to the floor via her feet. She experiences a force R at her feet—the ground reaction force—equal and opposite to W. In this situation the ground reaction force is a vertical force. To move forward, the woman pushes downward and backward against the floor. Provided her foot does not slip, she experiences a ground reaction force equal and opposite to the push she exerts against the floor (figure 10.9*b*). In this case the ground reaction force has a vertical component R_V counteracting W, and a horizontal component R_H enabling her to move forward. R_H does not move the body forward directly; it prevents the foot from slipping backward so that the leg can extend against a fixed point (Watkins 1983).

The greater the magnitude of R_H, the greater the horizontal velocity of the body determined by the impulse of R_H. For example, in a sprint start the sprinter tries to produce a large R_H to drive her body rapidly forward (figure 10.9*c*). Similarly, to jump upward it is necessary to push down hard on the floor; in response to the push against the floor the jumper experiences an equal and opposite ground reaction force, which enables her to thrust her body upward. Her upward velocity at take-off is determined by the impulse of the vertical component of the ground reaction force.

In the above examples the magnitude and direction of the ground reaction force is determined by muscular activity under the conscious control of the individual. In these circumstances, when the ground reaction force or any other external load (apart from body weight) is completely controlled by conscious muscular activity, the load is called an **active load**. By definition, active loads are unlikely to be harmful under normal circumstances. In everyday situations the muscles respond to changes in external loading to ensure that the body is not subjected to harmful loads.

However, it takes a finite time for muscles to fully respond (in terms of appropriate changes in the magnitude and direction of muscle forces) to changes in external

Active load: any external load (other than body weight) completely controlled by conscious muscular activity

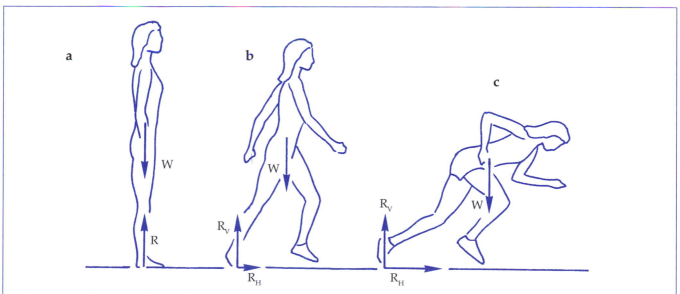

Figure 10.9   Horizontal (R_H) and vertical (R_V) components of the ground reaction force (R) in different activities; (*a*) standing upright; (*b*) pushing off during walking; (*c*) driving away from the blocks in a sprint start.

Muscle latency: the time muscles take to fully respond to changes in external loading

Passive load: any change in external load that occurs within the latency period of the muscles

loading; this time lag is referred to as **muscle latency**. Muscle latency varies between approximately 30 ms and 75 ms in adults (Nigg et al. 1984; Watt and Jones 1971). Consequently, muscles cannot fully respond to changes in external loading that occur in less than the latency period of the muscles. In these circumstances the body is forced to respond passively (by passive deformation) to the external load and, thus, this type of load is a **passive load**. By definition, the body is unable to control passive loads and is vulnerable to injury from high passive loads.

> *In everyday situations the muscles respond to changes in external loading to ensure that the body is not subjected to harmful loads. However, muscles cannot fully respond to changes in external loading that occur in less than the latency period of the muscles. In these circumstances the body responds passively and is more vulnerable to injury.*

The body is subjected to passive loading during, for example, the initial phase of contact of the foot with the ground during walking and running. Approximately 80% of runners and joggers are heel strikers; they contact the ground initially with the heel (Kerr et al. 1983). The remaining 20% of runners and joggers contact the ground initially with either the middle of the foot (midfoot strikers) or the front part of the foot (forefoot strikers).

Figure 10.10 shows the variation with time of the vertical component of the ground reaction force of a heel striker running at approximately 4.5 m/s (six-minute-mile pace). The period of ground contact (contact time) can be divided into component

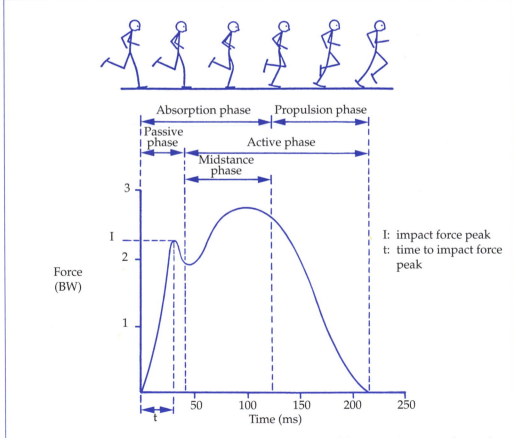

Figure 10.10   Force-time curve of the vertical component of the ground reaction force of a heel striker running at 4.5 m/s in running shoes. BW = body weight.

phases. In terms of changes in the gravitational potential and kinetic energy of the body, contact time can be divided into an absorption phase and a propulsion phase. During the absorption phase—55% to 60% of the contact time (Luethi and Stacoff 1987)—some of the body's gravitational potential and kinetic energy is transformed into strain energy in the form of deformation of the floor, shoe, sock, and the body (especially the musculoskeletal system). Lowering the body's center of gravity (due to leg flexion) reflects the decrease in gravitational potential energy. The decrease in kinetic energy is reflected in a decrease in the velocity of the center of gravity (due to the negative impulse—the braking effect of the ground reaction force; figure 10.11). For an average running velocity of 4.5 m/s the decrease in gravitational potential and kinetic energy during the absorption phase is about 12% of the average level of gravitational potential and kinetic energy (based on Alexander 1987).

During the propulsion phase, concentric contraction of the leg extensor muscles, assisted by the release of strain energy in the floor, shoe, sock, and musculoskeletal system, increases the body's gravitational potential and kinetic energy. Raising the center of gravity (due to leg extension) increases gravitational potential energy. The increase in the velocity of the center of gravity (due to the positive impulse of the ground reaction force) increases kinetic energy (figure 10.11). The ground reaction force is an active load in the propulsion phase. However, in the absorption phase the ground reaction force is initially a passive load and then an active load (see figure 10.10). The passive phase of the absorption phase—the period of muscle latency following heel strike—roughly corresponds to the period between heel strike and foot flat (the loading of the forefoot). The active phase of the absorption phase roughly corresponds to the period between foot flat and heel off—also called midstance phase (Luethi and Stacoff 1987). The propulsion phase corresponds to the period between heel off and toe off. Since the ground reaction force during midstance and propulsion is an active load, this phase of ground contact is called the active phase.

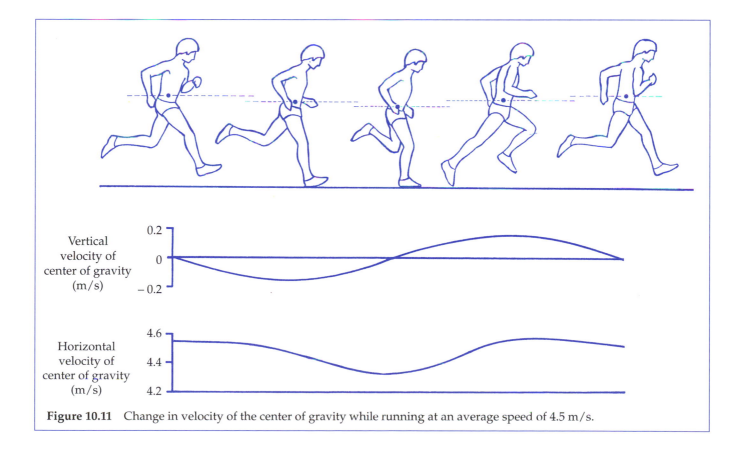

Figure 10.11   Change in velocity of the center of gravity while running at an average speed of 4.5 m/s.

Impact and Shock

In running, a high rate of loading immediately following heel strike, which culminates in a peak force, characterizes the ground reaction force-time curve (vertical component) during the passive phase of ground contact. The force then declines slightly before rising again at the start of the active phase (see figure 10.10). The slope of the force-time curve reflects the rate of loading; the steeper the slope, the higher the rate of loading. The higher the rate of loading, the higher the rate of strain on the system. In this context the system refers to the collection of materials subjected to the passive load—the support surface, shoe, sock, and human body.

In mechanics, **impact** refers to a collision, typically of short duration, between two objects (Goldsmith 1960). Since the heel of a runner collides with the support surface at heel strike, the peak force during the passive phase is called the **impact force peak** (Nigg, Cole, and Bruggemann 1995). The impact force peak reflects the amount of strain on the system; the higher the impact force peak, the greater the amount of strain. At the time of impact force peak the rate and amount of strain experienced by the system is highest in the part of the system at and around the impact site and least in the parts of the system farthest from the impact site. Following impact force peak the parts of the system closest to the impact site recoil and thus press on adjacent parts of the system subjecting them to a rate and amount of strain similar to that experienced around the impact site at impact. The adjacent parts of the system subsequently recoil such that the impact is transmitted to other parts of the system in the form of a **shock wave** (or **stress wave**). The amplitude (amount of strain experienced by the system) and frequency (rate of strain experienced by the system) of the shock wave decline with distance from the impact site due to damping (see later section on shock absorption).

When the initial shock wave passes through a particular part of the system, it is reflected toward as well as away from the impact site. Consequently, shortly after impact the system experiences a large number of shock waves that cause the system to vibrate. The vibration declines and eventually ceases as the energy contained in the shock waves dissipates as heat. In walking and running heel strike shock waves radiate upward throughout the body with the amplitude and frequency decreasing with increased distance from the impact site.

Shock is a transitory state in which the equilibrium of the system is disrupted, resulting in a nonuniform distribution of stress (and strain) in the system (Nigg, Cole, and Bruggemann 1995). Shock produces shock waves—spatial propagation of mechanical discontinuity in a system (Nigg, Cole, and Bruggemann 1995). Impact loading causes shock and the degree of shock produced by an impact depends on the rate of loading and the peak force during the impact; the greater the rate of loading and the greater the peak force, the greater the degree of shock produced and, consequently, the greater the amplitude and frequency of vibration.

When running at about 4.5 m/s, ground contact time is approximately 200 ms and the contributions of the passive, midstance, and propulsion phases to ground contact time are approximately 20%, 40%, and 40%, respectively (see figure 10.10) (Cavanagh and Lafortune 1980; Dickinson, Cook, and Leinhardt 1985; Luethi and Stacoff 1987). The body's center of gravity—reflecting the movement of the body as a whole—decelerates throughout the absorption phase; that is, transformation of gravitational potential and kinetic energy occurs throughout the absorption phase. However, the deceleration of individual body segments varies with distance from the point of impact; the closer the segment to the point of impact, the greater its deceleration, and the shorter the period over which its gravitational potential and kinetic energy are transformed. Consequently, the foot and lower leg experience high levels of deceleration and most of the gravitational potential and kinetic energy of these segments transformed during the absorption phase occurs between heel strike and impact force

Impact: a collision, typically of short duration, between two objects

Impact force peak: the peak force during the passive phase of impact of the body with another body or object

Shock wave: spatial propagation of mechanical discontinuity in a system as a result of shock

Shock: a transitory state in which the equilibrium of the system is disrupted resulting in a nonuniform distribution of stress (and strain) in the system

peak—a relatively small portion of the absorption phase. The higher the deceleration of the segment, the greater the rate and amount of loading experienced by the segment. In running, the impact force peak and rate of loading prior to impact force peak reflect the deceleration of the foot and lower leg (Bobbert, Schamhardt, and Nigg 1991).

The Power of Impact

The rate of loading and peak loading in an impact depend on the **power of the impact**—the rate at which gravitational potential and kinetic energy in the system transforms into other forms of energy such as strain energy during the impact. The power of the impact is given by E/t, where E is the amount of energy transformed and t is the time to impact force peak (see figure 10.10). The higher the power of the impact, the greater the rate of loading and impact force peak, and, consequently, the greater the degree of shock produced.

Power of the impact: the rate at which energy is transformed into other forms of energy in an impact

> *The body is subjected to passive loading following heel strike in walking and running; the greater the power of the impact, the greater the degree of shock produced.*

The amount of energy transformed in an impact depends considerably on the velocities of the colliding objects. The greater the relative velocity of the objects at impact—the velocity of the objects relative to each other—the greater the amount of energy transformed. For example, if a train moving at 50 k/h runs into the back of another train moving in the same direction at 40 k/h, the relative velocity of the trains at impact is 10 k/h. However, if the same trains were moving toward each other then their relative velocity at impact would be 90 k/h. Since the kinetic energy of the trains is proportional to the square of the relative velocity, the amount of energy transformed during a collision at a relative velocity of 90 k/h would be approximately 80 times greater than in a collision at a relative velocity of 10 k/h.

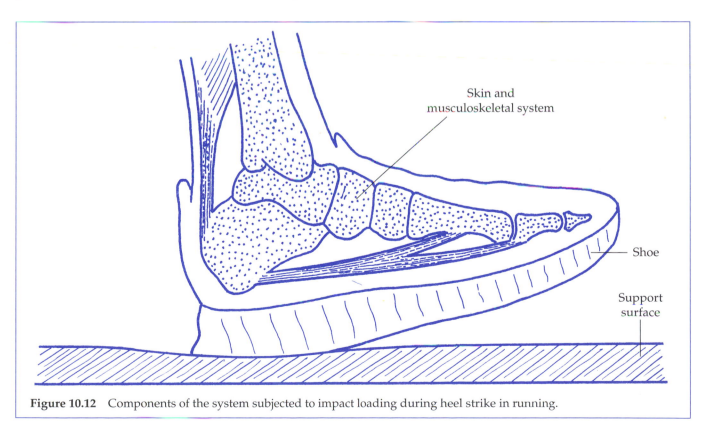

Skin and musculoskeletal system

Shoe

Support surface

Figure 10.12   Components of the system subjected to impact loading during heel strike in running.

The time to impact force peak depends on the stiffness of the system; the more stiff, the shorter the time, and, therefore, the higher the power of the impact. A system's stiffness depends on the stiffness of its materials and the arrangement of the materials relative to the line of action of the impact load. Figure 10.12 shows the composition of the system subjected to impact loading during heel strike in running. The system's components are the support surface, the shoe (including insole and sock), and the foot and ankle (representing the body as a whole). The foot and ankle consist of a combination of fairly stiff material (bone) and fairly compliant material (soft tissues including skin, adipose tissue, muscle-tendon units, ligaments, and cartilage).

> *The stiffness of a system depends on the stiffness of the materials that comprise the system and the arrangement of the materials in relation to the line of action of the impact load.*

At heel strike the calcaneus is protected naturally by a fairly thick layer of adipose tissue called the heel pad. The 2-cm-thick heel pad is underneath the calcaneus (Steinbach and Russell 1964). It has been shown that the heel pad and overlying skin are deformed (compressed) by approximately 1 cm by impact loads similar to those following heel strike when running at 3.6 m/s (Cavanagh, Valiant, and Misevich 1984). The deformation of the skin and heel pad cushions the impact load and, thus,

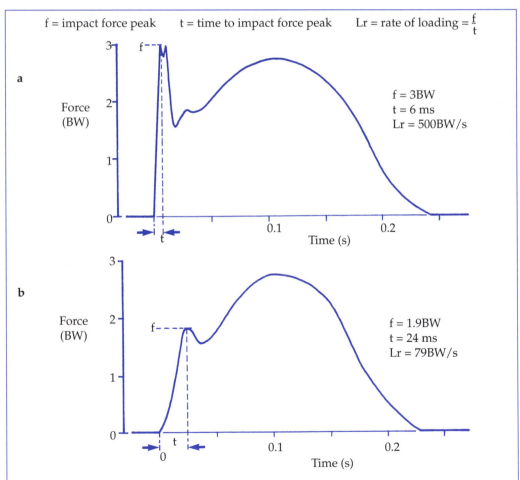

Figure 10.13   Force-time curve of the vertical component of the ground reaction force of a heel striker running at 4 m/s; (*a*) barefoot; (*b*) while wearing running shoes. BW = body weight.
Adapted from *Journal of Biomechanics* 18, J. Dickinson, S. Cook, and T. Leinhardt, "The measurement of shock waves following heel strike while running," 417, © 1985, with permission from Elsevier Science.

increases the time to impact force peak compared to if they were not present. The increase in time to impact force peak decreases the power of the impact and, consequently, decreases the rate of loading and impact force peak, which, in turn, reduces the level of shock produced.

Effect of Shoes and Surfaces on Time to Impact Force Peak

The compliant soles of running shoes and compliant surfaces also tend to increase time to impact force peak in the same way as the heel pad. Figure 10.13 shows the ground reaction force-time curve (vertical component) for the same subject running at approximately 4 m/s barefoot (10.13a) and in running shoes (10.13b) (Dickinson, Cook, and Leinhardt 1985). The ground contact time is similar in both conditions, as is the force-time curve in the active phase, but the force-time curve during the passive phase is different. The impact force peak in the barefoot condition is approximately 3BW compared to approximately 1.9BW with shoes. However, the most noticeable differences between the two conditions are the time to impact force peak and the rate of loading prior to impact force peak. The time to impact force peak when barefoot is approximately 6 ms compared to 24 ms when wearing shoes. The large difference in time to impact force peak is mainly responsible for the large difference in rate of loading: 500BW/s barefoot and 79BW/s with shoes. The rate of loading in the barefoot condition may result in a high level of shock, whereas the rate of loading with shoes is likely to result in relatively little shock.

> *Deformation of the skin and heel pad at heel strike cushions the impact load and decreases the power of the impact. The compliant soles of running shoes and compliant surfaces have a similar effect.*

Figure 10.14 shows the ground reaction force-time curves for the same individual running at 5 m/s in two types of running shoes, one with a hard (low compliance) sole and one with a soft (moderate compliance) sole (Nigg, Denoth, and Neukomm 1981). The hard-soled shoe results in a higher impact force peak, a higher rate of loading,

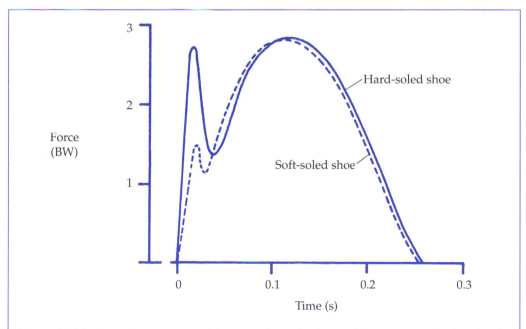

Figure 10.14   Vertical component of the ground reaction force while running at 5 m/s in hard-soled and soft-soled shoes. BW = body weight.
Adapted, with permission, from Nigg, Denoth, and Neukomm 1981.

and, consequently, a higher level of shock than the softer-soled shoe. Whereas soft-soled shoes produce less shock than harder soled shoes, there is a limit to how soft a shoe can be if it is to be effective in increasing time to impact force peak. If the shoe material is too soft it can bottom out quickly, such that the time to impact force peak is largely determined by other parts of the system.

Segmental Alignment and Time to Impact Force Peak

At foot strike the alignment of the upper and lower leg and foot in relation to the line of action of the ground reaction force affects the leg's stiffness. The closer the line of action of the ground reaction force to the hip, knee, and ankle joints, the more likely the leg will respond to the impact load like a rod subjected to a compression load at one end in line with the rod. Consequently, the system is stiff; most of the gravitational potential and kinetic energy transformed during the impact occurs in the form of strain energy in bones and joints such that the system experiences a high level of shock. This situation is typical of running heel strike, where the line of action of the ground reaction force passes close to the joint centers of the ankle and knee (figure 10.15a).

The greater the moment arm of the ground reaction force about the hip, knee, and ankle joints at foot strike, the more likely the leg will respond to the impact load like a spring, resulting in flexion of the hip, knee, and ankle. This type of action is typical of a forefoot-striking runner, where the line of action of the ground reaction force dorsiflexes the ankle and flexes the knee (figure 10.15b). The stiffness of the spring depends considerably on the degree of activation of the muscles controlling the joints. Whereas muscle latency limits the speed with which a muscle can fully respond to a change in external loading, the muscles can still influence the time to impact force peak by setting the rotational stiffness of the joints prior to foot strike. The greater the degree of tension in the muscles, the greater the degree of rotational stiffness of the joints. A certain amount of rotational stiffness of the joints is necessary to prevent the spring from bottoming out too quickly. In general, the more the leg behaves like a spring at foot strike, the longer the time to impact force peak. Furthermore, forced lengthening of the muscle-tendon units results in them absorbing some of the gravitational potential and kinetic energy transformed during the impact as strain energy. The more energy the muscle-tendon units absorb, the lower the stress (and strain) on other components of the musculoskeletal system, especially bones and joints. Striking the ground with the midfoot or forefoot is generally effective in preventing a high rate of loading following foot strike, such that most midfoot and forefoot strikers do not exhibit a discernible impact force peak (Cavanagh, Valiant, and Misevich 1984) (figure 10.15c).

> In heel-strike running the leg is fairly stiff at impact, resulting in a high level of shock. In contrast, the stiffness of the leg in forefoot-strike running is relatively low resulting in little if any shock.

The shape of the heel of a running shoe can affect the point of application of the ground reaction force and, as such, the line of action of the ground reaction force in relation to the hip, knee, and ankle joints. The center of pressure is the application point of the ground reaction force. Figure 10.16 shows the effect of a flared heel on the line of action of the ground reaction force at heel strike. The flared heel increases the moment arm of the ground reaction force about the ankle joint and thereby promotes springlike activity at the ankle. Flared heels have been shown to decrease the magnitude of impact force peak in heel strikers (Nigg and Bahlsen 1988; Nigg and Morlock 1987). However, since flared soles increase the moment arm of the ground reaction force about the ankle joint, they also increase the load on the supinators (the muscles controlling the amount and rate of pronation [figure 10.16, a and b]) and the dorsiflexors (the muscles that control the amount and rate of plantar flexion [figure 10.16, c and d]).

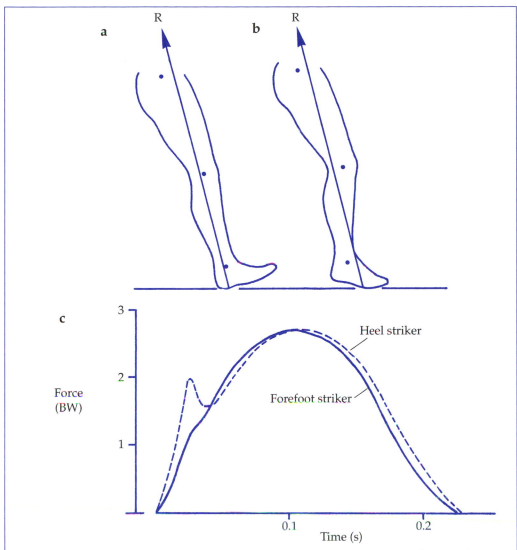

Figure 10.15   Alignment of the upper leg, lower leg, and foot in relation to the line of action of the ground reaction force (R) at (*a*) heel strike and (*b*) forefoot strike; (*c*) ground reaction force-time curves (vertical component) for a heel striker and a forefoot striker. BW = body weight.

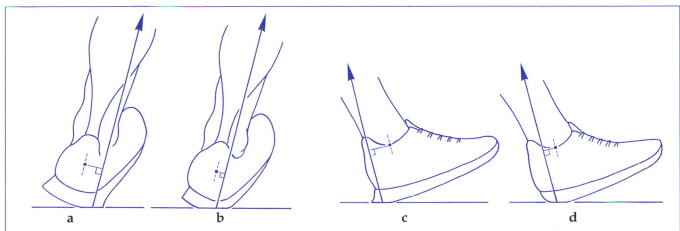

Figure 10.16   Effect of flared heels on the line of action of the ground reaction force; (*a* and *c*) flared heels; (*b* and *d*) nonflared heels.

Shock Absorption

Shock absorption refers to the damping of vibrations generated in a system. With regard to the human body, all of the tissues are viscoelastic to a certain extent and dissipate a certain amount of energy as heat by hysteresis during each cycle of vibration. Consequently, the amplitude of vibration progressively decreases and vibration usually ceases fairly quickly. The speed with which vibrations are absorbed and, therefore, the extent to which vibrations are transmitted to other regions of the body, depends on the degree of damping afforded by relative motion between bone and soft tissues and joints. Bones are much stiffer than the soft tissues that surround them. Due to the difference in stiffness, the frequency of vibration of bone is greater than that of the surrounding soft tissues following impact loading. Consequently, the bone and soft tissues move relative to each other, and some of the energy contained in the vibration of the bone is absorbed as strain energy by the soft tissues and subsequently dissipated as heat. Since hysteresis in the soft tissues is greater than in bone, the vibrational energy of bone dissipates fairly quickly.

> *All of the tissues of the body are viscoelastic to a certain extent and, as such, dissipate a certain amount of energy as heat by hysteresis during each cycle of vibration.*

Just as bones are much stiffer than the soft tissues surrounding them, bones are stiffer than the joints linking the bones together. Synovial and cartilaginous joints deform in response to loading and in so doing absorb energy, including vibrational energy of the bones. In symphysis joints, such as the intervertebral joints and pubic symphysis, the amount of fibrocartilage is large compared to the amount of articular (hyaline) cartilage in synovial joints. Consequently, symphysis joints absorb shock better than synovial joints. Figure 10.17 shows a section through a typical synovial joint. In response to loading, both epiphyses deform to a certain extent. Cancellous bone is stiffer than subchondral bone, which is stiffer than articular cartilage. Consequently, articular cartilage and subchondral bone absorb shock better than an equivalent amount of cancellous bone. However, in a typical synovial joint the volume of cancellous bone is greater than the volume of articular cartilage and subchondral bone, such that in healthy synovial joints the cancellous bone may be the major contributor to shock absorption (Radin and Paul 1971; Hoshino and Wallace 1987).

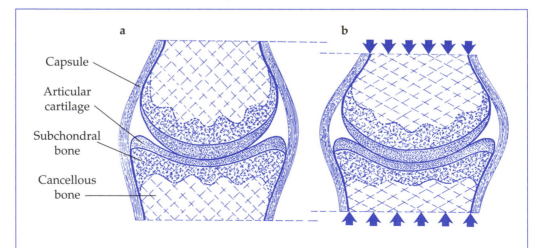

Figure 10.17   Response of a synovial joint to compression loading; (*a*) unloaded; (*b*) response to compression loading.

Effect of Impact Loading on Synovial Joints

With regard to synovial joints, the greater the power of the impact, the greater the likelihood of microfractures to trabeculae in cancellous bone. Healing of fractured trabeculae increases the stiffness of the cancellous bone and, consequently, decreases its shock-absorbing capacity (Radin and Paul 1971; Simon et al. 1972; Radin et al. 1973). Repeated microfracture and healing of cancellous bone progressively increases stiffness and progressively decreases shock-absorbing capacity. As the stiffness of cancellous bone increases, the subchondral bone and articular cartilage are subjected to increased strain (rate and amount), thereby increasing the likelihood of microtrauma—crushing or tearing of the matrix in subchondral bone and articular cartilage. Such microtrauma decreases the shock-absorbing capacity of the subchondral bone and articular cartilage. Not surprisingly, it has been shown that the shock-absorbing capacity of healthy joints is greater than that of degenerated joints (Voloshin and Wosk 1982). Progressive damage to articular cartilage can culminate in osteoarthritis (figure 10.18).

> *Repeated microfracture and healing of cancellous bone results in a progressive increase in stiffness, which, in turn, increases the strain on subchondral bone and articular cartilage.*

Heel strike during walking is the most common source of impact loading on the human body, and repeated exposure to high levels of shock resulting from walking heel strike appears to be a major cause of joint degeneration (Radin et al. 1973; Wosk and Voloshin 1985). Viscoelastic inserts in shoes have been shown to reduce the amplitude of heel-strike-induced shock waves and may compensate some for the decreased shock-absorbing capacity of degenerated joints (Voloshin and Wosk 1981; Johnson 1988). Voloshin and Wosk (1981) showed that prolonged use (18 months) of viscoelastic heel inserts can significantly reduce clinical symptoms, especially pain, in adults suffering from chronic degenerative joint conditions such as osteoarthritis and intervertebral disc damage.

In another study, the same researchers investigated the effect of prolonged use of viscoelastic heel inserts on a group of 254 males and 128 females, ages 14 years to 75 years, suffering from chronic low-back pain (Wosk and Voloshin 1985). The subjects rated the effect of the treatment at one month, three months, and one year, on a four-point scale. As shown in figure 10.19, the results reflect a progressive increase in the number of subjects who reported an excellent result (36%, 50%, 62%) and a progressive decrease in the number of subjects who reported a poor result (16%, 12%, 8%) over the one-year period of the study. The results suggest that the reduced level of shock from using the viscoelastic heel inserts helped the body to repair the damaged intervertebral joints. The extent of repair may depend on, among other factors, the age of the subject at the start of treatment. This is reflected in the number of young (14 to 25 years), middle-aged (26 to 45 years) and older-aged (46 to 75 years) subjects reporting excellent and good results (figure 10.20). The trend of decreased pain was similar in the three age groups, but the amount of improvement decreased with age.

> *Heel strike during walking is the most common source of impact loading on the human body; repeated exposure to high levels of shock resulting from heel strike may be a major cause of joint degeneration. Viscoelastic shoe inserts can compensate to a certain extent for the decreased shock-absorbing capacity of degenerated joints.*

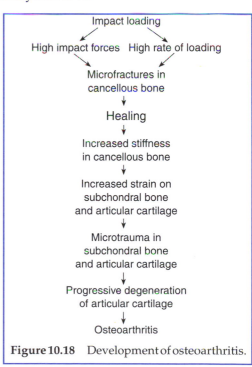

Figure 10.18   Development of osteoarthritis.

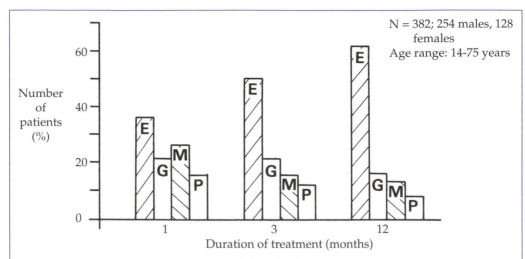

Figure 10.19   Effect of continuous use of viscoelastic shoe inserts on the severity of chronic lower back pain: E—excellent result, no pain and no restriction on habitual physical activity; G—good result, only slight pain periodically and no restriction on habitual physical activity; M—moderate result, significant decrease in pain and little or no restriction on habitual physical activity; P—poor result, either no decrease in pain or decrease not maintained. Adapted, with permission, from Wosk and Voloshin 1985.

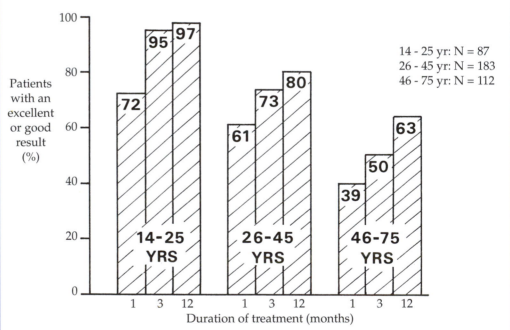

Figure 10.20   The effects of age at the start of treatment (use of viscoelastic shoe inserts) and the duration of treatment on reduction in chronic lower back pain.
Adapted, with permission, from Wosk and Voloshin 1985.

Energy Absorption During Active Loading

The amount of strain energy absorbed by the different components of a system during deformation depends on the toughness and volume of the components. The tougher the component and the greater its volume, the more strain energy it is likely to absorb. Of the various musculoskeletal tissues and related tissues such as the heel pad, muscle has the least capacity for absorbing strain energy, tendon has the highest capacity (approximately 65 times that of muscle), and the capacity of bone (approximately 55 times that of muscle) is between that of muscle and tendon (Evans 1971). Whereas the energy-absorbing capacity of muscle is very low compared to tendon, the muscle-tendon unit has the highest energy-absorbing capacity of all musculoskeletal and associated tissues (Evans 1971). During the midstance phase in running, the system of support surface, shoe (including sock and insole), and musculoskeletal system (including associated tissues such as heel pad and skin) deforms in response to the impulse of the (active) ground reaction force. The leg is forcibly flexed and the arch of the foot is forcibly flattened such that strain energy is stored in the bones (bending of long bones and compression of short bones), joints (compression of epiphyses), muscle-tendon units (eccentric contraction of leg extensor muscles and muscles supporting the arch of the foot), and ligaments supporting the arch of the foot. Figure 10.21 shows a model of the movement of the lower leg and foot from foot flat (figure 10.21, *a* and *b*) to peak force (figure 10.21, *c* and *d*) during the active phase, which illustrates the strain on the calf muscles and arch support mechanisms.

When running at a middle-distance pace (7 to 8 m/s), the gravitational potential and kinetic energy of a runner is estimated to decrease by about 100 J during the absorption phase and then increase by about 100 J during the propulsion phase (Ker et al. 1987). The same researchers also estimate that during the absorption phase approximately 35 J, 17 J, and 5 J is stored in the calf muscle-tendon units, arch support mechanisms, and shoe, respectively. The heel pad is estimated to store approximately 3 J (Cavanagh, Valiant, and Misevich 1984) and a specially designed indoor running track can store in the region of 7 J (McMahon and Greene 1978). These findings are summarized in the first column of table 10.2, while the resilience of the materials (reported in the same sources) is shown in the second column of the table. The third column shows the theoretical potential for the release of strain energy in the form of useful mechanical work during the propulsion phase. Whereas the estimate for the release of useful strain energy is likely higher than that which occurs in the real situation, the figures do indicate the considerable energy-saving potential of the system.

Clearly, the energy-absorbing capacity and resilience of muscle-tendon units is a major determinant of an individual's performance quality in games and sports. However, muscle-tendon units can only function efficiently when the muscles are not fatigued. A fatigued muscle is limited in its ability to apply tension to its tendons, which, in turn, limits the ability of the tendons to absorb strain energy. Consequently, muscle fatigue reduces the energy-absorbing capacity of a muscle-tendon unit; the greater the degree of muscle fatigue, the greater the reduction in energy-absorbing capacity. In any movement involving the absorption of gravitational potential and

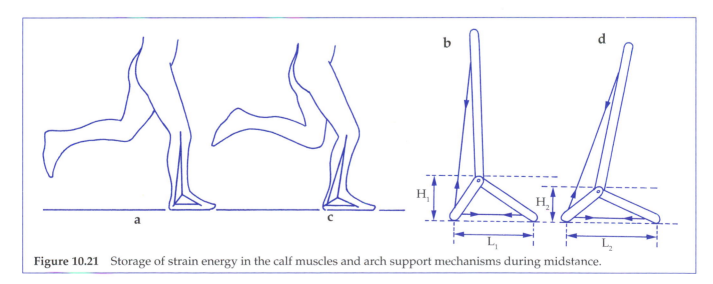

Figure 10.21 Storage of strain energy in the calf muscles and arch support mechanisms during midstance.

Table 10.2   Energy Storage and Resilience in System Components During the Ground Contact Phase in Middle-Distance Running

Material	Energy absorbed (J)	Resilience (%)	Energy returned (J)
1. Calf muscles	35	93*	32.5
2. Arch support mechanism	17	78	13.2
3. Heel pad	3	20	0.6
4. Shoe	5	60	3.0
5. Track	7	90	6.3
6. Other[†]	33	64**	21.1
Total	100		76.7

* Resilience of Achilles tendon
** average of 1, 2, and 3
[†] Energy stored in other parts of the body

kinetic energy, such as in the absorption phase of running, or landing from a jump, the extent to which a particular muscle-tendon unit has to lengthen to make its contribution to energy absorption is determined by the force the muscle can produce. The greater the force of the muscle, the shorter the distance it must lengthen to absorb a given amount of energy. Fatigue reduces the amount of contractile force a muscle can produce; a fatigued muscle has to lengthen farther than a nonfatigued muscle to absorb a given amount of energy.

Since it has been shown that a muscle yields at the same degree of strain irrespective of its level of fatigue, it follows that a fatigued muscle is more likely to be stretched beyond its yield point than a nonfatigued muscle (Mair et al. 1996). Furthermore, when the level of effort remains fairly constant, as when running at a constant speed, the gradual reduction in energy-absorbing capacity of muscle-tendon units due to fatigue inevitably results in increased strain on associated structures—especially bones and joints—since the amount of energy absorbed in each absorption phase of the activity is likely to remain fairly constant. Consequently, the greater the degree of muscular fatigue, the greater the risk of injury to all components of the musculoskeletal system (Grimston and Zernicke 1993).

The muscle-tendon unit has the highest energy-absorbing capacity of all the musculoskeletal tissues, but the capacity to absorb energy decreases with increasing fatigue. The gradual reduction in energy-absorbing capacity of muscle-tendon units due to fatigue results in increased strain on associated structures, especially bones and joints.

Summary

This chapter described the mechanical characteristics of musculoskeletal components and their response to loading. The musculoskeletal system responds to changes in loading in two ways: actively, in the form of muscular control of joint movements, and passively, by passive deformation. The combined effect of these responses is to prevent or reduce potentially harmful loads. The effectiveness of the active response is limited by muscle latency such that high rates of loading result in a completely passive response in the initial phase of loading. Furthermore, high passive loads tend to produce shock and repeated exposure to shock may cause joint degeneration. The musculoskeletal system normally adapts its structure to more readily withstand the time-averaged loads exerted on it; this is the subject of the next chapter.

Review Questions

1. Describe the stress-strain characteristics of bone and ligament.
2. Differentiate toughness, fragility, and brittleness.
3. Differentiate hysteresis, resilience, and damping.
4. Describe the effects of strain rate dependence on the mechanical characteristics of viscoelastic materials.
5. Differentiate between passive and active loading.
6. Differentiate between impact and shock.
7. Describe the effects of heel strike and forefoot strike on the level of shock produced and mechanism of energy absorption.
8. Describe the effects of muscle fatigue on the stress experienced by other musculoskeletal components.

Chapter 11

Structural Adaptation of the Musculoskeletal System

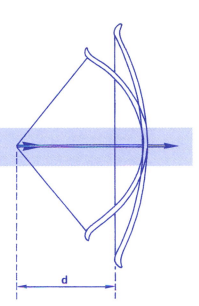

Musculoskeletal components experience strain in response to loading. A certain level of strain is necessary to ensure normal growth and development, but excessive strain results in abnormal growth and development and injury. Under normal circumstances the musculoskeletal components adapt their size, shape, and structure to the time-averaged strain. This chapter describes the effects of changes in time-averaged loading on the musculoskeletal system.

Adaptation

Genotype: the pattern of growth and development determined by the genes

Adaptation: the changes in structure and function that occur as a result of a change in environmental conditions

Phenotype: the appearance and form (size, shape, and structure) of the body that results from the interaction of the genotype and environmental influences

Two types of factors influence growth and development of the human body: genetic and environmental. The genes determine the basic pattern of growth and development in a set of "genetic instructions" called the **genotype**. However, the body is constantly subjected to a variety of environmental influences including nutritional state, changes in body temperature, and the physical stress imposed by movement of the body. These environmental influences affect the genotype by modifying the timing, rate, extent, and type of growth and development that occurs, such that the body becomes adapted to function effectively in relation to the environmental conditions (Malina and Bouchard 1991). The process of **adaptation** to environmental conditions is continuous throughout life.

The size, shape, and structure of the body that results from the effect of environmental conditions on the genotype is called the **phenotype**. The phenotype reflects both the general effects of the genotype and the specific effects of the environmental influences. We can actually see and measure the phenotype. Identical twins have identical genotypes and, consequently, tend to look similar to each other throughout life. However, they obviously have different phenotypes in that they are not subjected to exactly the same environmental influences. This is usually more apparent during adulthood when differences in lifestyle, especially in terms of nutrition and physical exercise, can result in marked differences in phenotype. Figure 11.1 illustrates the concept of adaptation.

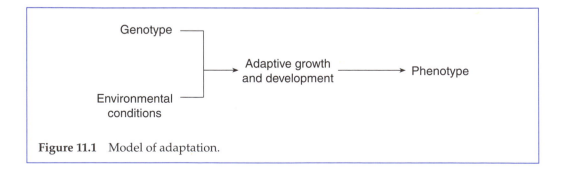

Figure 11.1   Model of adaptation.

Biopositive and Bionegative Effects of Loading

The effects of loading on the musculoskeletal system tend to be either **bionegative** or **biopositive** (Nigg, Denoth, and Neukomm 1981). Bionegative effects result from insufficient loading and excessive loading (figure 11.2). Prior to maturity, insufficient loading can result in abnormal growth and development of the musculoskeletal system, and in adults, insufficient loading can result in decreased functional capacity of musculoskeletal components. Like insufficient loading, excessive loading also may result in abnormal growth and development of the musculoskeletal system prior to maturity. In addition, excessive loading may result in injury to musculoskeletal components at any age. Biopositive effects—including normal growth and development of the musculoskeletal system prior to maturity and increased functional capacity of musculoskeletal components due to physical training—result from moderate loading.

Effects of Insufficient Loading

A certain amount of loading is essential to promote normal growth and development of the musculoskeletal system (Zernicke and Loitz 1992; Lieber 1992). This is clearly demonstrated in children with diseases of the nervous system, such as polio, which prevent normal muscular activity and, therefore, normal patterns of loading. The muscles of such children are usually poorly developed and the bones are usually abnormal in shape and much smaller than in a normal child (Gray 1964). To grow and develop normally the musculoskeletal system needs the mechanical stimulation provided by a normal pattern of motor skill development—the postures, movements, and levels of physical activity associated with an active childhood.

After maturity, the maintenance of the musculoskeletal system still depends on the mechanical stimulation provided by regular exercise. Lack of use results in **atrophy**—decrease in mass—of musculoskeletal components and, consequently, decreased functional capacity. With regard to muscle, loss of strength is rapid in totally inactive muscles. For example, bed rest studies have shown that a strength loss is evident after

Bionegative effects: changes in the structure and function of the musculoskeletal system resulting in abnormal growth and development and decreased functional capacity

Biopositive effects: changes in the structure and function of the musculoskeletal system resulting in normal growth and development and increased functional capacity

Atrophy: decrease in mass of musculoskeletal components, especially in reference to muscles, tendons, and ligaments

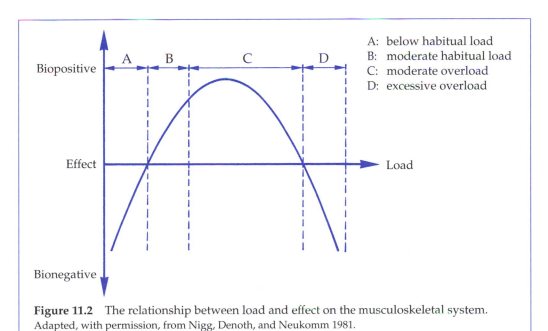

A: below habitual load
B: moderate habitual load
C: moderate overload
D: excessive overload

Figure 11.2   The relationship between load and effect on the musculoskeletal system. Adapted, with permission, from Nigg, Denoth, and Neukomm 1981.

only one day of inactivity and that loss in strength proceeds at the rate of 1% to 1.5% per day (Muller 1970). Immobilization of a limb in a plaster cast results in a loss of strength of about 22% in the first seven days (Muller 1970; Booth 1987). Similarly, lack of bearing one's weight due to bed rest (Donaldson et al. 1970), immobilization in a plaster cast (Anderson and Nilsson 1979), denervation of muscle (Brighton et al. 1985), and space flight (Whedon 1984) have been found to substantially decrease bone mass and, therefore, decrease bone strength and toughness.

> *To grow and develop normally, the musculoskeletal system needs the mechanical stimulation provided by a normal pattern of motor skill development. After maturity, the maintenance of normal musculoskeletal function depends on the mechanical stimulation provided by regular physical activity.*

Effects of Moderate Loading

Moderate habitual range: the range of loading on the musculoskeletal system due to normal daily activity

Moderate overload range: the range of loading on the musculoskeletal system due to physical training

Hypertrophy: increase in mass of musculoskeletal components, especially in reference to muscles, tendons, and ligaments

Injury: structural damage to one or more components of the musculoskeletal system resulting in decreased functional capacity

In general, moderate loading has a biopositive effect on growth, development, and maintenance of the musculoskeletal system. The moderate range can be subdivided into a **moderate habitual range**—loading associated with normal daily activity (see figure 11.2)—and a **moderate overload range**—loading associated with physical training. The moderate habitual range promotes normal growth and development of the musculoskeletal system prior to maturity and maintains a healthy level of musculoskeletal function consistent with normal daily activity after maturity. Loading within the moderate overload range has a biopositive effect on the musculoskeletal system resulting in **hypertrophy**—increase in mass—of musculoskeletal components and, consequently, increased functional capacity in terms of strength and toughness of muscle-tendon units, ligaments, cartilage, and bone (Zernicke and Loitz 1992; Lieber 1992). These changes occur in response to moderate overloading at all ages, but they are far more noticeable in adults than in children.

Effects of Excessive Loading

The upper part of the moderate overload range is associated with a progressive decrease in biopositive effect, and the excessive overload range is associated with a progressive increase in bionegative effect (see figure 11.2). Excessive overload in the form of continuous or highly repetitive nonimpact loading may result in severe angular deformities in joints or reduced bone lengths prior to maturity and degenerative joint disease in adults (Frost 1979). Excessive overload in the form of impact loading may result in **injury**—structural damage to one or more components of the musculoskeletal system resulting in decreased functional capacity.

The basic relationship between load and effect on the musculoskeletal system, shown in figure 11.2, is similar for everyone. However, the actual amount (magnitude and frequency) of loading corresponding to insufficient loading, moderate habitual loading, moderate overloading, and excessive overloading for a particular individual depends on her stage of maturity and level of physical conditioning. In other words, a particular amount of loading that corresponds to moderate overload for one individual may correspond to insufficient loading or excessive loading for someone else.

> *Bionegative effects result from insufficient or excessive loading. Biopositive effects result from moderate loading.*

Physical Training and Injury

In sport, an individual's performance largely depends on her physical fitness and motor ability. The need for a certain level of physical fitness—cardiorespiratory endurance, local muscular endurance, strength, and flexibility—reflects the physiological demands of the sport. Similarly, the need for a certain level of motor ability—ability to perform specific skills—reflects the technical demands of the sport. A high level of motor ability (skill) requires a high level of neuromuscular coordination in the form of a combination of speed, agility, balance, and power.

Clearly, the physiological and technical demands of different sports vary considerably. For example, performance in long-distance running events requires cardiorespiratory endurance, whereas performance in springboard diving requires more technical ability. A training program—exercises designed to improve an individual's performance in a sport—should reflect the balance between the physiological and technical demands of the sport. However, whether a particular training exercise is designed to improve physical fitness or technique, the effect on the musculoskeletal system is basically the same—the musculoskeletal system experiences overload (figure 11.3). The effect of training depends on the level of overload.

A fairly high level of physical effort is necessary to generate a large enough physiological stimulus to improve cardiorespiratory endurance in an already well trained endurance athlete. The intensity of such effort almost certainly subjects the musculoskeletal system to loads at the upper end of the moderate overload range and probably to loads in the lower end of the excessive overload range. This level of loading inevitably results in minor damage (microtrauma) to the musculoskeletal tissues, especially muscle and connective tissue. Given adequate rest, these tissues not only heal, but also adapt their structures over time to more readily withstand loads imposed on them during training. However, when rest periods are inadequate, the rate at which microtrauma occurs outpaces the processes of repair and structural adaptation such that microtrauma gradually accumulates and eventually results in a **chronic injury**—an injury that develops over a period of time (cumulative microtrauma) and is characterized by a gradual increase in pain and functional impairment (Watkins and Peabody 1996).

The main cause of chronic injuries appears to be **overtraining**, a training and competition schedule involving training sessions that are too long or rest periods between training sessions that are too short (Jones, Cowan, and Knapik 1994). Clearly, proper progression in the amount of exercise and, therefore, the level of loading on the musculoskeletal system, is essential to avoid injury. As a general rule, the amount of exercise should increase gradually so the musculoskeletal and cardiorespiratory systems have sufficient time to adapt to the gradual increase in the intensity of the training exercises. While it is possible to significantly increase

Chronic injury: an injury that develops over a period of time (cumulative microtrauma) characterized by a gradual increase in pain and functional impairment

Overtraining: a training and competition schedule involving training sessions that are too long or rest periods between training sessions that are too short resulting in decreased functional capacity

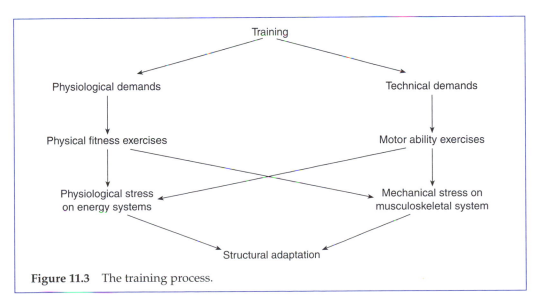

Figure 11.3   The training process.

Table 11.1 Classification of Jumper's Knee According to Symptoms

Stage 1 Pain after practice or after a game

Stage 2 Pain at the beginning of activity that disappears after warm-up and reappears after completion of activity

Stage 3 Pain present before, during, and after activity such that the individual is unable to participate

Adapted, with permission, from Blazina, Karlan, and Jobe (1973) and Roels et al. (1978).

cardiorespiratory fitness quickly, similar changes in the musculoskeletal system take much longer, especially if the individual has been sedentary for a number of years (Micheli 1982).

Jumper's knee is a common chronic injury in sports such as volleyball and basketball, which involve high-frequency jumping and landing. The condition is characterized by pain at one or more of three sites: the insertions of the quadriceps tendon to the upper pole of the patella and the insertion of the patellar ligament to the lower pole of the patella and tibial tuberosity (Ferretti, Papandrea, and Conteduca 1990). As described in table 11.1, the third and final stage of jumper's knee is characterized by fairly persistent pain usually severe enough to prevent participation in games and sports. Continuing to train or play at this stage may result in complete failure of the patellar ligament (Ferretti, Papandrea, and Conteduca 1990). This type of injury is usually referred to as an **acute injury**—an injury that occurs suddenly (sudden macrotrauma) and is severe enough in terms of pain or functional disablement to prevent further participation, at least temporarily (Watkins and Peabody 1996). Acute injuries range in severity from minor—a minor muscle tear—to severe—a complete ligament rupture or complete bone fracture. It has been shown that many acute injures, especially severe muscle, tendon, and ligament injuries, are the sudden end result of progressive degeneration of the structure concerned (Ferretti, Papandrea, and Conteduca 1990). Not surprisingly, Ferretti, Papandrea, and Conteduca (1990) suggest that continuing to train or play when suffering from a minor injury significantly increases the risk of a severe acute injury.

Acute injury: an injury that occurs suddenly (sudden macrotrauma) and is severe enough in terms of pain or functional disablement to prevent further participation, at least temporarily

Response and Adaptation of Musculoskeletal Components to Loading

Response to loading: the immediate changes in stress and strain experienced by musculoskeletal components following a change in loading

The **response** of a musculoskeletal component to loading refers to the immediate changes in stress and strain experienced by the component. Provided the load (or change in load) is not prolonged, the stress and strain experienced by the component during loading is unlikely to result in any permanent change in the external form (size and shape) or internal architecture of the component, even if the load is within the moderate overload range. For example, it is unlikely that the changes in stress and strain experienced by the elbow extensor muscle-tendon units during the performance of a single press-up result in a permanent change in the external form or internal architecture of the elbow extensor muscle-tendon units. However, if the press-up exercise is performed more frequently, for example, as part of a fitness training program, and the load on the elbow extensor muscle-tendon units is within the moderate overload range during each repetition, the elbow extensor muscle-tendon units may experience permanent changes in their external form or internal architecture in a way that enables them to more readily withstand the time-averaged increase in load. The permanent change in the external form or internal architecture of musculoskeletal components that occurs as a result of changes in the time-averaged loads exerted on them is referred to as **structural adaptation** (Carter, Wong, and Orr 1991).

Optimum Strain Environment

The period of growth and development prior to maturity is associated with a gradual increase in the range and complexity of movement patterns and an almost continuous increase in body weight. Consequently, the total load on the musculoskeletal system also increases during this period. To a certain extent this increase in loading is balanced by an increase in the size of the musculoskeletal system. However, the period of growth and development is also associated with considerable changes in the relative size and shape of the different body segments (figure 11.4). For example, up to about two years of age the length of the lower limbs is about the same length as the upper limbs. Thereafter, the lower limbs increase in length and weight more rapidly than the upper limbs. The change in the complexity of movement patterns, change in relative size of body segments, and gradual increase in body weight combine to produce an almost continuous change in the pattern of loading on the musculoskeletal system.

The musculoskeletal system adapts to these changes and continues to adapt to such changes throughout life. It is well established that all components of the musculoskeletal system adapt (or try to adapt) their external form and internal architecture to the time-averaged loads exerted on them to maintain an **optimum strain environment** (Frost 1973; Rubin 1984). The optimum strain environment

Structural adaptation: the permanent change in the external form or internal architecture of musculoskeletal components that occurs as a result of changes in the time-averaged loads exerted on them

Optimum strain environment: the genetically predetermined strain range within which each musculoskeletal component normally functions; structural adaptation maintains the optimum strain environment of each component

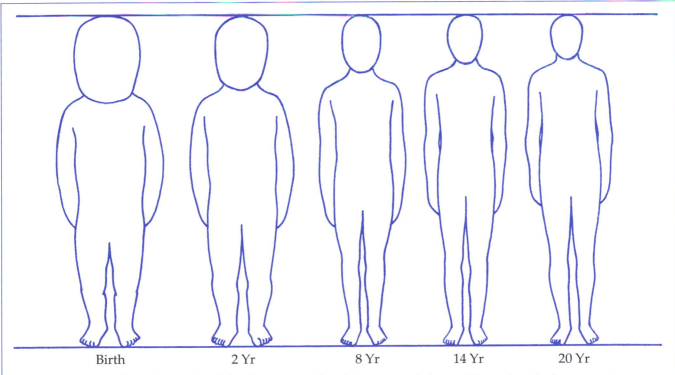

| Birth | 2 Yr | 8 Yr | 14 Yr | 20 Yr |

Figure 11.4   The effects of growth and development on the relative size and shape of the various body segments.

Genetic baseline: the genetically predetermined rate of growth and development of musculoskeletal components in the absence of loading

Wolff's law: bone adapts its external form and internal architecture to the time-averaged load exerted on it in an ordered and predictable manner to provide optimal strength with minimal bone mass

(OSE) of each component appears to be maintained by a negative-feedback system similar in operation to that of a thermostat; a change in time-averaged load producing a level of strain outside the strain limits of the OSE results in a change in the external form or internal architecture of the component so that the OSE is restored. For example, in bone, an increase in the time-averaged load above the upper limit of the OSE range will result in an increase in the cross-sectional area of the bone shaft or a realignment of the trabeculae in one or both epiphyses such that OSE is restored. Similarly, a decrease in the time-averaged load below the lower limit of the OSE range will result, for example, in a decrease in the cross-sectional area of the shaft of the bone such that OSE is restored. Prolonged absence of load or minimal load, for example, due to paralysis or prolonged immobilization following injury, will result in fairly rapid adaptation in the form of reduced mass. However, the loss of mass eventually ceases at a level referred to as the **genetic baseline**—the genetically predetermined mass due to growth and development in the absence of loading (Rubin 1984).

Structural Adaptation in Bone

The last 30 years have produced much of the present knowledge concerning the adaptation of musculoskeletal components to changes in time-averaged load (Frost 1988a, 1988b, 1990). However, the fundamental concepts concerning the adaptation of bone were established over 100 years ago (Gross and Bain 1993). In 1892 Julius Wolff (1836–1902) summarized the contemporary views of bone adaptation to changes in time-averaged load in what came to be known as **Wolff's law** (Wolff 1988). Wolff's law, shown to be more or less correct, hypothesized that

- bone adapts its external form and internal architecture to the load exerted on it,
- adaptation in bone provides optimum strength with minimum bone mass, and
- adaptation in bone occurs in an ordered and predictable manner.

Stereotypical Loading and Optimum Bone Mass

As described in chapter 4, the adaptation of bone to environmental influences, in particular, to time-averaged load, is referred to as modeling. In normal growth and development modeling has been estimated to account for 20% to 50% of the dimensions of mature bones (Frost 1988b). Some of the load experienced by bone is due to the weight of body segments. However, this source of loading is small relative to the loads exerted by muscles. Consequently, it is reasonable to assume that modeling in a particular bone reflects the time-averaged loads exerted by the muscles controlling the movement of the bone and, as such, modeling strengthens the bone to withstand the habitual load exerted by the muscles. Since bones, especially long bones, are stronger in relation to axial compression than to any other form of loading (Frost 1973), the habitual form of loading exerted by muscles on long bones is mainly axial compression—compression along the long axis of the bones. It is generally thought that the muscles controlling the movement of a particular long bone act synergistically stereotypically—they exert an axial compression load even though the magnitude of the load may vary (Rubin 1984; Frost 1973). Stereotypical loading produces a restricted strain environment in which the type of strain experienced by a bone is similar whenever it is loaded, even though the amount of strain may vary.

In a restricted strain environment a bone requires less bone mass than if it were required to function in a broad strain environment. It is generally agreed that all bones function in restricted strain environments and that modeling optimizes a bone's strength—provides maximum strength with minimum bone mass—for its particular restricted strain environment (Frost 1973; Gross and Bain 1993).

In normal growth and development modeling accounts for an esti-mated 20% to 50% of the dimensions of mature bones.

Modeling Processes in Bone

From birth to maturity, bone has the capacity to model external form and internal architecture. However, the capacity to model external form gradually decreases and virtually ceases at maturity. The capacity to model internal architecture also de-creases with age, but is retained to some extent throughout life. In general, bone adapts to changes in time-averaged loads in the same way as other musculoskeletal components—by increasing or decreasing bone mass to maintain an optimum strain environment. In bone, the optimum strain environment is characterized by mini-mum flexure (or bending) strain and an even distribution of stress (usually compres-sion stress) across articular areas. Minimum flexure strain is maintained by modeling in accordance with the phenomenon of **flexure-drift**. An even distribution of stress across articular areas is maintained by modeling in accordance with the phenom-enon of **chondral modeling** (see later in this chapter) (Frost 1973).

Flexure-drift: modeling that maintains minimum flexure strain in bones

Chondral modeling: model-ing of endochondral regions of the skeleton

> *The capacity of bone to model external form gradually decreases from birth and virtually ceases at maturity. The capacity to model internal architecture also decreases with age, but is retained to some extent throughout life.*

The Flexure-Drift Phenomenon

At birth, long bone shafts and the interarticular regions of many other bones such as the vertebrae are straight. However, prior to skeletal maturity the bone shafts adopt a narrow-waisted shape, such that the diameter of the shaft is smaller at the middle than at the ends. The change from a straight to a narrow-waisted shape reflects the shaft's modeling to minimize flexure strain. This is illustrated in figure 11.5 with respect to the development of a vertebra.

Figure 11.5a represents a frontal cross-section through the body of a vertebra of a young child. In an unloaded state, as in figure 11.5a, the sides of the body of the vertebra are basically straight. Figure 11.5b shows the vertebra subjected to a vertical compression load. The compression load reduces the vertical height of the vertebra (from h_1 to h_2) and the increased pressure on the cancellous bone and surrounding marrow bends the side walls of the vertebra outward. The greater the compression load, the greater the pressure exerted by the cancellous bone and marrow on the side walls. As the side walls bend in response to the pressure of the cancellous bone and marrow, the compression load adds to the bending load exerted on the side walls. The type of load exerted on the side walls by the compression load (which occurs after the side walls have started to bend) is referred to as a cantilever (indirect) bending load. The type of bending load exerted on the side walls by the pressure of the cancellous bone and marrow is referred to as a static (direct) bending load (figure 11.5c).

Whenever a bone (or part of a bone) is subjected to flexure strain above the upper limit of the optimum strain environment, flexure-drift occurs—bone is absorbed (removed) from the convex-tending surface of the bone and new bone is deposited on the concave-tending surface of the bone. The flexure-drift phenomenon is sometimes referred to as the flexure-drift law (Frost 1973). In the bodies of vertebrae the flexure-drift phenomenon results in bone absorption from the periosteal surfaces of the side walls and deposition of new bone on the endosteal surfaces such that each vertebra adopts a narrow-waisted shape (figure 11.5, *d* and *e*). When this shape is subjected to a vertical compression load, the side walls experience a cantilever bending load that bends them farther inward and a static bending load exerted by the cancellous bone and marrow that bends them outward. Thus, the cantilever and

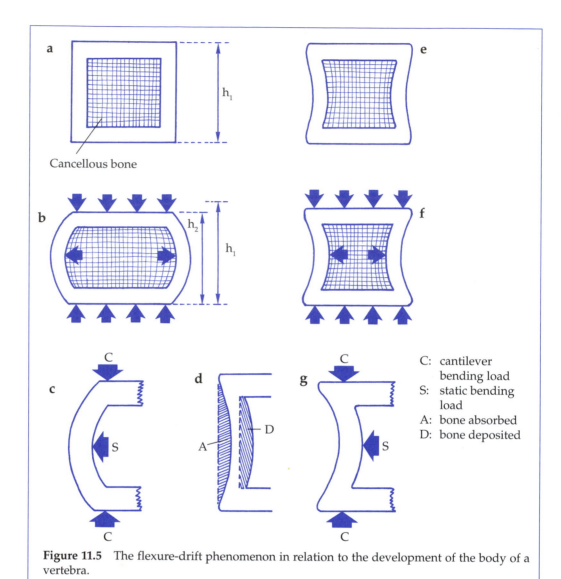

Figure 11.5   The flexure-drift phenomenon in relation to the development of the body of a vertebra.

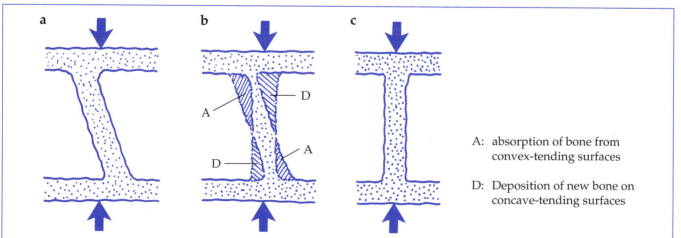

Figure 11.6   Adaptation of trabeculae to a change in the time averaged loading. The normal loading on trabeculae is axial loading: (a) change in load on the bone resulting in nonaxial load on the trabecula; (b) adaptation of the trabecula to restore axial loading; (c) axial load on the trabecula restored.

static bending loads oppose each other and, as such, minimize the amount of flexure strain on the side walls (figure 11.5, *f* and *g*). Consequently, whereas the thickness of the compact bone in the side walls of the vertebrae reflects the time-averaged compression load exerted on them, the shape of the walls reflects the time-averaged bending load exerted on them.

Trabeculae in cancellous bone adapt to axial loads (loads in line with the trabeculae) and bending loads in the same way as compact bone. The thickness of trabeculae reflects the magnitude of the time-averaged compression or tension loads experienced by the trabeculae, and the orientation of the trabeculae reflects the normal line of action of the compression or tension loads. A change in time-averaged loading on trabeculae will result in a change in the thickness or orientation of the trabeculae, as shown in figure 11.6.

In accordance with the flexure-drift phenomenon, the shafts of long bones and the interarticular regions of many other bones such as the vertebrae develop narrow-waisted shapes, minimizing the flexure strain on the bones. Trabeculae in cancellous bone also model in accordance with the flexure-drift phenomenon.

The Chondral Modeling Phenomenon

All bones that develop from hyaline cartilage via endochondral ossification (chapter 4) experience chondral modeling—the effect of loading on the rate and amount of new bone formed by hyaline cartilage. Chondral modeling applies to the following regions of bones:

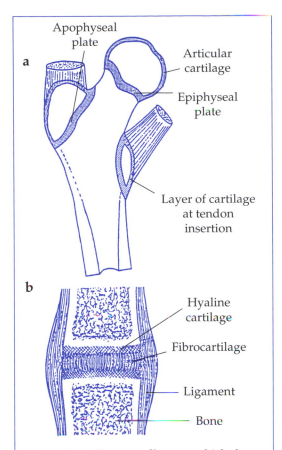

Figure 11.7 Regions of bone at which chondral modeling occurs; (*a*) vertical section through the proximal end of the femur; (*b*) vertical section through a typical intervertebral joint.

1. Articular cartilage: growth of epiphyses (figure 11.7*a*)
2. Epiphyseal plates: growth of metaphyses (figure 11.7*a*)
3. The layers of hyaline cartilage at the insertion of tendons and ligaments: growth of apophyses, nonapophyseal insertions of tendons, insertions of ligaments (figure 11.7*a*)
4. Apophyseal plates: growth of shaft adjacent to apophyseal plate (figure 11.7*a*)
5. The layers of hyaline cartilage at the bone-cartilage interfaces in symphysis joints: growth of bony end plates (figure 11.7*b*)
6. The hyaline cartilage surrounding sesamoid bones: growth of sesamoid bones

The effect of change of load on rate of growth in these regions is shown in the **chondral growth-force relationship** (Frost 1979) (figure 11.8). The relationship has the following main features:

1. Genetic baseline rate of growth: a certain amount of growth occurs in the absence of load
2. Ascending tension limb: increasing tension (within the moderate range) tends to increase rate of growth
3. Ascending compression limb: increasing compression (within the moderate range) tends to increase rate of growth
4. Descending compression limb: excessive compression tends to decrease the rate of growth and may result in the cessation of growth

Chondral growth-force relationship: the relationship between load (type and magnitude) and rate of growth in endochondral regions of the skeleton

Growth of Epiphyses and Metaphyses in Bones Forming Synovial Joints   In a long bone the size and shape of the epiphyses and metaphyses, and, consequently, the orientation of the epiphyses of a bone to its shaft, are determined by chondral modeling in articular cartilage and epiphyseal plates. Epiphyseal plates normally load to the left of the peak of the compression component of the chondral growth-force relationship, whereas articular cartilage normally loads around the peak (see figure 11.8). Consequently, the rate of growth of epiphyses is greater than that of metaphyses when congruence between the articular surfaces is maximum.

However, incongruence tends to have a more marked effect on rate of growth of the epiphyses than on the associated metaphyses. Incongruence decreases the rate of growth of the epiphyses below that of the associated metaphyses such that the orientation of the epiphyses to the shaft is largely determined by metaphyseal

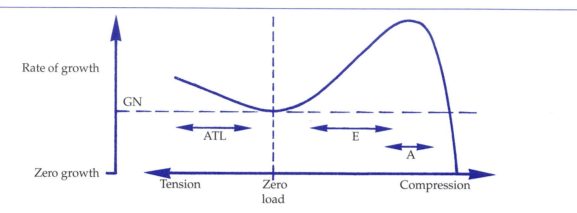

Figure 11.8   The chondral growth-force relationship; GN—genetic baseline rate of growth; ATL—normal range of loading on apophyseal plates, tendon insertions, and ligament insertions; E—normal range of loading on epiphyseal plates; A—normal range of loading on articular cartilage.

Adapted, with permission, from H.M. Frost, 1979, "A chondral modeling theory," *Calcified Tissue International* 28: 184.

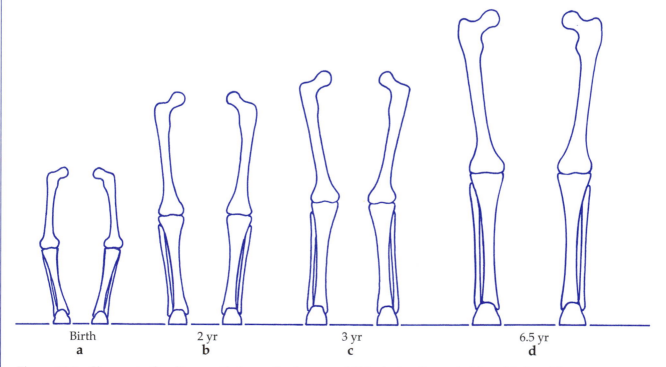

Birth	2 yr	3 yr	6.5 yr
a	**b**	**c**	**d**

Figure 11.9   Changes in the alignment between the femur and tibia during the period from birth to 6.5 yr.

growth. Metaphyseal growth not only determines the orientation of the epiphyses of a bone to its shaft, but it also affects the alignment of bones at joints.

Changes in the alignment of the long bones of the legs are particularly noticeable from birth to about 6.5 years of age (Salenius and Vankka 1975) (figure 11.9). At birth, most children have genu varum—bowlegs—with an angle between the femur and tibia of about 15° (figure 11.9a). Between birth and about two years of age the genu varum gradually decreases to zero; the femur and tibia become directly in line (figure 11.9b). During the third year, genu valgum—knock-knees—develop. This reaches a maximum of about 12° of valgus between the femur and tibia at about three years of age (figure 11.9c). During the next 3.5 years the genu valgum gradually decreases to about 5° of valgus at about 6.5 years of age (figure 11.9d). The changes in the degree of genu varum and genu valgum from birth to approximately 6.5 years of age presumably reflect the considerable changes in body weight, relative size of body segments, and complexity of movement patterns occurring during this period.

When a synovial joint is maximally congruent the loading on articular cartilage and epiphyseal plates tends to be evenly distributed. Incongruence results in an unequal distribution of load across articular cartilage and epiphyseal plates. If prolonged, such unequal loading results in modeling to restore maximum congruence. However, the actual changes that occur depend on the extent of the changes in the patterns of loading on the articular cartilage and epiphyseal plates. If the changes in loading remain within the normal range, as indicated by the chondral growth-force relationship, then the negative-feedback mode of modeling is invoked—maximum congruence is usually restored with normal alignment of the bones (see later discussion). However, if the changes in loading are outside the normal range then the positive-feedback mode of modeling is invoked, which aggravates the condition, resulting in progressively worsening misalignment.

Modeling of Metaphyses   A functionally normal joint is a congruent joint that transmits loads across the articulating surfaces in a normal manner. An anatomical misalignment at the knee, or any other joint, will be functionally normal if the misalignment stabilizes (does not get progressively worse). In these cases, the anatomical misalignments represent normal modeling in response to abnormal patterns of loading imposed on the skeleton by the combined effects of muscle forces and body weight. The skeletal adaptations ensure normal transmission of loads across the joints. The anatomical misalignments must be normal functionally, otherwise the joints become painful. For example, figure 11.10 illustrates the effect of the negative-feedback mode in relation to abductor-adductor muscle imbalance at the knee.

Figure 11.10a represents a knee with normal balance between the abductor and adductor muscles, that is, the resultant horizontal force at the knee is zero. This situation is associated with normal alignment between the femur and tibia and an even distribution of load across the epiphyseal plates (figure 11.10b). Figure 11.10c shows the loading (even distribution) on the epiphyseal plates in relation to the chondral growth-force curve. Figure 11.10d shows the same knee with an abductor-adductor imbalance such that there is a net medially directed horizontal force at the knee increasing the degree of genu valgum. Figure 11.10e shows the unequal pattern of loading on the epiphyseal plates associated with the muscle imbalance. Figure 11.10f shows the range of loading on the epiphyseal plates in relation to the chondral growth-force curve.

Since the unequal range of loading is within the normal range, the negative-feedback mode is invoked. The rate of growth of the lateral aspects of the metaphyses is increased and the rate of growth of the medial aspects of the metaphyses is decreased such that normal congruence is restored (with net zero horizontal force at the knee) at the expense of an abnormal alignment between the femur and tibia, that is, much reduced genu valgum or even slight genu varum relative to most individuals (figure 11.10g).

Whether or not a particular joint is anatomically misaligned during childhood, the only time it may become painful (excluding injuries and pathological conditions not due to loading) is during adulthood, when the bones are no longer capable of modeling in response to abnormal loading. In most adults, abnormal patterns of loading are the result of an increasingly sedentary lifestyle in which body weight gradually increases and muscle strength gradually decreases (figure 11.11).

Some of the world's best athletes have marked genu varum or genu valgum, but they train and perform without injury (Micheli 1982). Similarly, some of the world's best ballet dancers have severely pronated feet, but they practice and perform without injury (Micheli 1982). In these cases, it must be assumed that the abnormalities are functionally normal and, as such, arose during childhood. Anatomical misalignments that develop after maturity are the result of a combination of muscle weakness, ligament laxity, and progressive bone collapse (Frost 1979).

> *Anatomically misaligned but functionally normal joints represent normal modeling in response to abnormal patterns of loading imposed on the skeleton by the combined effects of muscle forces and body weight.*

Modeling of Articular Surfaces   Since articular surfaces in synovial joints are not physically attached to each other, even minor incongruences result in large changes in the load experienced by different parts of the articular surfaces. This is especially the case in joints with pulley-shaped articular surfaces such as the ankle joint (figure

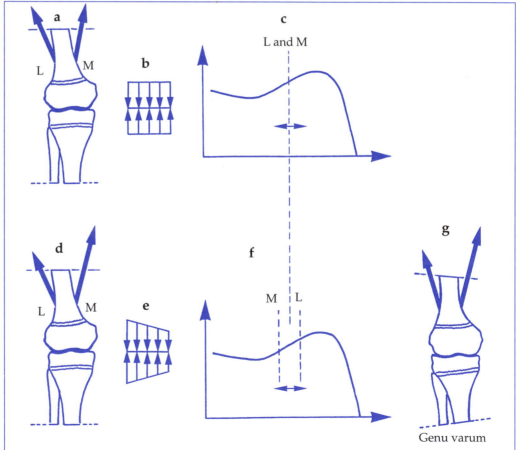

Figure 11.10   Chondral growth-force relationship: effect of negative-feedback mode in relation to abductor-adductor imbalance at the knee; L—lateral and M—medial.

11.12). Under normal circumstances inversion and eversion of the foot take place about the subtalar joint (figure 11.12, *a* and *b*). However, if the subtalar joint does not form properly, such that movement at the joint is absent or limited, inversion and eversion of the foot twists the talus in the tibiofibula mortise resulting in excessive loading on those parts of the articular surfaces that remain in contact (figure 11.12*c*). The excessive loading on the impinging areas reduces or halts growth in these areas, while growth of the unloaded areas proceeds at the genetic baseline rate (see figure 11.8). Consequently, the shapes of the articular surfaces adapt to the abnormal loading conditions by forming a rounded surface in the frontal plane rather than a trochlea-shaped surface, and the ankle joint as a whole resembles a ball and socket joint rather than a hinge joint (figure 11.12*d*) (Frost 1979).

Modeling of Vertebral Bodies   The load experienced by the articular regions of vertebral bodies in intervertebral joints in upright postures tends to be evenly distributed compression as shown in figure 11.13, *a* and *b*. Bending the vertebral column in any direction produces a pattern of loading on the discs and, therefore, on the hyaline cartilage interfaces between the end plates of the vertebrae and the fibrocartilage of the discs, ranging from relatively high compression on the concave-tending side of the column to relatively high tension on the convex-tending side of the column (figure 11.13, *c* and *d*). This loading pattern in response to bending the column is normal. However, if part of the column is subjected to higher than normal, unidirectional, bending stress during the period of growth and development, the vertebral bodies in the affected region may become wedge shaped to restore a more normal pattern of loading (figure 11.13, *e* and *f*). Wedging of vertebrae occurs in the thoracic and lumbar regions in the median and frontal planes (see chapter 6).

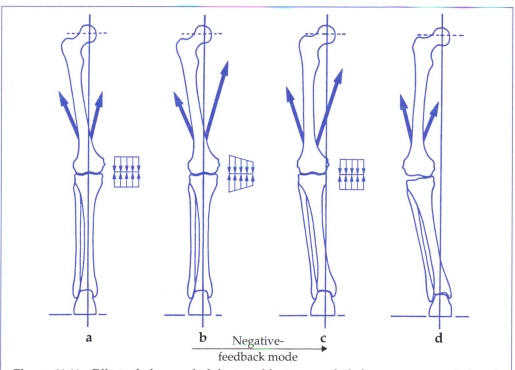

Figure 11.11   Effect of abnormal abductor-adductor muscle balance on anatomical and functional alignments in the knee joint; (*a*) normal muscle balance and normal anatomical and functional alignments; (*b*) abnormal muscle balance and functional alignment and normal anatomical alignment; (*c*) abnormal muscle balance and anatomical alignment and normal functional alignment; (*d*) change in muscle balance postmaturity resulting in worsening anatomical and functional alignments.

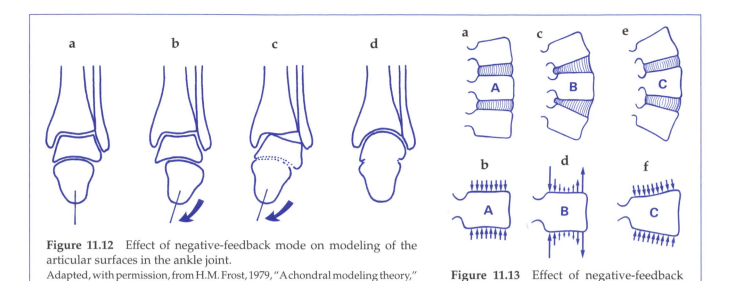

Figure 11.12   Effect of negative-feedback mode on modeling of the articular surfaces in the ankle joint.

Adapted, with permission, from H.M. Frost, 1979, "A chondral modeling theory," *Calcified Tissue International* 28: 187.

Figure 11.13   Effect of negative-feedback mode on modeling of vertebrae.

Effects of Age on Modeling Capacity   The modeling capacity of bone decreases with age. Whereas the ability to adapt internal architecture to changes in patterns of loading is somewhat retained throughout life, the ability to adapt size and shape decreases more rapidly and is negligible in the adult (Frost 1979; Lanyon 1981). Consequently, during childhood a change in the normal pattern of loading on bones usually results in adaptive bone growth to restore normal loading. However, in an adult, adaptive bone growth cannot accommodate a change in the normal pattern of loading due, for example, to an increase in body weight or a change in the muscle strength balance about a joint. Unless corrected by specific forms of treatment such as physiotherapy or orthotics, the abnormal pattern of loading increases the likelihood of degenerative joint disease. An orthosis is a musculoskeletal support—an external appliance such as leg calipers, a foot arch support, or heel raise designed to improve musculoskeletal function by helping to maintain normal transmission of loads in joints (Redford 1987). Orthotics is the theory and practice of development, manufacture, and the use of orthoses (Redford 1987).

The following examples illustrate the rate at which the modeling capacity of bone decreases with age and the possible effects of a lack of modeling capacity in the adult.

Tibial torsion. Tibial torsion is an abnormality of the tibia in which the lower end of the tibia is excessively rotated internally or externally with respect to the upper end about the long axis of the shaft. Internal and external tibial torsion are characterized by abnormal toeing in and toeing out, respectively (figure 11.14). The normal degree of rotation of the lower end of the tibia with respect to the upper end is 15° to 20° of external rotation (the foot is turned out 15° to 20° with respect to the knee; figure 11.14b). Tibial torsion can be corrected easily in an infant by applying a constant torsion load to the tibia by means of special splints (Frost 1979). The torsion load is applied in the direction opposite to that of the abnormality. The bone adapts to the torsion stress by modeling to relieve the torsion stress. In doing so, the tibial torsion is gradually reduced, provided the torsion load and, therefore, the torsion stress, is maintained. However, the effect of this type of treatment decreases fairly rapidly with age. For example, 30° to 40° of correction of an internal tibial torsion can be produced in 10 weeks of treatment when applied to a five-month-old child. The same amount and duration of treatment has a negligible effect in a five-year-old child (Frost 1979).

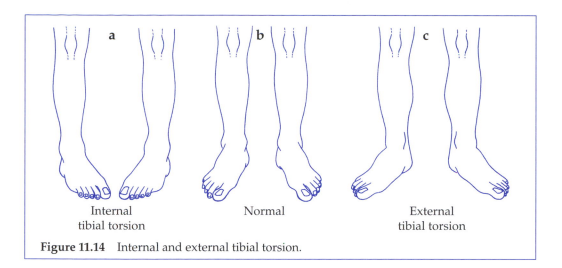

Internal
tibial torsion

Normal

External
tibial torsion

Figure 11.14   Internal and external tibial torsion.

Bone fracture. Following an injury to a bone involving two or more fractures, the chances of the broken pieces knitting together in exactly the same orientation to each other as existed prior to the injury are remote. Consequently, after healing, the line of action of the muscles crossing the joints formed by the two ends of the bone is altered. This may result in abnormal loading on the articulating surfaces of the joints. In a child the articulating surfaces usually restore normal loading by modeling. In an adult the abnormal loading increases the likelihood of degenerative joint disease, especially in weight-bearing joints (Radin 1986).

> *Anatomically misaligned but functionally normal joints that develop in childhood may become painful during adulthood when the bones are no longer capable of modeling in response to abnormal loading (and also due to injuries and pathological conditions not associated with loading). In most adults, abnormal patterns of loading result from an increasingly sedentary lifestyle in which body weight gradually increases and muscle strength gradually decreases.*

Structural Adaptation in Regular Fibrous Tissues

Regular fibrous tissues, and particularly ligaments and tendons, are designed to provide strength in tension. The effect of changes in time-averaged tension load on regular fibrous tissues depends on the change in magnitude and type of load. When subjected to continuous tension at a level exceeding the optimum strain environment, regular fibrous tissues experience creep (chapter 10). When subjected to intermittent tension at a level exceeding the optimum strain environment, regular fibrous tissues hypertrophy to restore the optimum strain environment. For example, increasing the strength of a muscle increases the time-averaged tension load exerted on the associated tendons, which experience an increase in strain. The tendons adapt to the increased strain by increasing their cross-sectional areas to a level that restores the optimum strain environment. The capacity of regular fibrous tissues to adapt to increases in time-averaged intermittent tension load is called the **stretch-hypertrophy rule** (Frost 1979).

Just as an increase in time-averaged intermittent tension on regular fibrous tissue results in hypertrophy, a decrease in time-averaged tension results in atrophy to restore the optimum strain environment. Atrophy in regular fibrous tissue involves a decrease in cross-sectional area or shortening (Akeson et al. 1977). Immobilization, which may occur due to paralysis or injury, results in considerable atrophy in

Stretch-hypertrophy rule: the capacity of regular fibrous tissues to hypertrophy when subjected to a time-averaged increase in intermittent tension that exceeds the optimum strain environment

ligaments. One of the most often quoted studies in this area is that of Noyes (1977). Noyes studied the effects of immobilization and subsequent rehabilitation on the strength of the anterior cruciate ligament (ACL) in rhesus monkeys. The results of the study showed that immobilization in a whole-body cast for eight weeks produced an average decrease of 39% in the strength of the ACL and after 5 months and 12 months of rehabilitation the average strength of the ACL was still only 79% and 91%, respectively, of the preimmobilization level. It is generally agreed that, in humans, rehabilitation of ligaments following immobilization and injury is a lengthy process (Frost 1979, 1990).

Under normal circumstances the load experienced by ligaments is markedly reduced by ligament-muscle stretch reflex mechanisms; stretching a ligament tends to result in reflex contraction of associated muscles, which relieves the load on the ligaments (Cohen and Cohen 1956). For example, hyperextension of the knee results in reflex contraction of the hamstrings triggered, at least in part, by stretch reflex loops between the knee ligaments (cruciate and collateral) and the hamstrings. Injury to ligaments may impair the stretch reflex mechanism by delaying the elicitation of muscular contraction. This may occur directly in the form of damage to nerves or indirectly in the form of laxity in the ligaments such that greater movement is necessary in the joint to stretch the ligaments sufficiently to elicit the stretch reflex. The greater the delay in the elicitation of the stretch reflex, the greater the likelihood of ligaments being excessively overloaded or associated joints being dislocated. Injury to ligaments may also impair their blood supply, which, in turn, may reduce the rate of hypertrophy.

It is, perhaps, not surprising that many ligament reconstructions based on grafts from other parts of the body fail because it takes time for an adequate blood supply to develop, which is essential for hypertrophy, and for stretch reflex loops to develop, which protect the ligaments from overload (Frost 1973).

> *Under normal circumstances the load experienced by ligaments is markedly reduced by ligament-muscle stretch reflex mechanisms. Injury to ligaments can impair the functioning of the stretch reflex mechanisms by delaying the elicitation of muscular contraction, which, in turn, increases the likelihood of excessively overloading ligaments or dislocating associated joints.*

Structural Adaptation at Ligament and Tendon Insertions

All ligaments and tendons insert onto bone via a hyaline cartilage interface and, thus, the bone adapts to the tension load exerted via the tendons and ligaments in accordance with the chondral modeling phenomenon (figure 11.15). Just before entering the layer of hyaline cartilage the ligament or tendon separates into bundles of collagen fibers. Each bundle of fibers passes through and is embedded within all three layers of the region of insertion: hyaline cartilage, calcified cartilage, and subchondral bone. In the layer of subchondral bone the collagen fiber bundles become encrusted with mineral salts. At this stage each bundle is referred to as a **Sharpey fiber** (Frost 1973).

The ultimate shear stress between the Sharpey fiber and the surrounding bone determines each Sharpey fiber's resistance to being pulled out of the insertion. The deeper the penetration of the fiber, the greater the area of fiber presented to the surrounding bone, and, therefore, the greater the shear strength of the fiber. The greater the shear strength of the fiber, the greater the shear load required to pull out the fiber. It is, therefore, not surprising that each ligament and tendon insertion consists of a very large number of Sharpey fibers; the larger the number of fibers, the

Sharpey fiber: a terminal bundle of collagen fibers in a ligament or tendon insertion encrusted with mineral salts

lower the shear stress on each fiber. Furthermore, the fibers are distributed over a relatively large area (much larger than that generally indicated on bony specimens and models) and merge with adjacent fibrous tissues. This is referred to as **fiber fan-out** (Frost 1973). The larger the area of insertion, the lower the tension stress on the region of insertion.

The combination of fiber fan-out and large number of Sharpey fibers produces strong attachments that, in the case of apophyses, are stronger than the apophyseal plate. Avulsion fractures occur along the apophyseal plate—the tendon stays intact and the apophysis is displaced. This can reasonably be compared to pulling up a plant having an extensive root system; there is usually a lot of soil still attached to the roots. In this sense a tendon or ligament can be described as having a very extensive root system. The insertions of ligaments and tendons adapt to changes in time-averaged tension loads by varying the degree of fiber fan-out and the number of Sharpey fibers. Increases in loading increase the area of fiber fan-out and the number of fibers, and decreases in loading reduce the area of fan-out and number of fibers.

Structural Adaptation in Muscle

Muscle tissue adapts to changes in time-averaged tension load more quickly than other musculoskeletal components. Time-averaged increases in tension that exceed the optimum strain environment result in increases in strength, and time-averaged decreases in tension that fall below the optimum strain environment result in decreases in strength. However, the changes in strength accompanying changes in time-averaged tension are likely due to a combination of changes in muscle mass and motor unit recruitment.

Each motor unit has an optimal rate of stimulation resulting in maximal tension. Strength training appears to develop the ability of the individual to stimulate muscles optimally during voluntary effort (Sale 1992). Furthermore, some motor units have a higher stimulus threshold than others such that for activities requiring less than maximum strength only the motor units with low to moderate thresholds are used. Strength training based on high-resistance, low-repetition exercises appears to enhance the voluntary recruitment of higher threshold motor units (Sale 1992). Consequently, strength training can bring about increases in muscle strength by changes in motor unit activation as well as by hypertrophy. Changes in motor unit activation involving optimal stimulation of motor units and recruitment of higher threshold motor units account, at least in part, for the increases in strength exhibited by women and men during the initial stages of a strength-training program, that is, before hypertrophy occurs.

Fiber fan-out: the area of distribution of Sharpey fibers in a ligament or tendon insertion

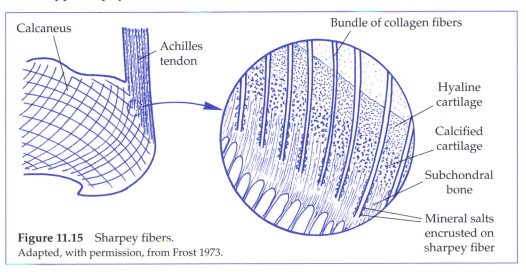

Figure 11.15   Sharpey fibers.
Adapted, with permission, from Frost 1973.

Changes in strength are likely due to changes in muscle mass and motor unit activation.

Muscle fiber length determines a muscle's extensibility, and the number of sarcomeres in each myofibril determines the length of each muscle fiber. Changes in the length of muscle fibers and, therefore, in the extensibility of muscles occur due to changes in time-averaged range of movement of the muscles (Goldspink 1992). Muscles adapt to changes in time-averaged range of movement by increasing or decreasing the number of sarcomeres in each myofibril in order to maintain optimum function of muscle fibers in terms of length-tension characteristics. For example, immobilizing the ankles of cats in full dorsiflexion—with the soleus muscles in a stretched position—has been found to result in about a 20% increase in the number of sarcomeres in the fibers of the soleus muscle in four weeks. The same researchers found that immobilizing the ankles of cats in full plantar flexion—the soleus muscles in a shortened position—resulted in about a 40% decrease in the number of sarcomeres in the fibers of the soleus muscle in four weeks (Williams and Goldspink 1973, 1978).

Muscles adapt to changes in time-averaged range of movement by increasing or decreasing the length of muscle fibers to maintain optimum function in terms of length-tension characteristics.

Summary

This chapter described the effects of changes in time-averaged loading on the musculoskeletal system. A certain level of loading is necessary to ensure normal growth and development; however, excessive loading, like insufficient loading, is associated with abnormal growth and development. Each musculoskeletal component has an optimum strain environment and changes in time-averaged loading result in structural adaptation to restore the optimum strain environment. Decreases in time-averaged loading result in atrophy and decreased functional capacity, and moderate increases in time-averaged loading result in hypertrophy and increased functional capacity. Injury occurs due to excessive loading of an acute or chronic nature and in order to prevent or minimize the risk of injury in a particular activity it is necessary to understand the factors that contribute to loading; this is the subject of the final chapter.

Review Questions

1. Differentiate between biopositive and bionegative effects of loading.
2. Differentiate between moderate habitual load and moderate overload.
3. Differentiate between acute and chronic injury.
4. Differentiate between response to loading and adaptation to loading.
5. Describe the chondral growth-force relationship.
6. Differentiate between cantilever and static bending loads.
7. Differentiate between the negative- and the positive-feedback mode in chondral modeling.
8. Differentiate between anatomical and functional alignment in normal joints.
9. Describe the effects of age on the modeling capacity of bone.
10. Describe the structure of ligament and tendon insertions.
11. Describe the effects of changes in loading and range of movement on muscle.

Chapter 12

Etiology of Musculoskeletal Disorders and Injuries

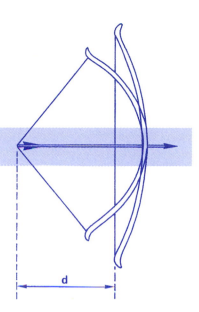

Musculoskeletal disorders affect most people at some time in their lives. The underlying cause of many musculoskeletal disorders is excessive overload of a chronic or acute nature. In any particular situation the load on the musculoskeletal system depends on the influence of a number of risk factors and their interaction. Without a clear understanding of the relevant risk factors it is difficult to determine the cause of a disorder and treatment may be confined to dealing with signs and symptoms. This chapter describes the three main types of risk factors that influence the level of loading on the musculoskeletal system.

Kinetic Chain

Kinetic chain: the transmission of forces between body segments in any particular posture or movement

All postures and movements involve the transmission of forces across a number of joints. For example, even in a sitting position the weight of the upright trunk is transmitted to the chair seat by the joints of the vertebral column and hips. Similarly, bending over a bath to turn on a tap involves the transmission of forces across a large number of joints. The term **kinetic chain** describes the transmission of forces between body segments in any particular posture or movement (Steindler 1955). Since the joints in any particular kinetic chain are linked together, the movement of each separate joint affects the movement and, therefore, the loading on some or all of the other joints in the chain. It follows that abnormal movement in one joint may cause overloading in the other joints in the chain. Consequently, in any particular movement or sequence of movements, the movement of the various joints must be properly coordinated so none of the separate joints, or the segmental links between them, are excessively overloaded.

Compensatory Movements

All joints have one to six degrees of freedom (chapter 5). Consequently, the number of possible movement patterns that can accomplish a given task is, for most whole-body movements, infinite. It follows that when movement in one or more joints in a particular kinetic chain is restricted, perhaps due to muscle weakness or anatomical abnormalities, the other joints in the chain can usually compensate by altering their own ranges of movement so that the task can still be accomplished. For example, stiffness in one knee is often compensated in walking by tilting the pelvis upward on the affected side to allow the leg to swing forward without the foot catching on the floor. However, for any given task there is one most efficient movement pattern (or range of movement patterns within fairly strict limits) that minimizes the amount of muscle force required and, therefore, minimizes the load on the musculoskeletal system as a whole. Even a slight amount of **compensatory movement** in a joint alters the normal pattern of loading in the joint (Riegger-Krugh and Keysor 1996). Furthermore, the greater the degree of compensatory movement in a joint, the greater the loading on the joint and its supporting structures. If compensatory movements are excessive and prolonged (continuous or repetitive) the joints concerned and their supporting structures may be excessively overloaded, resulting in a musculoskeletal disorder.

Compensatory movement: the change in the movement of one or more joints in a kinetic chain in response to restricted movement in other joints so that a given task can still be carried out

Even a slight amount of compensatory movement in a joint can alter the normal pattern of loading in the joint and, in time, may result in some kind of musculoskeletal disorder.

Risk Factors

A **musculoskeletal disorder** is any temporary or permanent abnormality of the musculosketal system resulting in pain or discomfort. Disorders of the musculoskeletal system affect most people at some stage in their lives. Some disorders occur dramatically and their cause is obvious, for example, a broken leg caused by the impact of a motor vehicle accident or a hard tackle in football. In these contexts injury may be the more appropriate term. Many other disorders of the musculoskeletal system, such as chronic low-back pain and osteoarthritis of the hip joint, develop over time, and their causes are usually less obvious.

Without a thorough understanding of the underlying cause of a musculoskeletal disorder, treatment may be confined to dealing with the signs and symptoms of the disorder. However, if the cause of the disorder is not removed, the disorder may worsen and result in more severe pain or discomfort. The underlying cause of many musculoskeletal disorders is excessive overload of an acute or chronic nature. The actual level of overload depends on the influence of associated risk factors. There are three main categories of risk factors: **movement factors**, **intrinsic factors**, and **extrinsic factors** (Lysens et al. 1984; Lorenzton 1988; Nigg 1988).

Movement factors concern the characteristics of the movements comprising the activity, in particular, the speed of movement and the degree of physical contact between the individual and her surroundings. The latter includes, for example, impacts with other participants, playing surfaces, walls, and implements such as sticks, bats, and balls. In general, the greater the movement speed and the greater the degree of physical contact, the higher the risk of injury (Backx et al. 1991). Consequently, some activities are inherently more dangerous than others. Intrinsic risk factors are personal, physical, and psychological characteristics that distinguish individuals from each other, and extrinsic risk factors concern environmental conditions and the manner in which activities are administered (figure 12.1). Whereas the main cause of a particular disorder or injury may be apparent, in most cases the cause of the disorder or injury is the result of a complex interaction of movement factors and intrinsic and extrinsic factors.

> *Without a thorough understanding of the underlying cause of a musculoskeletal disorder, treatment is confined to dealing with the signs and symptoms of the disorder. In most cases the cause of the disorder or injury is the result of a complex interaction of movement, intrinsic, and extrinsic factors.*

Musculoskeletal disorder: any temporary or permanent abnormality of the musculoskeletal system resulting in pain or discomfort

Movement risk factors: the char-acteristics of a movement that influence the level of loading on the musculoskeletal system

Intrinsic risk factors: the personal, physical, and psychological characteristics of an individual that influence the level of loading on the musculoskeletal system

Extrinsic risk factors: the environmental conditions and administrative procedures that influence the level of loading on the musculoskeletal system

Injury and Public Health

The benefits of a physically active lifestyle to enhance health are well documented (Blair, Kohl, and Gordon 1992). These benefits have been the basis of numerous campaigns in many countries to encourage people to adopt more physically active lifestyles (Mechelen 1993). The success of these campaigns is partly reflected in the significant increase in sport and physical activity participation in most Western countries over the past two decades (Stephens 1987; Roberts and Brodie 1989). Not surprisingly, research indicates that the increase in the number of sport and physical activity participants is accompanied by an increase in the number of injuries. Sport and physically active leisure-related injuries are part of the total number of injuries sustained by the population as a whole, and injuries are a major health problem (in terms of financial cost and personal suffering) in most Western countries (Guyer and Ellers 1990; Towner et al. 1994). Sport and physically active leisure pursuit injuries account for 16% to 32% of total injuries in studies based on injuries treated in local health care systems (Watkins 1995). It is not surprising that injury prevention has

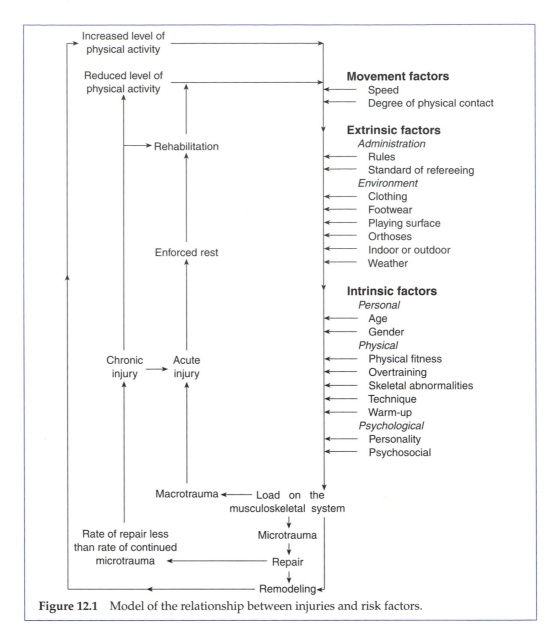

Figure 12.1   Model of the relationship between injuries and risk factors.

become a public health priority in some of these countries (Rivara et al. 1993; Department of Health 1992).

Intrinsic Risk Factors

Most epidemiological surveys of sport-related injuries focus on extrinsic factors in specific sports (Lysens et al. 1984). However, with the increase in sport and active leisure-related injuries in recent years in some Western countries there is now more interest in intrinsic risk factors, particularly personal factors (age and gender), physical factors (physical fitness, overtraining, skeletal abnormalities, technique, and warm-up), and psychological factors (personality and psychosocial).

Age and Gender

Age-related injuries fall into two major categories: those occurring prior to maturity and those occurring after maturity. Compared to adults, skeletally immature individuals have more pliable bone and softer cartilage, and their ligaments are stronger than the associated centers of bone growth (Micheli 1983). Consequently, loading that causes a ligament or tendon tear in an adult may cause an epiphyseal or apophyseal avulsion in a child or adolescent. Unless properly diagnosed and treated, such damage to epiphyseal or apophyseal regions may result in abnormal bone growth (Peterson and Renström 1986).

Knee injuries and knee pain are the most common orthopedic complaints in children and adolescents (Kannus, Nittymaki, and Jarvinen 1988). For example, Osgood-Schlatter's disease (inflammation and pain at the insertion of the patellar ligament to the tibial tuberosity) is a frequent complaint of young athletes, especially volleyball and basketball players who experience a high frequency of high-intensity jumping and landing. Just as rupture of the patellar ligament may follow the final stage of chronic jumper's knee in an adult, avulsion of the tibial tuberosity may follow chronic Osgood-Schlatter's disease in an immature athlete. Unless properly treated, avulsion of the tibial tuberosity may result in patella alta, which, in turn, may result in disorders of the patellofemoral joint (Kujala et al. 1989).

Epidemiological data suggests that the incidence of knee disorders, in particular anterior cruciate ligament injuries and patellofemoral syndrome, is higher in females than in males, especially in those who participate in sports involving high-frequency jumping and landing such as volleyball, basketball, and soccer (Arendt and Dick 1995). The etiology of these injuries is not clear, but suggested risk factors include knee alignment (Q angle), muscle balance, and muscle strength (Sommer 1988; Arendt 1996).

Physical Fitness

The muscles control joint movements and, as such, not only determine the pattern of movement in a given situation, but also protect the joints and joint-supporting structures by helping maintain normal load transmission in joints. The muscles' level of conditioning—the endurance, strength, and extensibility characteristics of the muscles—affects the risk of injury to all components of the musculoskeletal system. In general, increases in the endurance, strength, and extensibility of muscles will reduce the risk of injury.

Zohar (1973) compares the process of conditioning the body for a particular physical activity to the process of constructing a building or a bridge. When constructing a bridge, it is important to ensure that every part of the structure, irrespective of its size and role, has the required strength and compliance to withstand the forces to which it is subjected. Furthermore, the engineer builds in a suitable safety margin for each component to allow for unusually high loads. Zohar suggests that in many physical conditioning programs certain parts of the musculoskeletal system, especially those not readily targeted by training equipment, are frequently neglected and, thus, result in weak links in the kinetic chain. The neglected areas include the axial rotator muscles of the shoulders and hips and the muscles responsible for pronation and supination of the ankles. Zohar also suggests that those parts of the musculoskeletal system that are conditioned are frequently not conditioned to a level to incorporate an appropriate safety margin.

For example, in a study of the muscular control of the ankle joint in running, Reber, Perry, and Pink (1993) found that the tibialis anterior muscle had a higher sustained level of activity than the other muscles making it more susceptible to fatigue, and that there was a significant increase in the activity of the peroneus brevis muscle as the pace of running increased. Inappropriate conditioning of these muscles may account for the relatively high incidence of fatigue-related lower-leg injuries in running.

> *The level of conditioning of the muscles—their endurance, strength, and extensibility—considerably affects the risk of injury to all components of the musculoskeletal system. Increases in the endurance, strength, and extensibility of muscles will reduce the risk of injury.*

Imbalances in Strength and Extensibility Normal joint movement involves a high degree of coordination between the members of the various antagonistic pairs of muscles that control joint movement. The coordination between the opposing groups in each antagonistic pair depends on a functional balance between the groups in terms of strength and extensibility. A strength imbalance between the groups may occur if one group is trained more than the other (such as in a strength-training program in which the hamstrings are neglected relative to the quadriceps) or experiences significantly greater relative loading than the other during sport performance (such as jumping and landing in volleyball, which load the quadriceps more than the hamstrings) (Grace 1985).

Strength imbalances may also occur between similar groups of muscles on opposite sides of the body, such as the right and left quadriceps; this may occur, for example, in athletes who favor one leg over the other in jumping and landing (Knapik et al. 1991). The normal balance of strength between similar groups of muscles on opposite sides of the body may be viewed in absolute terms. However, the normal balance in strength between the opposing muscle groups in each antagonistic pair must be viewed in relative terms. For example, the cross-sectional area and, therefore, potential strength of the plantar flexors of the ankle joint is approximately seven times greater than that of the dorsiflexors (Silver, de la Garza, and Rang 1985; Wickiewicz et al. 1983). Even allowing for differences in leverage, there could not be a strength balance between the two muscle groups in absolute terms. The main function of the dorsiflexors is to provide toe-clearance during the swing phase in walking, whereas the plantar flexors are responsible for supporting body weight in the upright posture and for propelling the body forward and upward in activities such as walking, running, and jumping. Consequently, the plantar flexors and dorsiflexors are different functionally and the strength ratio of 7:1 in favor of the plantar flexors is probably functionally normal. Therefore, strength imbalance between the plantar and dorsiflexors may be reflected in a strength ratio markedly different from 7:1. Similarly, the cross-sectional area of the quadriceps is approximately two and a half times that of the hamstrings (Wickiewicz et al. 1983). Consequently, strength imbalance between the quadriceps and hamstrings may be reflected in a strength ratio markedly different from 2.5:1.

Just as differences in the (time-averaged) relative load on antagonistic muscle groups may result in strength imbalances, highly repetitive and exclusive movement patterns may produce extensibility imbalances between antagonistic muscle groups (Bach et al. 1985). Strength imbalances combined with extensibility imbalances may result in functional imbalances (also referred to as muscle imbalances), which predispose the individual to musculoskeletal disorder or injury (Grace 1985). Functional imbalance is manifest in a number of ways, including

- imbalances between left and right sides of the body (Knapik et al. 1991);
- imbalances between antagonistic muscle groups (Bach et al. 1985; Francis and Bryce 1987); and
- imbalances between muscle groups of different joints that work together to bring about particular movements (Sommer 1988).

Sommer (1988) investigated the effect of fatigue on the movement of the hips, knees, and ankles in jumping and landing in volleyball and basketball players. The results showed that as fatigue increased, the tendency for the knee joints to be forced into abduction during the powerful phases of leg extension (jumping) and leg flexion (landing) increased. Sommer suggested that this was due to functional imbalances between the muscles controlling the hip joints (the gluteal muscles), and that the effects of the imbalances became more pronounced as fatigue increased. The likely effect of knee abduction during intense activity of the knee extensor muscles would be

- lateral tracking of the patella,
- strain on the medial ligaments and other medial supporting structures, and
- asymmetric loading on the quadriceps tendon and patellar ligament, especially at their insertions on the patella.

Lateral tracking may result in patellar chondromalacia, and asymmetric loading on the quadriceps tendon and patellar ligament may result in jumper's knee—inflammation and pain at the upper and lower poles of the patella (Watkins 1994).

A strength imbalance between antagonistic muscle groups may occur if one group is trained more than the other or experiences significantly greater relative loading than the other during sports performance. Highly repetitive and exclusive movement patterns may produce extensibility imbalances between antagonistic muscle groups. Strength imbalances combined with extensibility imbalances result in functional imbalances (muscle imbalances) that predispose the individual to injury.

Growth-Related Imbalances in Strength and Extensibility   Imbalances in strength and extensibility can occur at any age, but some imbalances seem to be growth related. Longitudinal growth in the bones of the limbs and in the vertebrae is always more advanced than that of the associated soft tissues–muscle-tendon units and fibrous supporting structures (Micheli 1983). Consequently, during the growth period the soft tissues have to continually adapt their length to the length of the bones. For most of the growth period, soft tissue adaptation can keep pace with the rate of bone growth. However, during a period of rapid bone growth, such as the adolescent growth spurt, the rate of bone growth may outpace that of soft tissue adaptation. In these circumstances, there may be a significant increase in the tightness of the soft tissues resulting in a loss of joint flexibility, increased compression loading on articular surfaces, and increased tension loading on ligament and tendon insertions. All of these factors may increase the risk of injury to the tissues concerned, especially when the loss of flexibility is asymmetric. For example, adolescents tend to develop hollowback during the adolescent growth spurt. The hollowback seems to be caused by asymmetric tightness in the fibrous structures supporting the lumbar region. The fibrous supporting structures on the posterior aspect of the lumbar region (the lumbodorsal fascia) are thicker and more extensive than those on the anterior aspect. Consequently, in response to increased growth in the vertebrae—an increase in the height of the vertebral bodies—the posterior soft tissues may take longer to adapt (by structural elongation) than the anterior soft tissues, resulting in temporary hollowback. As long as the hollowback persists, the lumbar region is, in effect, in a permanent state of extension, which results in an increase in the compression loading on the posterior aspects of the vertebrae and discs, and changes the form of loading on the anterior aspects of the vertebrae and discs from compression to tension.

Since hollowback restricts forward bending of the lumbar region (lumbar flexion), there is a tendency to develop a compensatory roundback (round shoulders) in the thoracic region. In effect, the vertebral column in the region of roundback is flexed, increasing the compression loading on the anterior aspects of the vertebrae and discs, and changing the form of loading on the posterior aspects from compression to tension. If prolonged or excessive, the abnormal loading in the hollowback and roundback regions may result in abnormal growth of the affected vertebrae and discs, resulting in permanent hollowback and roundback (Scheuermann's disease). However, in most adolescents the normal shape of the vertebral column and, consequently, normal loading on the vertebrae and discs, is gradually restored as the posterior soft tissues gradually elongate. Nevertheless, during the period of hollowback and roundback the vertebrae and discs, as well as the soft tissue-supporting structures, are more likely to be excessively overloaded by activities that would not normally cause excessive overload. For example, sport training involving repetitive flexion, extension, or rotation of the trunk may result in a variety of injuries such as ruptured discs and fractures of the vertebrae (Micheli 1983).

Another example of a growth-related soft tissue imbalance is genu valgum (knock-knees), which occurs during the adolescent growth spurt. As the bones of the upper and lower legs grow in length, the soft tissues that support and move the knee joints must adapt their lengths to accommodate the increase in bone length. During the adolescent growth spurt the rate of bone growth exceeds that of the soft tissues such that they become tighter, especially when the knees are in the extended position. The tightness is, however, likely to be asymmetric since the fibrous supporting structures on the lateral aspect of the knee (in particular, the iliotibial tract) are thicker and more extensive than those on the medial aspect. Consequently, the lower leg is pulled outward (abduction of the knee joint) resulting in temporary genu valgum. Genu valgum increases the compression load on the articular surfaces of the lateral condyles of the femur and tibia and unloads the medial condyles. In addition, the patella is displaced laterally, increasing the compression load on the lateral articular surfaces of the patella and patellar surface of the femur and unloading the medial articular surfaces.

In most adolescents a growth-spurt-related genu valgum gradually disappears as the lateral soft tissues elongate. However, if the genu valgum is prolonged or excessive, the abnormal loading is likely to result in abnormal growth of the metaphyses or epiphyses of the femur and tibia causing permanent genu valgum or abnormal growth of the articular surfaces between the patella and patellar surfaces of the femur, resulting in recurrent partial or complete lateral dislocation of the patella.

Persistent postural defects including hollowback, roundback, genu valgum, and genu varum have been shown to be associated with increased risk of injury, especially muscle and ligament injuries of the low back and lower limbs, in football and basketball (Shambaugh, Klein, and Herbert 1991; Watson 1995; Winslow and Yoder 1995). It is reasonable to assume that proper conditioning, especially muscle strength and extensibility, may compensate to a certain extent for the increased risk of injury associated with persistent postural defects.

> *Imbalances in strength and extensibility can occur at any age, but some imbalances seem to be growth related. During a period of rapid bone growth, such as an adolescent growth spurt, the rate of growth of bone may outpace that of the associated soft tissues. In these circumstances there may be a significant increase in the tightness of the soft tissues, increasing the risk of injury or disorder.*

Overtraining

Normal daily activities such as walking, carrying a shopping bag, pushing a child's buggy, or climbing a flight of stairs involve a relatively low level of physical effort; the load imposed on the musculoskeletal system by these activities is well within the moderate habitual loading range (see figure 11.2). In sharp contrast, the intensity of effort required to improve cardiorespiratory fitness may excessively overload the musculoskeletal system, especially in a previously sedentary individual. Consequently, the first few weeks of a training program designed to improve the cardiorespiratory fitness of an inactive individual should be mainly concerned with conditioning the musculoskeletal system. This involves a level of effort less than that required to produce a cardiorespiratory training effect.

There have been a number of studies concerning the effects of fitness training on previously sedentary adults. Even in studies in which the amount of exercise was carefully controlled to minimize the risk of injury, the injury rate of the participants was still 50% to 90%, and all of the injuries occurred in the first six weeks of training (Jones, Cowan, and Knapik 1994). In these studies, injury was defined as a conditioning-related musculoskeletal disorder severe enough to necessitate reducing training intensity, changing exercise mode, or discontinuing training for at least a week. It is, therefore, easy to overtrain (or overuse) the musculoskeletal system when it is in a deconditioned state.

However, overtraining is not confined to the previously inactive individual. In a review of injuries to recreational joggers and runners, it was estimated that 60% of all injuries were due to overtraining (Krissoff and Ferris 1979). Even well-trained athletes can suffer from overtraining, especially if they train too frequently or abruptly increase their amount of training (Alfred and Bergfeld 1987). For example, in a study of knee injuries in elite volleyball players, the incidence of injury was found to be directly related to the frequency of play (including training sessions and games); the incidence of injury in those who played once or twice a week was 3.3%, whereas the incidence of injury in those who played more than four times a week was 41.8% (Ferretti et al. 1984) (figure 12.2). In ballet and all forms of modern dance the major sites of injury are the knee, ankle, and lower back, and the most frequently reported cause of injury is overtraining due to long practice sessions (Solomon and Micheli 1986; Reid 1987).

Skeletal Abnormalities

Abnormalities in the length or shape of bones can cause excessive overloading of the musculoskeletal system by causing compensatory movement in associated joints, for example, as with leg length inequality, femoral anteversion, and tibial varum.

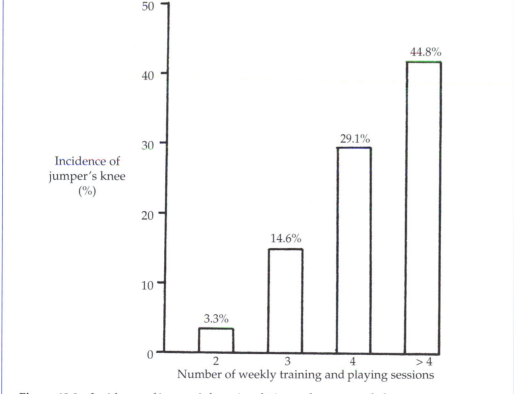

Figure 12.2   Incidence of jumper's knee in relation to frequency of play.
From Ferretti et al., 1984, "Jumper's knee," *Physician and Sportsmedicine* 12:98. Adapted with permission of McGraw-Hill.

Leg Length Inequality   Ideally, the right and left sides of the body should be equally loaded during standing, walking, and other daily activities. However, most individuals have some degree of leg length inequality (LLI), which, unless compensated, results in abnormal loading of the hips, pelvis, and lower back (Gibson, Papaigannou, and Kenwright 1983). In most individuals LLI is less than 5 mm. This discrepancy can be compensated for easily by making minor adjustments in the movement of the feet to effectively shorten a long leg (via pronation of the foot) and lengthen a short leg (via supination of the foot) to minimize the LLI and bring about normal loading (Gibson, Papaigannou, and Kenwright 1983). However, when the degree of LLI is too large to be completely compensated by foot movements, the pelvis invariably tilts downward on the shorter side when standing and walking (figure 12.3). The pelvic tilt considerably reduces the load-bearing articular surface area in the hip joint of the longer leg, which considerably increases the compression stress on the articular surfaces. To compensate for the abnormally tilted pelvis, the lower part of the vertebral column bends toward the longer leg so that the trunk can remain fairly upright.

The compensatory lateral bending of the lower part of the vertebral column is called functional scoliosis. The functional scoliosis is brought about by asymmetric activity in the muscles on the left and right sides of the lower back. The degree of asymmetry in muscle activity depends on the extent of the LLI; it has been shown that an LLI as small as 10 mm leads to a considerable asymmetric increase in the activity of several muscle groups, making it impossible for the individual to maintain a relaxed standing position. The increased muscular activity causes greater energy expenditure and increased fatigue (Friberg 1983; Blustein and D'Amico 1985). The functional scoliosis not only bends the lower back toward the longer leg, but also rotates the lower back about a vertical axis. The combination of lateral bending and axial rotation results in the intervertebral discs being compressed on the concave side of the column, stretched on the other side, and twisted about a vertical axis. This is thought to be the pattern of loading most likely to damage the intervertebral discs (Friberg 1983). Furthermore, this pattern of loading causes the discs to bulge outward at the rear on the long leg side with the likelihood of impingement of the adjacent spinal nerves. The impingement of spinal nerves may result in pain such as sciatica and, subsequently, muscle spasm and more pain.

The most common symptom associated with LLI is low-back pain (Giles and Taylor 1981). The extent of abnormal loading on low-back muscles, intervertebral discs, and hip joint articular surfaces, which result from LLI and the subsequent functional scoliosis, depends on the extent of LLI. However, the extent of abnormal loading also depends on the type of movement being performed; the more dynamic the activity, the greater the load. For example, a relatively small LLI may not result in excessive loading when walking, but may result in excessive loading when running due to the greater muscle and joint reaction forces.

The long-term effects of LLI on the musculoskeletal system depend on the age of the individual at the time the LLI occurs. If LLI occurs during childhood the initial functional scoliosis becomes a structural scoliosis; the vertebrae, pelvic bones, and upper ends of both femurs (and their supporting structures) adapt by chondral modeling to create functionally normal joints at the hips, within the pelvis, and between the vertebrae (Frost 1973). Wedge-shaped vertebrae (Scheuermann's disease) are a well-recognized feature of structural scoliosis. Even though a structural scoliosis minimizes the effects of LLI, the loading imposed by dynamic activities is more likely to become excessive in a scoliotic

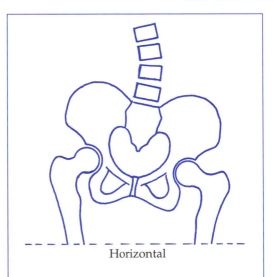

Figure 12.3   Effect of leg length inequality on the orientation of the articular surfaces in the hip joints.

Horizontal

individual than in an individual without scoliosis. LLI occurring after maturity, or to be more precise, becoming excessive after maturity, is usually the result of fracture of a femur or tibia (Gibson, Papaigannou, and Kenwright 1983). Since the bones are no longer able to model in response to the abnormal loads imposed by the LLI, the long-term effects may be degenerative joint disease, especially in the hip joint of the longer leg, and damage to intervertebral discs causing spinal nerve impingement and muscle spasm.

Femoral Anteversion   Turnout—the ability to turn both feet out so that they are in line with each other—is a fundamental posture in ballet and other forms of theatrical dance (Ende and Wickstrom 1982; Hardaker, Margello, and Goldner 1985) (figure 12.4a and b). In normal upright posture most people have about 10° to 20° of out-toeing so that the feet have only to be turned out through another 70° to 80° in order to achieve a full 90° turnout (figure 12.4, a and b). Since the knee and ankle joints are not designed to rotate about a vertical axis, turnout of each foot should, ideally, be

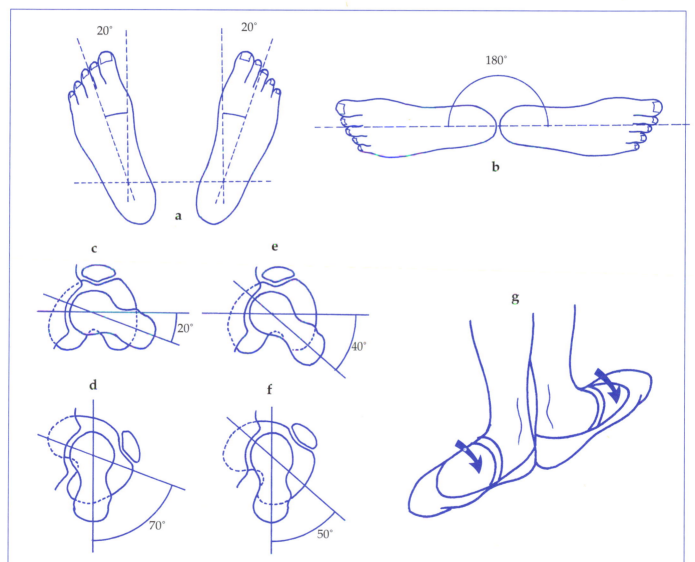

Figure 12.4   Effect of femoral anteversion on turnout; (*a*) natural out-toeing in normal upright standing posture; (*b*) full turnout; (*c*) normal upright posture with femoral anteversion of right leg of 20°; (*d*) with femoral anteversion of 20°, outward rotation of the hip is limited to approximately 70°; (*e*) normal upright posture with femoral anteversion of right leg of 40°; (*f*) with femoral anteversion of 40°, outward rotation of the hip is limited to approximately 50°; (*g*) rolling in.

achieved entirely by outward rotation of the whole leg about the hip joint. However, the normal range of outward rotation of the hip is 60° to 70° (Hardaker, Margello, and Goldner 1985). It follows, therefore, that most dancers can only achieve full turnout by twisting the knee and ankle joints outward (up to 20°). Whereas the knee and ankle joints may be able to tolerate small amounts of twisting, excessive twisting may cause excessive overloading of the articular surfaces and supporting structures of these joints. This situation is likely to arise if outward rotation of the hip is restricted.

Movement in the hip, as in other joints, may be restricted by tight ligaments or bony abnormalities. For example, the range of outward rotation of the hip joint is largely determined by the orientation of the neck of the femur to the shaft of the femur. In most people the neck of the femur is directed slightly backward from the hip joint such that it makes an angle of about 20° with the frontal plane (figure 12.4c). This angle is called the angle of femoral anteversion. The smaller the anteversion angle, the greater the amount of outward rotation of the hip, and vice versa.

An anteversion angle of about 20° allows 60° to 70° of natural rotation of the hip (figure 12.4, c and d). However, an anteversion angle of 40° limits outward rotation of the hip to 40° to 50° (figure 12.4, e and f). In this case the only way full turnout can be achieved is by forcibly twisting the knees and ankles outward, which will sublux the tibiofemoral and patellofemoral joints and may excessively overload parts of the articular surfaces and supporting structures. Twisting the foot (forced eversion) will sublux the ankle and subtalar joints and may excessively overload parts of the articular surfaces as well as the arch support structures; this movement of the foot is called rolling-in (figure 12.4g).

Tibial Varum   Downhill skiing is essentially a series of linked turns used to control the rate of descent and direction of travel (Matheson and Macintyre 1987). The skill of turning depends largely on the skier's ability to control the degree of edging of the skis—the angle of the skis to the surface of the snow—by varying degrees of pronation, dorsiflexion, knee and hip flexion, and torso counterrotation. The skier's ability to perform these movements effectively and efficiently depends considerably on the degree to which the ski boots accommodate the natural alignment of the lower legs and feet. In normal upright standing posture most people's lower legs are vertical with the feet flat on the floor and the ankle and subtalar joints in a neutral position (figure 12.5a). Most ski boots are designed to accommodate this type of alignment (figure 12.5b). However, there is considerable individual variation in the alignment of the lower legs and feet, especially in the degree of tibial varum—when the lower leg is angled medially from the knee (figure 12.5c). In many ski boots the angle between the cuff and the boot is adjustable. However, if the boot is not adjustable, it may not properly accommodate a marked tibial varum such that in upright standing posture the weight of the skier is transmitted to the floor via the lateral edges of the boots (figure 12.5d). When skiing in these boots the skier may find it difficult to keep the skis flat on snow when traversing, which will predispose him to falling as a result of catching the outside edges of the skis. The skier may also find it difficult to press the medial edges of the skis into the snow—essential for turning. To try to keep the skis flat and medially edge the skis, the skier is likely to abduct the knees or adopt a wide-tracked stance. These movements may be strenuous and result in rapid fatigue with a consequent increase in the risk of injury.

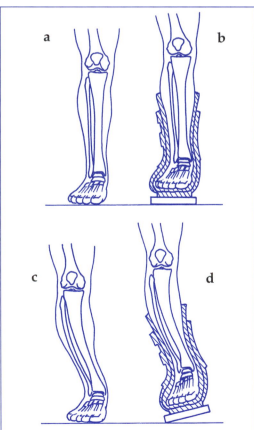

Figure 12.5   The inability of a nonadjustable ski boot to accommodate a marked tibial varum.

Skeletal abnormalities can cause excessive overloading of the musculoskeletal system by causing compensatory movement in associated joints.

Technique

In sports and physically active leisure pursuits there are basically two types of movements:

1. Those that involve generating or increasing speed of movement of the whole body (in running and jumping activities) and of an arm or leg to propel an implement (in throwing and striking activities)

2. Those that involve decreasing speed of movement of the whole body (in landing from a jump) and of an implement (in catching a ball)

The movement pattern—the way the body segments move in relation to each other—of an individual during the performance of a particular movement or sequence of movements is usually referred to as the individual's technique. Good technique is a movement pattern that is not only effective in terms of performance (the desired outcome such as height jumped, sprinting speed, accuracy of a tennis serve, or number of points scored in gymnastics) but one that also minimizes the risk of injury by appropriate distribution of the overall load throughout the kinetic chain. In terms of performance, the quality of technique is far more important in some sports than in others. For example, in gymnastics, diving, and figure skating, the quality of the technique is the performance criterion (Watkins 1987a). In other sports, such as track and field athletics and swimming, the performance criterion is the end result of a particular technique, for example, the time to run 100 m (Watkins 1987b). However, in terms of minimizing the risk of injury, good technique is essential, irrespective of how performance is actually judged. An individual is unlikely to achieve full potential in a sport if she can occasionally perform well, but is frequently injured due to a poor technique that excessively overloads certain muscles and joints. Inappropriate utilization and summation of muscular effort and abnormal joint movements characterize poor technique, result in localized overload, and, thus, increase risk of injury.

Inappropriate Utilization and Summation of Muscular Effort   To generate maximum speed in the body as a whole or in a body part it is necessary to involve as much of the body's musculature as possible; the more muscles and, therefore, joints used, the greater the force application due to increased range of movement. In propulsive actions, the impulse of the thrust is reduced if some muscles (and joints) that could contribute to the action are not used, or not used maximally. Furthermore, some joints in the kinetic chain may be excessively overloaded in an effort to compensate for the lack of involvement of the underused joints. For example, in the volleyball spiking action, the objective is to hit the ball downward into the opponent's court. The speed of the ball on leaving the hand depends on the force applied to the ball and the time of force application during ball contact. To apply a large force to the ball the hand must be moving as quickly as possible on impact. Ideally—with good technique—hand speed is generated largely by the hip and trunk flexor muscles, with little movement in the shoulder (figure 12.6, a and b). In other words, the large hip and trunk flexor muscles are the sources of power in the spiking action. This minimizes the load on the shoulder and arm muscles, and enables the arm to exert control over the hand movement prior to ball contact. Lack of involvement of the hip and trunk muscles is usually compensated by excessive shoulder activity, which, with repeated use, can cause a variety of disorders of the shoulder joint and its supporting structures (figure 12.6, c and d).

Figure 12.6   Spiking actions in volleyball; (*a* and *b*) good technique—emphasis is on hip and trunk flexion; (*c* and *d*) poor technique—emphasis is on shoulder flexion.

A tennis serve is similar to the volleyball spiking action in that the power of the movement is largely produced by the hips and trunk (as well as the legs), rather than the shoulder and arm muscles, which are responsible for controlling the racket. Forehand and backhand strokes in tennis are both similar to the spike and tennis serve in the way speed of movement is generated in the racket head (hips and trunk) and the way the racket head is controlled (shoulder and arm muscles). Tennis elbow refers to a range of disorders of the elbow and its supporting structures that occur as a result of lack of involvement of the hips or the trunk. To compensate and generate normal racket head speed, the shoulder and arm muscles have to exert large forces that excessively overload the insertions of muscles and ligaments, especially those around the elbow. Repeated use leads to progressive microtrauma and tennis elbow syndrome (Groppel 1986).

The previous examples from volleyball and tennis demonstrate the importance of the legs and trunk in movements involving the development of large forces. This is not surprising since the largest and, therefore, potentially strongest muscles in the body are those of the legs and trunk. Because the speed of movement resulting from a propulsive action depends on the amount of force produced throughout the whole action, it is important to involve as many muscle groups as possible, not just the strongest groups. This is achieved by sequential and overlapping involvement of the various muscle groups starting with the strongest muscles and finishing with the weaker muscles. In this way the summation of force throughout the kinetic chain is maximized, resulting in maximum speed of movement (Hopper 1973). This mode of operation can be compared to a multistage rocket in which the first stage (most powerful stage) is responsible for initiating upward movement of the rocket. The thrust produced by the first stage decreases as its fuel starts to run out, so the first stage is eventually replaced by the second stage (next most powerful stage) to maintain upward thrust on the rocket. As the thrust produced by the second stage decreases, the third stage takes over and so on through the remaining stages. In a human kinetic chain there is considerable overlap between the stages, that is, two or more muscle groups are likely to be contributing to propulsion simultaneously.

Utilizing and summating muscular effort is as important in decreasing the speed of movement of the body or an implement such as a cricket ball as it is in generating or increasing speed of movement. The greater the deceleration of the body (or implement), the greater the load on the body, and, therefore, the greater the risk of injury, especially to bones and joints. For example, actions such as landing from vertical jump with the legs too straight, plant-and-cut (sudden deceleration on a straight or near straight leg with intent to change direction in one step), and one-step-stop (attempting to stop and reverse direction in one step), which occur in basketball are major risk factors for anterior cruciate ligament injury (Henning et al. 1994). Controlling joint flexion can reduce deceleration in these movements and, consequently, reduce the risk of injury.

Abnormal Joint Movements   Abnormal joint movements occur when a joint is forced to do the following:

1. Move about axes abnormal for the joint. For example, abduction of the knee joint during the leg kick in breaststroke may result in breaststroker's knee.

2. Move beyond the normal maximum range of movement about a normal axis. For example, hyperextension of the knee joint resulting from a tackle in football can damage the knee ligaments, and combined adduction and internal rotation of the shoulder joint can result in shoulder impingement in a swimmer or thrower.

- **Breaststroker's knee.** The incidence of knee pain in swimmers who specialize in breaststroke is high and has given rise to the term breaststroker's knee (Vizsolyi et al. 1987; Johnson et al. 1987). In the early stages of its development the condition is characterized by pain and tenderness in the supporting structures on the medial side of the knee. Later, the front and lateral side of the knee also become involved, resulting in generalized knee pain.

Breaststroker's knee is caused by a particular kind of kicking action called the whip kick (figure 12.7). At the end of the recovery phase, as the heels approach the buttocks, the knees separate (abduction of the hips), the lower legs turn outward, and the ankles dorsiflex to present as large an area as possible to the water at the start of the kicking phase (figure 12.7a). The whip kick is initiated by inward rotation of the hips, which rotates the lower legs farther outward. This is closely followed by the propulsion phase involving simultaneous hip extension, hip adduction, and knee extension. These actions drive the feet backward along semicircular paths. During this action the lower legs are forcibly abducted and outwardly rotated at the knees, largely by the pressure of water against the lower legs and feet. Abduction and outward rotation of the lower legs at the knees are abnormal movements and result in the symptoms described above. The load on the knees is greatest at the start of the propulsion phase when the knees experience a whiplike action, and during the middle phase of the kick when the lower legs present a large surface area to the water (figure 12.7b). By reducing the amount of hip abduction and internal rotation of the hips at the start of the propulsion phase the feet move directly backward rather than in a semicircular path, which considerably reduces the thrust of the leg kick, but also reduces the amount of forced abduction and outward rotation of the knees.

The abnormal pattern of loading on the knees caused by the whip kick is similar to that produced by "screwing the knee" in ballet dancing. This action is caused by inadequate outward rotation of the hips (see earlier section on femoral anteversion and turnout) and occurs when the knees are straightened from a position in which the knees are bent and the feet fully turned out (Ende and Wickstrom 1982). The situation may also arise in the propulsion phase of throwing a javelin; if the foot of the rear leg is turned outward as the hip and knee joints extend, the knee will be forcibly abducted and outwardly rotated (figure 12.8).

- **Shoulder impingement syndrome.** Overhead movements of the arms in swimming, throwing, and racket games are brought about by movements in three joints: the shoulder,

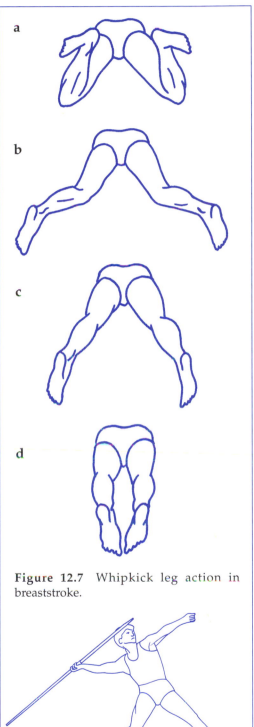

Figure 12.7 Whipkick leg action in breaststroke.

Figure 12.8 Rear leg action in throwing the javelin.

sternoclavicular, and acromioclavicular joints (as described in chapter 7). In abducting the arm directly overhead, 180° from the resting position, the shoulder, sternoclavicular, and acromioclavicular joints contribute approximately 120°, 40°, and 20°, respectively (Peat 1986). Consequently, when range of movement in the sternoclavicular and acromioclavicular joints is restricted, the shoulder joint must be hyperabducted to achieve full abduction of the arm. In doing so, the supporting structures of the shoulder joint come into contact with the coracoacromial arch, resulting in a range of painful disorders referred to as shoulder impingement syndrome (Miniaci and Fowler 1993).

Even with normal ranges of movement in the shoulder, sternoclavicular, and acromioclavicular joints, impingement between the shoulder joint structures and the coracoacromial arch is still likely to occur when shoulder abduction is accompanied by internal rotation. This situation can arise during the arm recovery phase in butterfly and freestyle swimming due to a high elbow position (shoulder abduction) and the hand lagging behind the elbow (internal rotation of shoulder) (figure 12.9a). The likelihood of impingement can be reduced by leading with the hand (external rotation of shoulder; figure 12.9b) and increasing body roll (decreases shoulder abduction; figure 12.9, c and d) during arm recovery (Penny and Smith 1980).

Technique refers to the movement pattern of an individual during a particular movement or sequence of movements. Good technique is a movement pattern not only effective in performance, but also one that minimizes the risk of injury by appropriately distributing the overall load throughout the kinetic chain. Poor technique is characterized by inappropriate utilization and summation of muscular effort and abnormal joint movements, both of which result in localized overload and, therefore, increased risk of injury.

Warm-Up

Warm-up refers to preliminary exercise that prepares the body physically and mentally for the more strenuous activity that is to follow (Watkins 1982). The rationale underlying warm-up is that the cardiorespiratory and musculoskeletal systems need a certain amount of time to fully respond to an increase in the level of physical activity. A gradual increase in the intensity of effort enables the cardiorespiratory system to gradually increase the supply of blood, and, therefore, oxygen, to the heart and muscles. The increase in muscle temperature accompanying the increased level of muscular activity decreases the viscosity of the muscles, which increases their mechanical efficiency. The joints (synovial and symphysis) also benefit from a warm-up by becoming more stable due to an increase in the thickness of cartilage as a result of an increase in the circulation of fluid through the cartilage.

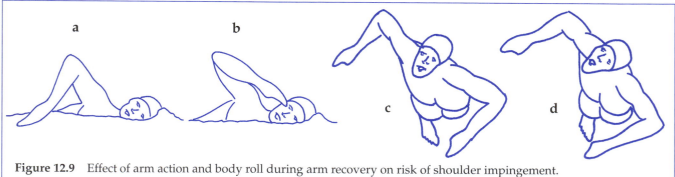

Figure 12.9   Effect of arm action and body roll during arm recovery on risk of shoulder impingement.

For individuals engaging in physically active leisure, a 10- to 15-minute warm-up consisting of light to moderate whole-body exercise, such as jogging and stretching the joints, is recommended (Wilmore and Costill 1994).

> *Warm-up is preliminary exercise that prepares the body both physically and mentally for the more strenuous activity to follow. A warm-up can decrease the risk of injury.*

Psychological Factors

Psychological factors can be described as either personality variables or psychosocial variables (Kerr and Fowler 1988). Personality variables refer to fairly stable characteristics such as locus of control and level of self esteem. Psychosocial variables refer to the effect that changes in the personal and social life of the individual may have on the response of the individual in a given situation. In the context of sport there is good evidence that psychological stress affects attentional and physiological responses in the form of, for example, above normal tension in muscles, defects in fine motor control, distraction, and reduction in peripheral vision, all of which may make the individual more prone to injury (Petrie 1993; Kelley 1990).

> *There is good evidence that psychological stress affects attentional and physiological responses, which may make the individual more prone to injury.*

External Risk Factors

In the sense that the size of the internal loads on the musculoskeletal system reflects the size of the external loads acting on the body it is, perhaps, not surprising that many studies of injury risk factors associated with particular sports and physical activity have focused on extrinsic factors, that is, factors concerning environmental conditions and the administration of the activity (Lysens et al. 1984).

Environmental Conditions

The main environmental risk factors appear to be clothing, footwear, playing surfaces, and orthoses.

Clothing, Footwear, and Playing Surfaces Protective clothing designed to reduce impact loading is a feature of many sports. These include, for example, helmets (e.g., football, hockey, skiing, horse riding, rock climbing, cricket), headguards (amateur boxing, rugby), shoulder pads (football), elbow pads (volleyball), gloves (boxing, cricket), torso padding (hockey), knee pads (volleyball), and leg pads (hockey, soccer, cricket).

Footwear is important in many sports, not just for protection but also to enhance performance. With regard to protection, developments in footwear have been concerned with reducing impact loading in activities such as running and those involving high-frequency jumping and landing such as basketball and volleyball, and providing support for the ankle to reduce the risk of ankle sprains. Softer-soled shoes reduce impact force peak and the rate of loading (chapter 10). Ankle sprains, especially inversion sprains, are common in sport and can cause significant, chronic disability (Smith and Reischl 1986; Boruta et al. 1990). In recent years a lot of research has been carried out regarding the effect of high-top shoes on the risk of inversion sprains. Whereas high-top shoes appear to increase ankle stability in laboratory tests, it is not clear whether they reduce the risk of ankle sprain in normal training and playing conditions (Barrett and Bilisko 1995).

With regard to playing surfaces, there are two main mechanical characteristics that influence the risk of injury: surface hardness and frictional properties (Nigg and Segesser 1988). Just as softer-soled shoes reduce impact force peak and rate of loading compared to harder-soled shoes, softer playing surfaces reduce impact force peak and rate of loading compared to harder surfaces. Ferretti et al. (1984) and Watkins and Green (1992) found a clear positive relationship between the hardness of the playing surface and the incidence of injuries in volleyball. Evidence regarding the effects of friction on injury risk is limited, but the frequency of football and tennis injuries is reported to be higher on high-friction surfaces than on lower-friction surfaces (Nigg and Segesser 1988). The risk of injury increases when friction is too low because players slip too much. The risk of injury also increases when friction is too high since the foot experiences high deceleration at foot strike, resulting in high muscle forces and joint reaction forces at the knee and ankle, and, consequently, a higher risk of injury.

> Clothing, footwear, and playing surfaces have a major influence on injury risk. Developments in footwear have been concerned with reducing impact loading and providing support for the ankle to reduce the risk of ankle sprains. Playing surfaces with moderate levels of hardness and friction are associated with reduced risk of injury.

Orthoses   Orthoses are used in sport and nonsport activities to assist the joint-supporting structures in maintaining normal joint transmission of loads by reducing the risk of abnormal movements and restricting the amount of abnormal movement when it occurs. Not surprisingly, most research on sports orthoses has been concerned with knee and ankle orthoses, reflecting the relatively high incidence of chronically debilitating sports-related knee and ankle injuries. There are basically two types of sports orthoses: those that support joints directly and those that support joints indirectly. Orthoses providing direct support range from well-established taping and strapping methods used to support the ankle, knee, and elbow (Anderson, Woiyts, and Lambert 1992) to jointed semirigid appliances such as knee braces (Zachazewski and Geissler 1992). Orthoses providing indirect support commonly come in the form of forefoot or rearfoot valgus or varus wedges, which are used to maintain the correct alignment of the forefoot or rearfoot to the ankle and subtalar joints to maintain normal pronation and supination (McKenzie, Clement, and Taunton 1985; Matheson and Macintyre 1987; Petrov et al. 1988).

Administration

The rules of competition and the extent to which the rules are enforced—the standard of refereeing—clearly affect the risk of injury to participants in any given sport. The rules of competition are normally designed to ensure fair play and safe competition (Sperryn 1988), and, as such, should reflect the overall risk of injury in a particular sport or activity. Consequently, rule changes may be necessary to reduce the risk of injury to an acceptable level. For example, in American football, after it became clear that spearing—the action of a player driving his head (protected by a helmet) into an opponent's body—was associated with a high incidence of severe neck injuries, the practice of spearing was banned in 1975. Over the following nine years the frequency of injuries in the National Football League (NFL) resulting in permanent cervical quadriplegia decreased from 28 per season to 5 (Torg et al. 1985).

Summary

This chapter described the risk factors that influence the level of load exerted on the musculoskeletal system. There are three main groups of risk factors—movement related factors, intrinsic factors, and extrinsic factors. Excessive overload is likely to be the result of a complex interaction of factors from the three groups. To determine the cause of excessive overload in a particular situation it is necessary to identify the relevant risk factors and discern their interaction.

Review Questions

1. Differentiate causes, signs, and symptoms of injury.
2. Identify and classify the main intrinsic risk factors.
3. Identify and classify the main extrinsic risk factors.
4. Describe how strength and extensibility imbalances may occur.
5. Differentiate between absolute and relative strength balance.
6. Describe what is meant by good and bad technique in sports.
7. Differentiate between personality and psychosocial risk factors.
8. Describe the two main types of orthoses used in sports.

References

Adams, M.A. and W.C. Hutton. 1980. The effect of posture on the role of the apophysial joints in resisting intervertebral compressive forces. *Journal of Bone and Joint Surgery* 62B (3):358–62.

———. 1982. Prolapsed intervertebral disc: A hyperflexion injury. *Spine* 7 (3):184–91.

———. 1983. The mechanical function of the lumbar apophysial joints. *Spine* 8 (3) 327–30.

———. 1985. The effect of posture on the lumbar spine. *Journal of Bone and Joint Surgery* 67B (4):625–29.

Aglietti, P., J.L. Insall, and G. Cerulli. 1983. Patellar pain and incongruence. *Clinical Orthopaedics* 176:217–24.

Akeson, W.H., D. Amiel, G.L. Mechanic, S.L.Y. Woo, F.L. Harwood, and M.L. Hamer. 1977. Collagen cross-linking alterations in joint contractures: Changes in the reducible cross-links in periarticular connective tissue collagen after nine weeks of immobilization. *Connective Tissue Research* 5:15–19.

Akeson, W.H., C.B. Frank, D. Amiel, and S.L. Woo. 1985. Ligament biology and biomechanics. In *Symposium on sports medicine: The knee*, ed. G. Finerman. St Louis: Mosby. 111-15

Alexander, R. McN. 1968. *Animal mechanics*. London: Sidgwick and Jackson.

———. 1975. *Biomechanics*. London: Chapman and Hall.

———. 1987. The spring in your step. *New Scientist* 114 (1558):42–44.

———. 1989. Muscles for the job. *New Scientist* 122 (1660):50–53.

———. 1992. *The human machine*. London: Natural History Museum Publications.

Alfred, R.H., and J.A. Bergfeld. 1987. Diagnosis and management of stress fractures of the foot. *The Physician and Sportsmedicine* 15 (8):83–85, 88–89.

An, K-N., F.C. Hui, B.F. Morrey, R.L. Linscheid, and E.Y. Chao. 1981. Muscles across the elbow joint: A biomechanical analysis. *Journal of Biomechanics* 14 (10):659–69.

Anderson, F.C., and M.G. Pandy. 1993. Storage and utilization of elastic strain energy during jumping. *Journal of Biomechanics* 26 (12):1413–27.

Anderson, K., E.M. Woiyts, and P.V. Lambert. 1992. A biomechanical evaluation of taping in reducing knee joint translation and rotation. *American Journal of Sports Medicine* 20 (4):416–21.

Anderson, S.M., and B.E. Nilsson. 1979. Changes in bone mineral content following ligamentous knee injuries. *Medicine and Science in Sports* 11:351–53.

Arendt, E. 1996. Common musculoskeletal injuries in women. *The Physician and Sportsmedicine* 24 (7):39–42, 45-48.

Arendt, E., and R. Dick. 1995. Knee injury patterns among men and women in collegiate basketball and soccer. *American Journal of Sports Medicine* 23 (6):694–700.

Ascenzi, A., and G.H. Bell. 1971. Bone as a mechanical engineering problem. In *The biochemistry and physiology of bone*, ed. G. H. Boume. New York: Academic Press.

Aspden, R.M. 1987. Intra-abdominal pressure and its role in spinal mechanics. *Clinical Biomechanics* 2:168–74.

Bach, D.K., D.S. Green, G.M. Jensen, and E. Savinar. 1985. A comparison of muscular tightness in runners and nonrunners and the relation of muscular tightness to low back pain in runners. *Journal of Orthopaedic and Sports Physical Therapy* 6:315–23.

Backx, F.J.G., H.J.M. Berger, E. Bol, and W.B.M. Erich. 1991. Injuries in high-risk persons and high-risk sports: A longitudinal study of 1818 schoolchildren. *American Journal of Sports Medicine* 19 (2):124–30.

Bailey, D.A. 1995. The role of mechanical loading in the regulation of skeletal development during growth. In *New horizons in pediatric exercise science*, eds. C.J.R. Blimkie and O. Bar-Or, 97–108. Champaign, IL: Human Kinetics.

Bailey, D.A., A.D. Martin, C.S. Houston, and J.L. Howie. 1986. Physical activity, nutrition, bone density and osteoporosis. *Australian Journal of Science and Medicine in Sport* 18 (3): 3–8.

Baker, B., A.C. Peckham, F. Pupparo, and J.C. Sanborn. 1985. Review of meniscal injury and associated sports. *American Journal of Sports Medicine* 13 (1):1–4.

Barrett, J., and T. Bilisko. 1995. The role of shoes in the prevention of ankle sprains. *Sports Medicine* 20 (4):277–80.

Basmajian, J.V. 1965. Man's posture. *Archives of Physical Medicine and Rehabilitation* 46 (1):26–35.

———. 1970. *Primary anatomy*. Baltimore: Williams & Wilkins.

Bauer, G. 1960. Epidemiology of fracture in aged persons. *Clinical Orthopedics* 17:219–25.

Beard, D.J., C.A.F. Dodd, H.R. Trundle, and A.H.R.W. Simpson. 1994. Proprioception enhancement for anterior cruciate ligament deficiency. *Journal of Bone and Joint Surgery* 76B: 654–59.

Blair, S.N., H.W. Kohl, and N.F. Gordon. 1992. How much physical activity is good for health? *Annual Review of Public Health* 13:99–126.

Blazina, M.E., R.K. Karlan, and F.W. Jobe. 1973. Jumper's knee. *Orthopedic Clinics of North America* 4:665–73.

Blustein, S.M., and J.C. D'Amico. 1985. Limb length discrepancy: Identification, clinical significance, and management. *Journal of the American Podiatric Medical Association* 75 (4):200–206.

Bobbert, M.F., H.C. Schamhardt, and B.M. Nigg. 1991. Calculation of vertical ground reaction force estimates during running from positional data. *Journal of Biomechanics* 24:1095–105.

Booth, F.W. 1987. Physiologic and biochemical effects of immobilisation on muscle. *Clinical Orthopaedics and Related Research* 219:15–20.

Boruta, P.M., J.O. Bishop, G. Braly, and H.S. Tullos. 1990. Acute lateral ankle ligament injuries: A literature review. *Foot and Ankle* 11:107–13.

Bosco, C., P. Luhtanen, and P.V. Komi. 1983. A simple method for measurement of mechanical power in jumping. *European Journal of Applied Physiology* 50:273–82.

Bose, K., R. Kanagasuntheram, and M.B.H. Hosman. 1980. Vastus medialis: An anatomic and physiological study. *Orthopedics* 3:880–83.

Bowen, V., and J.D. Cassidy. 1981. Macroscopic and microscopic anatomy of the sacroiliac joint from embryonic life until eighth decade. *Spine* 6 (6):620–27.

Bradford, D.S., J.H. Moe, F.J. Montalvo, and R.B. Winter. 1974. Scheuermann's kyphosis and roundback deformity. *Journal of Bone and Joint Surgery* 56A (4):740–58.

Brighton, C.T., M.J. Katz, S.R. Goll, and C.E. Nicholls. 1985. Prevention and treatment of sciatic denervation disuse osteoporosis in a rat tibia with capacitatively coupled electrical stimulation. *Bone* 6:87–97.

Brinckmann, P. 1985. Pathology of the vertebral column. *Ergonomics* 28 (1):77–80.

Brody, D.M. 1980. Running injuries. *Clinical Symposia* 32 (4):1–36.

Brunet, M.E., R.J. Haddad, and E.B. Porche. 1982. Rotator cuff impingement syndrome in sports. *The Physician and Sportsmedicine* 10 (12):86–97.

Bullock-Saxton, J. 1988. Normal and abnormal postures in the sagittal plane and their relationship to low back pain. *Physiotherapy Practice* 4 (2):94–104.

Bunnell, W.P. 1986. The natural history of idiopathic scoliosis before skeletal maturity. *Spine* 11 (8):773–76.

Burstein, A.H., and T.M. Wright. 1994. *Fundamentals of orthopedic biomechanics*. Baltimore: Williams & Wilkins.

Callaghan, M.J., and J.A. Oldham. 1996. The role of quadriceps exercise in the treatment of patellofemoral pain syndrome. *Sports Medicine* 21 (5):384–91.

Caplan, A.I. 1984. Cartilage. *Scientific American* 251 (4):82–90.

Carter, D.R., M. Wong, and T.E. Orr. 1991. Musculoskeletal ontogeny, phylogeny, and functional adaptation. *Journal of Biomechanics* 24S1:3–16.

Cash, J.D., and J.C. Hughston. 1988. Treatment of acute patellar dislocation. *American Journal of Sports Medicine* 16 (3):244–48.

Casscells, W. 1982. Chondromalacia of the patella. *Journal of Pediatric Orthopaedics* 2 (5):560–64.

Castle, F. 1969. *Five-figure logarithmic and other tables.* London: Macmillan.

Cavanagh, P.R. 1980. *The running shoe book.* Mountain View, CA: Anderson World.

Cavanagh, P.R., and M.A. Lafortune. 1980. Ground reaction forces in distance running. *Journal of Biomechanics* 13:397–406.

Cavanagh, P.R., G.A. Valiant, and K.W. Misevich. 1984. Biological aspects of modeling shoe/foot interaction during running. In *Sport shoes and playing surfaces,* ed. E.C. Frederick, 24-46. Champaign, IL: Human Kinetics.

Chalmers, J., and K.C. Ho. 1970. Geographical variations in senile osteoporosis: The association with osteoporosis. *Journal of Bone and Joint Surgery* 52:667–75.

Clarkson, P.M., and I. Tremblay. 1988. Exercise-induced muscle damage, repair, and adaptation in humans. *Journal of Applied Physiology* 65 (1):1–6.

Clement, D.B., J.E. Taunton, G.W. Smart, and K.L. McNicol. 1981. A survey of overuse running injuries. *The Physician and Sportsmedicine* 9 (5):47–58.

Cohen, L.A., and M.L. Cohen. 1956. Arthrokinetic reflex of the knee. *American Journal of Physiology* 184:433–37.

Colachis, S.C., R.E. Worden, C.O. Bechtol, and B.R. Strohm. 1963. Movement of the sacroiliac joint in the adult male: A preliminary report. *Archives of Physical Medicine and Rehabilitation* 44:490–98.

Colliton, J. 1996. Back pain and pregnancy. *The Physician and Sportsmedicine* 24 (7):89–93.

Combs, J.A. 1994. Hip and pelvis avulsion fractures in adolescents. *The Physician and Sportsmedicine* 22 (7):41–44, 47–49.

Copeland, S. 1993. Throwing injuries of the shoulder. *British Journal of Sports Medicine* 27 (4):221–27.

Dempster, W.T. 1955. Space requirements of the seated operator. *WADC Technical Report 55159.* Wright-Patterson Air Force Base, Ohio: Wright Air Development Center.

———. 1965. Mechanisms of shoulder movement. *Archives of Physical Medicine and Rehabilitation* 46 (1):49–70.

Department of Health. 1992. *The health of the nation: A strategy for health in England.* London: Her Majesty's Stationery Office.

Dickenson, R.P., W.C. Hutton, and J.R.R. Stott. 1981. The mechanical properties of bone in osteoporosis. *Journal of Bone and Joint Surgery* 63B (2):233–38.

Dickinson, J.A., S.D. Cook, and T.M. Leinhardt. 1985. The measurement of shock waves following heel strike while running. *Journal of Biomechanics* 18 (6):415–22.

Don Tigny, R.L. 1985. Function and pathomechanics of the sacroiliac joint: A review. *Physical Therapy* 65 (1):35–44.

Donaldson, C., S. Hulley, J. Vogel, R. Hattner, J. Bayers, and D. McMillan. 1970. Effect of prolonged bed rest on bone mineral. *Metabolism* 19:1071–1084.

Drummond, D.S. 1987. Kyphosis in the growing child. *Spine: State of the Art Reviews* 1 (2):339–56.

Dunlop, R.B., M.A. Adams, and W.C. Hutton. 1984. Disc space narrowing and the lumbar facet joints. *Journal of Bone and Joint Surgery* 66B (5): 706–10.

During, J., H. Goudfrooij, W. Keessen, T.W. Beeker, and A. Crowe. 1985. Towards standards for posture: Postural characteristics of the lower back system in normal and pathologic conditions. *Spine* 10 (1): 83–87.

Dwyer, A.P. 1987. Backache and its prevention. *Clinical Orthopaedics and Related Research* 222: 5–43.

Edington, J., E.C. Frederick, and P.R. Cavanagh. 1990. Rearfoot motion in distance running. In *Biomechanics of distance running,* ed. P.R. Cavanagh, 135-164. Champaign, IL: Human Kinetics.

Edman, K.A.P. 1992. Contractile performance of skeletal muscle fibres. In *Strength and power in sport,* ed. P.V. Komi. Oxford, UK: Blackwell Scientific.

Elftman, H. 1966. Biomechanics of muscle. *Journal of Bone and Joint Surgery* 48A (2):363–77.

Ende, L.S., and J. Wickstrom. 1982. Ballet injuries. *The Physician and Sportsmedicine* 10 (7):101–103, 106-109, 113-115, 118.

Endresen, E.H. 1995. Pelvic pain and low back pain in pregnant women: An epidemiological study. *Scandinavian Journal of Rheumatology* 24 (3):135–41.

Enoka, R.M. 1994. *Neuromechanical basis of kinesiology*. Champaign, IL: Human Kinetics.

Evans, F.G. 1971. Biomechanical implications of anatomy. In *Biomechanics*, ed. J.M. Cooper. Chicago: The Athletic Institute.

Ferretti, A., P. Papandrea, and F. Conteduca. 1990. Knee injuries in volleyball. *Sports Medicine* 10 (2):132–38.

Ferretti, A., G. Puddu, P.P. Mariani, and M. Neri. 1984. Jumper's knee: An epidemiological study of volleyball players. *The Physician and Sportsmedicine* 12 (10):97–106.

Fleisig, G.S., S.W. Barrentine, R.F. Escamilla, and J.R. Andrews. 1996. Biomechanics of overhand throwing with implications for injuries. *Sports Medicine* 21 (6):421–37.

Ford, D.M., K.M. Bagnall, C.A. Clements, and K.D. McFadden. 1988. Muscle spindles in the paraspinal musculature of patients with adolescent idiopathic scoliosis. *Spine* 13 (5):461–65.

Francis, R.S., and G.R. Bryce. 1987. Relationships between lumbar lordosis, pelvic tilt, and abdominal muscle performance. *Physical Therapy* 67 (8):1221–225.

Freeman, M., and B. Wyke. 1967. Articular reflexes at the ankle joint: An electromyographic study of normal and abnormal influences of ankle joint mechanoreceptors upon reflex activity in the leg muscles. *British Journal of Surgery* 54:990–1001.

Freeman, W.H., and B. Bracegirdle. 1967. *An atlas of histology*. London: Heinemann.

Friberg, O. 1983. Clinical symptoms and biomechanics of lumbar spine and hip joint in leg length inequality. *Spine* 8 (6):643–51.

Frost, H.M. 1967. *An introduction to biomechanics*. Springfield, IL: Charles C Thomas.

———. 1973. *Orthopedic biomechanics*. Vol. 5. Springfield, IL: Charles C Thomas.

———. 1979. A chondral modeling theory. *Calcified Tissue International* 28:181–200.

———. 1988a. Structural adaptations to mechanical usage: A proposed three-way rule for bone modeling. Part I. *Veterinary and Comparative Orthopaedics and Traumatology* 1:7–17.

———. 1988b. Structural adaptations to mechanical usage: A proposed three-way rule for bone modeling. Part II. *Veterinary and Comparative Orthopaedics and Traumatology* 2:80–85.

———. 1990. Skeletal structural adaptations to mechanical usage: Four mechanical influences on intact fibrous tissues. *The Anatomical Record* 226:433–39.

Galea, A.M., and J.M. Albers. 1994. Patellofemoral pain: Beyond empirical diagnosis. *The Physician and Sportsmedicine* 22 (4):48, 53–54, 56, 58.

Gamble, J.G. 1988. *The musculoskeletal system: Physiological basics*. New York: Raven Press.

Gandevia, S.C., D.I. McClosky, and D. Burke. 1992. Kinesthetic signals and muscle contraction. *Trends in Neuroscience* 15:62–65.

Garn, S., and R. Newton. 1988. Kinesthetic awareness in subjects with multiple ankle sprains. *Physical Therapy* 68:1667–671.

Garrett, W. E., M. R. Safran, A. V. Seaber, R.R. Glisson, and B.M. Ribbeck. 1987. Biomechanical comparison of stimulated and nonstimulated skeletal muscle pulled to failure. *American Journal of Sports Medicine* 15 (5):448–54.

Gibson, P.H., T. Papaigannou, and J. Kenwright. 1983. The influence on the spine of leg-length discrepancy after femoral fracture. *Journal of Bone and Joint Surgery* 65B:584–87.

Giles, L.G.F., and J.R. Taylor. 1981. Low-back pain associated with leg length inequality. *Spine* 6 (5):510–21.

Goldsmith, W. 1960. *Impact: The theory and physical behaviour of colliding solids*. London: Edward Arnold.

Goldspink, G. 1992. Cellular and molecular aspects of adaptation in skeletal muscle. In *Strength and power in sport*, ed. P.V. Komi. Oxford, UK: Blackwell Scientific.

Gollehon, D.L., P.A. Torzilli, and R.F. Warren. 1987. The role of the posterolateral and cruciate ligaments in the stability of the human knee: A biomechanical study. *Journal of Bone and Joint Surgery* 69A:233–42.

Gordon, A.M., A.F. Huxley, and F.J. Julian. 1966. The variation in isometric tension with sarcomere length in vertebrate muscle fibres. *Journal of Physiology* (London) 184:170–92.

Gozna, E.R., and I.J. Harrington. 1982. *Biomechanics of musculoskeletal injury*. Baltimore: Williams & Wilkins.

Grace, T.G. 1985. Muscle imbalance and extremity injury: A perplexing relationship. *Sports Medicine* 2:77–82.

Grana, W.A., S. Holder, and E. Schelberg-Karnes. 1987. How I manage acute anterior shoulder dislocations. *The Physician and Sportsmedicine* 15 (4):88–93.

Gray, J.A. 1964. Organ systems in adaptation: the skeleton. Section 4: Adaptation to the environment. In *American Physiological Society: The handbook of physiology*, eds. D.B. Dill, E.F. Adolph, and C.G. Wilber, 133–51. Baltimore: Williams & Wilkins.

Greenspan, A., J.W. Pugh, A. Norman, and R.S. Norman. 1978. Scoliotic index: A comparative evaluation of methods for the measurement of scoliosis. *Bulletin of Hospital Joint Disorder* 39:117–25.

Gregor, R.J. 1993. Skeletal muscle mechanics and movement. In *Current issues in biomechanics*, ed. M.D. Grabiner, 171-211. Champaign, IL: Human Kinetics.

Grieve, G.P. 1976. The sacroiliac joint. *Physiotherapy* 62:384-400.

Grigg, P. 1994. Peripheral neural mechanisms in proprioception. *Journal of Sport Rehabilitation* 3:2–17.

Grimston, S.K., and R.F. Zernicke. 1993. Exercise-related stress responses in bone. *Journal of Applied Biomechanics* 9:2–14.

Groppel, J.L. 1986. The biomechanics of tennis: An overview. *International Journal of Sports Biomechanics* 2:141–55.

Gross, M.L., M. Flynn, and J.J. Sonzogni. 1994. Overworked shoulders: Managing injury of the proximal humeral physis. *The Physician and Sportsmedicine* 22 (3):81–82, 85–86.

Gross, T.S., and S.T. Bain. 1993. Skeletal adaptation to functional stimuli. In *Current issues in biomechanics*, ed. M.D. Grabiner, 151-169. Champaign, IL: Human Kinetics.

Guyer, B., and B. Ellers. 1990. Childhood injuries in the United States: Mortality, morbidity and cost. *American Journal of Diseases in Children* 144:649–52.

Haderspeck, K., and A. Schultz. 1981. Progression of scoliosis: An analysis of muscle actions and body weight influences. *Spine* 6 (5):447–55.

Hainline, B. 1994. Low-back pain in pregnancy. *Advances in Neurology* 64:65–76.

Hall, M.G., W.R. Ferrell, R.H. Baxendale, and D.L. Hamblen. 1994. Knee joint proprioception: Threshold detection levels in healthy young subjects. *Neuro-Orthopedics* 15:81–90.

Hall, M.G., W.R. Ferrell, R.D. Sturrock, D.L. Hamblen, and R.H. Baxendale. 1995. The effect of the hypermobility syndrome on knee joint proprioception. *British Journal of Rheumatology* 34:121–25.

Hanson, P.G., M. Angevine, and J.H. Juhl. 1978. Osteitis pubis in sports activities. *The Physician and Sportsmedicine* 6 (10):111–14.

Hardaker, W.T., S. Margello, and J.L. Goldner. 1985. Foot and ankle injuries in theatrical dancers. *Foot and Ankle* 6 (2):59–69.

Hawkins, R.J., and N. Mohtadi. 1994. Rotator cuff problems in athletes. In *Orthopaedic Sports Medicine: Principles and Practice*, eds. J.C. DeLee and D. Drez. Philadelphia: Saunders. 623–56.

Henning, C.E., N.D. Griffis, S.W. Vequist, K.M. Yearout, and K.A. Decker. 1994. Sport-specific knee injuries. In *Clinical practice of sports injury prevention and care*, ed. P.A. Renström. Oxford, UK: Blackwell Scientific.

Herbert, R. 1988. The passive mechanical properties of muscles and their adaptations to altered patterns of use. *Australian Journal of Physiotherapy* 34 (3):141–49.

Hollinshead, W.H. 1962. *Textbook of anatomy.* New York: Harper & Row.

Hontas, M.J., R.J. Haddad, and L.C. Schlesinger. 1986. Conditions of the talus in the runner. *American Journal of Sports Medicine* 14 (6):486–90.

Hopper, B.J. 1973. *The mechanics of human movement.* London: Crosby Lockwood Staples.

Hoshina, H. 1980. Spondylolysis in athletes. *The Physician and Sportsmedicine* 8 (9):75–79.

Hoshino, A., and A.W. Wallace. 1987. Impact absorbing properties of the human knee. *Journal of Bone and Joint Surgery* 69B (5):807–11.

Huberti, H.H., and W.C. Hayes. 1984. Patellofemoral contact pressures: The influence of Q angle and tendofemoral contact. *Journal of Bone and Joint Surgery* 66A:715–24.

Huijing, P.A. 1992. Mechanical muscle models. In *Strength and power in sport*, ed. P.V. Komi. Oxford, UK: Blackwell Scientific.

Hutton, W.C., and B.M. Cyron. 1978. Spondylolysis. *Acta Orthopedica Scandinavica* 49:604–609.

Huxley, H.E., and J. Hanson. 1954. Changes in the cross striations of muscle during contraction and stretch and their structural interpretation. *Nature* 173:973–77.

Indelicato, P.A. 1995. Isolated medial collateral ligament injuries in the knee. *Journal American Academy of Orthopedic Surgeons* 3 (1):9–14.

Inman, V.T. 1976. *Joints of the ankle*. Baltimore: Williams & Wilkins.

Insall, J.N., and E. Salvati. 1971. Patellar position in the normal knee joint. *Radiology* 101:101–9.

Jackson, D.W., L.L. Wiltse, and R.J. Cirincione. 1976. Spondylolysis in the female gymnast. *Clinical Orthopaedics* 117:68–73.

Johnson, G.R. 1988. The effectiveness of shock-absorbing insoles during normal walking. *Prosthetics and Orthotics International* 12 (1):91–95.

Johnson, J.E., F.H. Sim, and S.G. Scott. 1987. Musculoskeletal injuries in competitive swimmers. *Mayo Clinical Proceedings* 62:289–304.

Jones, B.H., D.N. Cowan, and J. Knapik. 1994. Exercise, training, and injuries. *Sports Medicine* 18 (3):202–14.

Kannus, P., S. Nittymaki, and M. Jarvinen. 1988. Athletic overuse injuries in children: A 30-month prospective follow-up study at an outpatient sports clinic. *Clinical Paediatrics* 27 (7):333–37.

Kapandji, I.A. 1970. *The physiology of the joints*. Vol. 2. Edinburgh: Churchill Livingstone.

———. 1974. *The physiology of the joints*. Vol. 3. Edinburgh: Churchill Livingstone.

Kaplan, F.S. 1983. Osteoporosis. Clinical Symposia. *Ciba-Geigy* 35 (4):32–42.

Keim, H.A. 1982. *The adolescent spine*. New York: Springer-Verlag.

Kelley, M.J. 1990. Psychological risk factors and sports injuries. *Journal of Sports Medicine and Physical Fitness* 30 (2):202–21.

Ker, R.F., M.B. Bennett, S.R. Bibby, R.C. Kester, and R. McN. Alexander. 1987. The spring in the arch of the human foot. *Nature* 325 (7000):147–49.

Kerr, B.A., L. Beauchamp, V. Fisher, and R. Neil. 1983. Foot-strike patterns in distance running. In *Biomechanical aspects of sport shoes and playing surfaces*, eds. B.M. Nigg and B.A. Kerr. Calgary: University Printing.

Kerr, G., and B. Fowler. 1988. The relationship between psychological factors and sports injuries. *Sports Medicine* 6:127–34.

Kleinrensink, G.J., R. Stoeckart, J. Meulstee, D.M.K.S. Kaulesar Sukul, A. Vleeming, C.J. Snijders, and A. Van Noort. 1994. Lowered motor conduction velocity of the peroneal nerve after inversion trauma. *Medicine and Science in Sports and Exercise* 26 (7):877–83.

Knapik, J.J., C.L. Bauman, B.H. Jones, J. M. Harris, and L. Vaughan. 1991. Preseason strength and flexibility imbalances associated with athletic injuries in female collegiate athletes. *The American Journal of Sports Medicine* 19 (1):76–81.

Komi, P.V. 1992. Stretch-shortening cycle. In *Strength and power in sport*, ed. P.V. Komi. Oxford, UK: Blackwell Scientific.

Komi, P.V., and C. Bosco. 1978. Utilization of stored elastic energy in leg extensor muscles by men and women. *Medicine and Science in Sports* 10 (4):261–65.

Krissoff, W.B., and W.D. Ferris. 1979. Runners' injuries. *The Physician and Sportsmedicine* 7 (12):54–64.

Krueger-Franke, M., C.H. Siebert, and W. Pfoerringer. 1992. Sports-related epiphyseal injuries of the lower extremity: an epidemiological study. *Journal of Sports Medicine and Physical Fitness* 32 (1):106–11.

Kujala, U.M., T. Aalto, K. Ostermann, and S. Dahlstrom. 1989. The effect of volleyball playing on the knee extensor mechanism. *American Journal of Sports Medicine* 17 (6):766–69.

Kujala, U.M., O. Friberg, T. Aalto, M. Kvist, and K. Osterman. 1987. Lower limb asymmetry and patellofemoral joint incongruence in the etiology of knee exertion injuries in athletes. *International Journal of Sports Medicine* 8 (3):214–20.

Kujala, U.M., M. Kvist, K. Osterman, O. Friberg, and T. Aalto. 1986. Factors predisposing army conscripts to knee exertion injuries in a physical training program. *Clinical Orthopaedics and Related Research* 210:203–11.

LaBrier, K., and D.B. O'Neill. 1993. Patellofemoral stress syndrome: Current concepts. *Sports Medicine* 16 (6):449–59.

Lancourt, J.E., and J.A. Cristini. 1975. Patella alta and patella infera. *Journal of Bone and Joint Surgery* 57A:1112–115.

Lanyon, L.E. 1981. Adaptive mechanics: The skeleton's response to mechanical stress. In *Mechanical factors and the skeleton*, ed. I.A. Stokes. London: John Libbey.

Larson, R.L. 1973. Physical activity and the growth and development of bone and joint structures. In *Physical activity: Human growth and development*, ed. G.L. Rarick. New York: Academic Press.

Larson, R.L., and R.O. McMahan. 1966. The epiphyses and the childhood athlete. *Journal of the American Medical Association* 196:607–12.

Lassiter, T., T. Malone, and J. Garrick. 1989. Injury to the lateral ligaments of the ankle. *Orthopedic Clinics of North America* 20:629–40.

Letts, M., T. Smallman, R. Afanasiev, and G. Gouw. 1986. Fracture of the pars interarticularis in adolescent athletes: A clinical-biomechanical analysis. *Journal of Paediatric Orthopaedics* 6 (1):40–46.

Lieber, R. L. 1992. *Skeletal muscle structure and function.* Baltimore: Williams & Wilkins.

Lieber, R.L., and S.C. Bodine-Fowler. 1993. Skeletal muscle mechanics: Implications for rehabilitation. *Physical Therapy* 73 (12):844–56.

Lieber, R.L., B.M. Fazeli, and M.J. Botte. 1990. Architecture of selected wrist flexor and extensor muscles. *Journal of Hand Surgery* 15 (2):244–50.

Lindh, M. 1989. Biomechanics of the lumbar spine. In *Basic biomechanics of the musculoskeletal system*, eds. M. Nordin and V.H. Frankel. Philadelphia: Lea & Febiger.

Lloyd-Smith, R., D.B. Clement, D.C. McKenzie, and J.E. Taunton. 1985. A survey of overuse and traumatic hip and pelvic injuries in athletes. *The Physician and Sportsmedicine* 13 (10):131–41.

Lonstein, M.B., and S.W. Wiesel. 1987. Standardized approaches to the evaluation and treatment of industrial low back pain. *Spine: State of the Art Reviews* 2 (1):147–67.

Lorenzton, R. 1988. Intrinsic factors. In *The Olympic Book of Sports Medicine*, eds. A. Ditrix, H.G. Knuttgen, and K. Tittel. Oxford, England: Blackwell Scientific.

Lowe, J., E. Libson, I. Ziv, M. Nyska, Y. Floman, R.A. Bloom, and G.C. Robin. 1986. Spondylolysis in the upper lumbar spine. *Journal of Bone and Joint Surgery* 69B (4):582–86.

Luethi, S.M., and A. Stacoff. 1987. The influence of the shoe on foot mechanics in running. In *Current research in biomechanics*, eds. B. Van Gheluwe and J. Atha. Basel, Switzerland: Karger.

Lundberg, A., and O.K. Svensson. 1988. The axes of rotation of the talocalcaneal and talonavicular joints. In *Patterns of motion of the ankle/foot complex*, ed. A. Lundberg. Gotab, Stockholm: Carolina Institute.

Luoto, S., M. Heliovaara, H.Hurri, and H. Alaranta. 1995. Static back endurance and the risk of low back pain. *Clinical Biomechanics* 10:325-30.

Lysens, R., A. Steverlynck, Y. van den Auweele, J. Lefevre, L. Renson, A. Claessens, and M. Ostyn. 1984. The predictability of sports injuries. *Sports Medicine* 1:6–10.

MacKinnon, J.L. 1988. Osteoporosis. *Physical Therapy* 10:1533–39.

Mair, S.D., A.V. Seaber, R.R. Glisson, and W.E. Garrett. 1996. The role of fatigue in susceptibility to acute muscle strain injury. *American Journal of Sports Medicine* 24 (2):137–43.

Malina, R.M., and C. Bouchard. 1991. *Growth, maturation, and physical activity.* Champaign, IL: Human Kinetics.

Martinez, S.F., M.A. Steingard, and P.M. Steingard. 1993. Thigh compartment syndrome: A limb threatening emergency. *The Physician and Sportsmedicine* 21 (3):94, 96, 99–100, 103–104.

Matheson, G.O., and J.G. Macintyre. 1987. Lower leg varum alignment in skiing: Relationship to foot pain and suboptimal performance. *The Physician and Sportsmedicine* 15 (9):162–69, 172, 176.

Matthews, P.B. 1988. Proprioceptors and their contribution to somatosensory mapping: Complex messages require complex processing. *Canadian Journal of Physiology and Pharmacology* 66:430–38.

McArdle, W.D., F.I. Katch, and V.L. Katch. 1996. *Exercise physiology: Energy, nutrition, and human performance.* Philadelphia: Lea & Febiger.

McConnell, J. 1986. The management of chondromalacia patellae: A long term solution. *Australian Journal of Physiotherapy* 32:215–22.

McGill, S.M., and R.W. Norman. 1987. Reassessment of the role of intra-abdominal pressure in spinal compression. *Ergonomics* 30 (11):1565–588.

McKenzie, D.C., D.B. Clement, and J.E. Taunton. 1985. Running shoes, orthotics, and injuries. *Sports Medicine* 2:334–47.

McMahon, T.A., and P.R. Greene. 1978. Fast running tracks. *Scientific American* 239 (6):148–63.

Mechelen, W. van. 1993. Incidence and severity of sports injuries. In *Sports injuries: Basic principles of prevention and care*, ed. P.A.F.H. Renström. Oxford, UK: Blackwell Scientific.

Micheli, L.J. 1982. Lower extremity injuries: Overuse injuries in the recreational adult. In *The exercising adult*, ed. R.C. Cantu. Massachusetts: Collamore Press.

———. 1983. Overuse injuries in children's sports: The growth factor. *Orthopedic Clinics of North America* 14 (2):337–60.

Micheli, L.J., and C.L. Stanitski. 1981. Lateral patellar retinacular release. *American Journal of Sports Medicine* 9 (5):330–36.

Miller, D.I. 1980. Body segment contributions to sport skill performance: Two contrasting approaches. *Research Quarterly for Exercise and Sport* 51:219–33.

Miller, J.A.A., A.B. Schultz, and G.B.J. Anderson. 1987. Load-displacement behaviour of sacroiliac joints. *Journal of Orthopaedic Research* 5:92-101.

Miniaci, A., and P.J. Fowler. 1993. Impingement in the athlete. *Clinics in Sports Medicine* 12 (1):91–110.

Morris, J.M., D.B. Lucas, and B. Bresler. 1961. Role of the trunk in the stability of the spine. *Journal of Bone and Joint Surgery* 43A (3):327–51.

Muller, E.A. 1970. Influence of training and of inactivity on muscle strength. *Archives of Physical Medicine and Rehabilitation* 51:449–62.

Nachemson, A. 1992. Newest knowledge of low back pain: A critical look. *Clinical Orthopaedics* 279:8–20.

Nelson, B.W., E. O'Reilly, M. Miller, M. Hogan, J.A. Wegner, and C. Kelly. 1995. The clinical effects of intensive, specific exercise on chronic low back pain: A controlled study of 895 consecutive patients with 1-year follow up. *Orthopedics* 18 (10):971–81.

Nicholas, J.A., and M. Marino. 1987. The relationship of injuries of the leg, foot, and ankle to proximal thigh strength in athletes. *Foot & Ankle* 7 (4):218–28.

Nigg, B.M. 1985. Biomechanics, load analysis, and sports injuries in the lower extremities. *Sports Medicine* 2:367–79.

———. 1988. Extrinsic factors. In *The Olympic book of sports medicine*, eds. A. Ditrix, H.G. Knuttgen, and K. Tittel. Oxford, UK: Blackwell Scientific.

Nigg, B.M., and H.A. Bahlsen. 1988. The influence of heel flare and midsole construction on pronation, supination, and impact forces for heel-toe running. *International Journal of Sport Biomechanics* 4:205–19.

Nigg, B.M., G.K. Cole, and G-P. Bruggemann. 1995. Impact forces during heel-toe running. *Journal of Applied Biomechanics* 11:407–32.

Nigg, B.M., J. Denoth, B. Kerr, S. Luethi, D. Smith, and A. Stacoff. 1984. Load sport shoes and playing surfaces. In *Sport shoes and playing surfaces*, ed. E.C. Frederick. Champaign, IL: Human Kinetics.

Nigg, B.M., J. Denoth, and P.A. Neukomm. 1981. Quantifying the load on the human body: problems and some possible solutions. In *Biomechanics*, Vol. VIIB, eds. A. Morecki, K. Fidelus, K. Kedzior, and I. Wit. Baltimore: University Park Press.

Nigg, B.M., and M. Morlock. 1987. The influence of lateral heel flare of running shoes on pronation and impact forces. *Medicine and Science in Sports and Exercise* 19 (3):294–302.

Nigg, B.M., and B. Segesser. 1988. The influence of playing surfaces on the load on the locomotor system and on football and tennis injuries. *Sports Medicine* 5:375–85.

Norkin, C.C. and P.K. Levangie. 1992. *Joint structure and function: a comprehensive analysis.* Philadelphia: F.A. Davis Company.

Noth, J. 1992. Motor units. In *Strength and power in sport*, ed. P.V. Komi. Oxford, UK: Blackwell Scientific.

Noyes, F.R. 1977. Functional properties of knee ligaments and alterations induced by immobilization. *Clinical Orthopedics* 123:210–42.

Pappas, A.M. 1983. Epiphyseal injuries in sports. *The Physician and Sportsmedicine* 11 (6): 140–48.

Peat, M. 1986. Functional anatomy of the shoulder. *Physical Therapy* 66:1855–865.

Penny, J.N., and C. Smith. 1980. The prevention and treatment of swimmer's shoulder. *Canadian Journal of Applied Sports Science* 5 (3):195–202.

Percy, E.C., and R.T. Strother. 1985. Patellagia. *The Physician and Sportsmedicine* 13 (7):43–59.

Perdriolle, R., and J. Vidal. 1985. Thoracic idiopathic scoliosis curve evolution and progression. *Spine* 10 (9):785–91.

Perlman, M., D. Leveille, J. De Leonibus, R. Hartman, J. Klein, R. Handelman, E. Schultz, and S. Wertheimer. 1987. Inversion lateral ankle trauma: Differential diagnosis, review of the literature, and prospective study. *Journal of Foot Surgery* 26:95–135.

Peters, J.W., S.G. Trevino, and P.A. Renström. 1991. Chronic lateral ankle instability. *Foot and Ankle* 12:182–91.

Peterson, L., and P.A. Renström. 1986. *Sports injuries: Their treatment and prevention*. London: Martin Dunitz.

Petrie, T.R. 1993. Coping skills, competitive trait anxiety, and playing status: Moderating effects on the life stress-injury relationship. *Journal of Sport and Exercise Psychology* 15: 261–74.

Petrov, O., D.W. Roth, W.W. Weis, and A.J. Rader. 1988. Nonprescription custom insoles for ski boots. *Journal of the American Podiatric Medical Association* 78 (8):422–28.

Plum, P., and T. Ofeldt. 1985. *Low back pain: New treatment*. Skaerup, Denmark: Brage.

Radin, E.L. 1984. Biomechanical considerations. In *Osteoarthritis: Diagnosis and management*, eds. R.W. Moskowitz, D.S. Howell, V.M. Goldberg, and H.J. Mankin. Philadelphia: Saunders.

———. 1986. Osteoarthrosis: What is known about prevention. *Clinical Orthopaedics and Related Research* 222:60–65.

Radin, E.L., H.G. Parker, J.W. Pugh, R.S. Steinberg, I.L. Paul, and R.M. Rose. 1973. Response of joints to impact loading III: Relationship between trabecular microfractures and cartilage degeneration. *Journal of Biomechanics* 6:51–57.

Radin, E.L., and I.L. Paul. 1971. Response of joints to impact loading I: In vitro wear. *Arthritis and Rheumatology* 14:356–62.

Reber, L., J. Perry, and M. Pink. 1993. Muscular control of the ankle in running. *American Journal of Sports Medicine* 21 (6):805–10.

Recht, M.P., D.L. Burk, and M.K. Dalinka. 1987. Radiology of wrist and hand injuries in athletes. *Clinics in Sports Medicine* 6 (4):811–28.

Redford, J.B. 1987. Orthotics: General principles. *Physical Medicine and Rehabilitation* 1 (1):1–10.

Reid, D.C. 1987. Preventing injuries to the young ballet dancer. *Physiotherapy Canada* 39 (4):231–35.

Reider, B. 1996. Medial collateral ligament injuries in athletes. *Sports Medicine* 21 (2):147–56.

Richter, D.E., C.L. Nash, R.W. Moskowitz, V.M. Goldberg, and I.A. Rosner. 1985. Idiopathic adolescent scoliosis: A prototype of degenerative joint disease. *Clinical Orthopaedics and Related Research* 193:221–29.

Riegger-Krugh, C., and J.L. Keysor. 1996. Skeletal malalignments of the lower quarter: Correlated and compensatory motions and postures. *Journal of Orthopaedic Sports Physical Therapy* 23 (2):164–70.

Risch, S.V., N.K. Norvell, M.L. Pollock, E. Rische, H. Langer, and M. Fulton. 1993. Lumbar strengthening in chronic low back pain patients: Physiologic and psychologic benefits. *Spine* 18:232–38.

Rivara, F.P., B. Alexander, A.B. Johnston, and R. Sodenberg. 1993. Population-based study of fall injuries in children and adolescents resulting in hospitalisation or death. *Paediatrics* 92 (1):61–63.

Roberts, K., and D.A. Brodie. 1989. The rise of sports participation in the United Kingdom. *Society and Leisure* 12:307–24.

Roberts, T.D.M. 1995. *Understanding balance: The mechanics of posture and locomotion*. London: Chapman and Hall.

Roels, J., M. Martins, J.C. Mulier, and A. Burssens. 1978. Patellar tendinitis (jumper's knee). *American Journal of Sports Medicine* 6:362–68.

Rubin, C.T. 1984. Skeletal strain and the functional significance of bone architecture. *Calcified Tissue International* 36:S11–S18.

Ryan, C., P.A. Fricker, and P.G.A. Hannaford. 1987. Muscle compartment pressure syndrome of the upper limb and shoulder: Two case studies. *Australian Journal of Science and Medicine in Sport* 19 (3):24–25.

Rydevik, B., M.D. Brown, and G. Lundborg. 1984. Pathoanatomy and pathophysiology of nerve root compression. *Spine* 9 (1):7-15.

Sale, D. G. 1992. Neural adaptation to strength training. In *Strength and power in sport*, ed. P.V. Komi. Oxford, UK: Blackwell Scientific.

Salenius, P., and E. Vankka. 1975. Development of the tibiofemoral angle of children at different ages. *Journal of Bone and Joint Surgery* 57A:259–61.

Sargeant, A.J. 1994. Human power output and muscle fatigue. *International Journal of Sports Medicine* 15 (3):116–21.

Schultz, A., K. Haderspeck, and S. Takashima. 1981. Correction of scoliosis by muscle stimulation: Biomechanical analyses. *Spine* 6 (5):468–76.

Seddon, J.H. 1972. *Surgical disorders of peripheral nerves*. Edinburgh: Livingstone.

Segan, D.J., E.C. Sladek, J. Gomez, H.J. McCoy, and D.A. Cairns. 1988. Weight lifting as a cause of bilateral upper extremity compartment syndrome. *The Physician and Sportsmedicine* 16 (10):73–75.

Shambaugh, J.P., A. Klein, and J.H. Herbert. 1991. Structural measures as predictors of injury in basketball players. *Medicine and Science in Sports and Exercise* 23 (3):522–27.

Siegler, S., J. Block, and C.D. Schneck. 1988. The mechanical characteristics of the collateral ligaments of the human ankle joint. *Foot and Ankle* 8:234–42.

Siffert, R.S. 1987. Lower limb-length discrepancy. *Journal of Bone and Joint Surgery* 67A:1100–106.

Silliman, J.F., and R.J. Hawkins. 1991. Current concepts and recent advances in the athlete's shoulder. *Clinics in Sports Medicine* 10 (4):693–705.

Silver, D.M., and P. Campbell. 1985. Arthroscopic assessment and treatment of dancers knee injuries. *The Physician and Sportsmedicine* 13 (11):75–82.

Silver, J.R., D.D. Silver, and J.J. Godfrey. 1986. Trampoline injuries of the spine. *Injury: British Journal of Accident Injury* 17 (2):117–24.

Silver, R.L., J. de la Garza, and M. Rang. 1985. The myth of muscle balance. *Journal of Bone and Joint Surgery* 67B (3):432–37.

Simon, S.R., E.L. Radin, I.L. Paul, and R.M. Rose. 1972. The response of joints to impact loading II: In vivo behavior of subchondral bone. *Journal of Biomechanics* 5:267–72.

Skinner, H.B., M.P. Wyatt, M.L. Stone, J.A. Hodgdon, and R.L. Barrack. 1986. Exercise-related knee joint laxity. *American Journal of Sports Medicine* 14:30–34.

Skoglund, L.B., and J.A. Miller. 1981. The length of the thoracolumbar spine in children with idiopathic scoliosis. *Acta Orthopaedica Scandinavica* 52:77–185.

Smith, E.L. 1982. Exercise for prevention of osteoporosis: A review. *The Physician and Sportsmedicine* 10 (3):72–83.

Smith, E.L., and C. Gilligan. 1987. Effects of inactivity and exercise on bone. *The Physician and Sportsmedicine* 15 (1):91-100.

Smith, R.W., and S.F. Reischl. 1986. Treatment of ankle sprains in young athletes. *American Journal of Sports Medicine* 14 (6):465–71.

Solomon, R.L., and L.J. Micheli. 1986. Technique as a consideration in modern dance injuries. *The Physician and Sportsmedicine* 14 (8):83–89, 92.

Sommer, H.M. 1988. Patellar chondropathy and apicitis: Muscle imbalances of the lower extremities in competitive sports. *Sports Medicine* 5:386–94.

Sparto, P.J., M. Parnianpour, T.E. Reinsel, and S. Simon. 1997. The effect of fatigue on multijoint kinematics, coordination, and postural stability during a repetitive lifting task. *Journal of Orthopaedic and Sports Physical Therapy* 25:3-12.

Speer, D.P., and J.K. Braun. 1985. The biomechanical basis of growth plate injuries. *The Physician and Sportsmedicine* 13 (2):72–78.

Sperryn, P.N. 1988. Safety and hygiene in sport. In *The Olympic book of sports medicine*, eds. A. Ditrix, H.G. Knuttgen, and K. Tittel. Oxford, UK: Blackwell Scientific.

Stacoff, A., X. Kalin, and E. Stussi. 1991. The effects of shoes on the torsion and rearfoot motion in running. *Medicine and Science in Sports and Exercise* 23:482–90.

Stanitski, C.L. 1982. Low back pain in young athletes. *The Physician and Sportsmedicine* 10 (1):77–91.

Steinbach, H.L., and W. Russell. 1964. Measurement of the heel pad as an aid to diagnosis of acromegaly. *Radiology* 82:418–23.

Steindler, A. 1955. *Kinesiology of the human body*. Springfield, Illinois: Charles C Thomas.

Steiner, M.E. 1987. Hypermobility and knee injuries. *The Physician and Sportsmedicine* 15 (6):159–65.

Steiner, M.E., W.A. Grana, K. Chillag, and E. Schelberg-Karnes. 1986. The effect of exercise on anterior-posterior knee laxity. *American Journal of Sports Medicine* 14:24–29.

Stephens, T. 1987. Secular trends in adult physical activity: Boom or bust? *Research Quarterly for Exercise and Sport* 58 (2):94-105.

Steyers, C.M., and P.H. Schelkun. 1995. Practical management of carpal tunnel syndrome. *The Physician and Sportsmedicine* 23 (1):83–88.

Stillman, R.J., T.G. Lohman, M.H. Slaughter, and B.H. Massey. 1986. Physical activity and bone mineral content in women aged 30 to 85 years. *Medicine and Science in Sports and Exercise* 18:576–80.

Taft, T.N., F.C. Wilson, and J.W. Oglesby. 1987. Dislocation of the acromioclavicular joint: an End result study. *Journal of Bone and Joint Surgery* 69:1045–1051.

Tall, R.L., and W. Devault. 1993. Spinal injury in sport: Epidemiological considerations in clinical sports medicine. *Clinics in Sports Medicine* 12 (3):441–48.

Talmage, R.V., and J.J.B. Anderson. 1984. Bone density loss in women: Effects of childhood activity, exercise, calcium intake, and oestrogen therapy. *Calcified Tissue International* 36:522–32.

Taylor, D.C., J.D. Dalton, A.V. Seaber, and W.E. Garrett. 1990. Viscoelastic properties of muscle-tendon units. *American Journal of Sports Medicine* 18:300–308.

Torg, J.S., J.J. Vegso, B. Sennet, and M. Dass. 1985. The National Football Head and Neck Injury Registry: 14 year report on cervical quadriplegia, 1971-1984. *Journal of the American Medical Association* 254:3439–443.

Tortora, G.J., and N.P. Anagnostakos. 1984. *Principles of anatomy and physiology*. New York: Harper & Row.

Towner, E.M.L., S.N. Jarvis, S.S.M. Walsh, and A. Aynsley-Green. 1994. Measuring exposure to injury risk in schoolchildren aged 11–14. *British Medical Journal* 308 (6926):449–52.

Tricker, R.A.R., and B.J.K. Tricker. 1967. *The science of movement*. London: Mills & Boon.

Troup, J.D.G. 1970. Functional anatomy of the spine. *British Journal of Sports Medicine* 5 (1):27–33.

Van Ingen Schenau, G.J., C.A. Pratt, and J.M. Macpherson. 1994. Differential use and control of mono- and biarticular muscles. *Human Movement Science* 13:495-517.

Vizsolyi, P., J. Taunton, G. Robertson, L. Filsinger, H.S. Shannon, D. Whittingham, and M. Gleave. 1987. Breaststroker's knee: An analysis of epidemiological and biomechanical factors. *American Journal of Sports Medicine* 15 (1):63–71.

Voloshin, A., and J. Wosk. 1981. Influence of artificial shock absorbers on human gait. *Clinical Orthopaedics and Related Research* 160:52–56.

———. 1982. In-vivo study of low back pain and shock absorption in human locomotor system. *Journal of Biomechanics* 15:21–27.

Walker, M.L., J.M. Rothstein, S.D. Finucane, and R.L. Lamb. 1987. Relationships between lumbar lordosis, pelvic tilt, and abdominal muscle performance. *Physical Therapy* 67 (4):512–16.

Watkins, J. 1982. Warming up and warming down. *New Zealand Journal of Sports Medicine* 10 (1):10–14.

———. 1983. *An introduction to mechanics of human movement*. Lancaster, England: MTP Press.

———. 1987a. Qualitative movement analysis. *British Journal of Physical Education* 18 (4):177–79.

———. 1987b. Quantitative movement analysis. *British Journal of Physical Education* 18 (6):271–75.

———. 1994. Injuries in volleyball. In *Clinical practice of sports injury prevention and care*, ed. P.A. Renström. Oxford, UK: Blackwell Scientific.

———. 1995. The need for agreed procedures in sports injury epidemiological research. *New Zealand Journal of Sports Medicine* 23 (4):34–37.

Watkins, J., and B.N. Green. 1992. Volleyball injuries: A survey of injuries of Scottish National League male players. *British Journal of Sports Medicine* 26 (2):135–37.

Watkins, J., and P. Peabody. 1996. Sports injuries in children and adolescents treated at a sports injury clinic. *Journal of Sports Medicine and Physical Fitness* 36 (1):43–48.

Watson, A.W.S. 1995. Sports injuries in footballers related to defects of posture and body mechanics. *Journal of Sports Medicine and Physical Fitness* 35 (4):289-94.

Watt, D.G.D., and G.M. Jones. 1971. Muscular control of landing from unexpected falls in man. *Journal of Physiology* 219:729–37.

Weiker, G.G., and F. Munnings. 1994. Selected hip and pelvis injuries. *The Physician and Sportsmedicine* 22 (2):96, 98–99, 103–106.

Whedon, G.D. 1984. Disuse osteoporosis: Physiological aspects. *Calcified Tissue International* 36:146–150.

Wickiewicz, T.L., R.R. Roy, P.L. Powell, and V.R. Edgerton. 1983. Muscle architecture of the human lower limb. *Clinical Orthopaedics and Related Research* 179:275–83.

Wilder, D.G., M.H. Pope, and J.W. Frymoyer. 1980. The functional topography of the sacroiliac joint. *Spine* 5 (6):575–79.

Wilkerson, G.B., and A.J. Nitz. 1994. Dynamic ankle stability: Mechanical and neuromuscular interrelationships. *Journal of Sport Rehabilitation* 3:43–57.

Wilkie, D.R. 1960. Man as a source of mechanical power. *Ergonomics* 3:1–8.

Williams, P., and G. Goldspink. 1973. The effect of immobilization on the longitudinal growth of striated muscle fibers. *Journal of Anatomy* 116:45–55.

———. 1978. Changes in sarcomere length and physiological properties in immobilized muscle. *Journal of Anatomy* 127:459–68.

Williams, P.L., L.H. Bannister, M.M. Berry, P. Collins, M. Dyson, J.E. Dussek, and M.W.J. Ferguson, eds. 1995. *Gray's anatomy*. Edinburgh: Longman.

Wilmore, J.H., and D.L. Costill. 1994. *Physiology of sport and exercise*. Champaign, IL: Human Kinetics.

Winslow, J., and E. Yoder. 1995. Patellofemoral pain in female ballet dancers: Correlation with iliotibial band tightness and tibial external rotation. *Journal of Orthopaedic and Sports Physical Therapy* 22 (1):18–21.

Wolff, J. 1988. *The law of bone modeling*. Translated by P. Maquet and R. Furlong. New York: Springer Verlag. Originally published as Wolff, J. 1892. *Das gesetz der transformation der knochen*. Berlin: A. Hirschwald.

Wosk, J., and A. Voloshin. 1985. Low back pain: Conservative treatment with artificial shock absorbers. *Archives of Physical Medicine and Rehabilitation* 66:145–48.

Yates, J.W., and D.W. Jackson. 1984. Current status of meniscus injury. *The Physician and Sportsmedicine* 12 (2):51–59.

Zachazewski, J.E., and G. Geissler. 1992 When to prescribe a knee brace. *The Physician and Sportsmedicine* 20 (1):91–99.

Zernicke, R.F., and B.J. Loitz. 1992. Exercise-related adaptations in connective tissue. In *Strength and power in sport*, ed. P.V. Komi. Oxford, UK: Blackwell Scientific.

Zohar, J. 1973. Preventive conditioning for maximum safety and performance. *Sports Coach* 42 (9):65, 113–15.

Index

Note: The letters *f* and *t* after page numbers refer to figures and tables, respectively.

About the Author

James Watkins, PhD, teaches functional anatomy and biomechanics in the Scottish School of Sport Studies at the University of Strathclyde in Glasgow, Scotland, where he served as department head from 1989 to 1994.

Dr. Watkins' publications include more than 70 academic journal papers and four books. He is an advisory board member of the *Journal of Sports Sciences* and an editorial board member of the *European Journal of Physical Education* and the *British Journal of Physical Education*. He was chair of the Biomechanics section of the British Association of Sport and Exercise Sciences from 1993 to 1996.

Dr. Watkins received his PhD in biomechanics from the University of Leeds, England, in 1975.

Related Books from Human Kinetics

Biomechanics of Musculoskeletal Injury

William C. Whiting, PhD, and Ronald F. Zernicke, PhD
1998 • Hardcover • Item BWHI0779
ISBN 0-87322-779-4 • $49.00 ($73.50 Canadian)

Explores the mechanical bases of musculoskeletal injury to better understand causal mechanisms, the effect of injury on musculoskeletal tissues, and how our current knowledge of biomechanics can contribute to injury prevention.

Kinematics of Human Motion

Vladimir M. Zatsiorsky, PhD
1998 • Hardcover • Item BZAT0676
ISBN 0-88011-676-5 • $49.00 ($73.50 Canadian)

The first major text on the kinematics of human motion, written by one of the world's leading authorities on the subject.

Neuromechanical Basis of Kinesiology (Second Edition)

Roger M. Enoka, PhD
1994 • Hardcover • Item BENO0665
ISBN 0-87322-665-8 • $52.00 ($77.95 Canadian)

Integrates biomechanics and neurophysiology to provide a unique theoretical framework for the study of human movement.

To request more information or to order, U.S. customers call 1-800-747-4457, e-mail us at humank@hkusa.com, or visit our Web site at www.humankinetics.com. Persons outside the U.S. can contact us via our Web site or use the appropriate telephone number, postal address, or e-mail address shown in the front of this book.

HUMAN KINETICS
The Information Leader in Physical Activity
P.O. Box 5076, Champaign, IL 61825-5076
2335

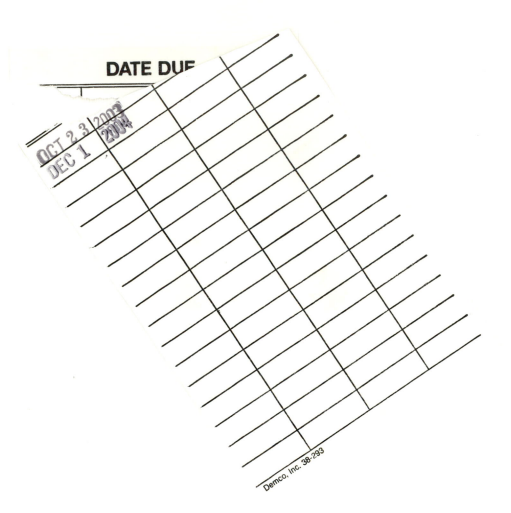

DATE DUE

OCT 2 3 2003
DEC 1 2004

Demco, Inc. 38-293